Applied Mathematical Sciences | Volume 12

L. D. Berkovitz

Optimal Control Theory

With 10 Illustrations

Springer-Verlag New York · Heidelberg · Berlin
1974

L. D. Berkovitz
Division of Mathematical Sciences
Purdue University
West Lafayette, Indiana

AMS Classifications
49A10, 49A35, 49A40, 49B10, 49B35, 49B40, 49C05

Library of Congress Cataloging in Publication Data

Berkovitz, Leonard David, 1924-
Optimal control theory.

(Applied mathematical sciences; v. 12)
Bibliography: p.
Includes index.
1. Control theory. 2. Mathematical optimization.
I. Title. II. Series.
QA1.A647 vol. 12 [QA402.3] 510'.8s [629.8'312] 74-20837

Printed in the United States of America.

ISBN 0-387-90106-X Springer-Verlag New York · Heidelberg · Berlin
ISBN 3-540-90106-X Springer-Verlag Berlin · Heidelberg · New York

PREFACE

This book is an introduction to the mathematical theory of optimal control of processes governed by ordinary differential equations. It is intended for students and professionals in mathematics and in areas of application who want a broad, yet relatively deep, concise and coherent introduction to the subject and to its relationship with applications. In order to accommodate a range of mathematical interests and backgrounds among readers, the material is arranged so that the more advanced mathematical sections can be omitted without loss of continuity. For readers primarily interested in applications a recommended minimum course consists of Chapter I, the sections of Chapters II, III, and IV so recommended in the introductory sections of those chapters, and all of Chapter V. The introductory section of each chapter should further guide the individual reader toward material that is of interest to him. A reader who has had a good course in advanced calculus should be able to understand the definitions and statements of the theorems and should be able to follow a substantial portion of the mathematical development. The entire book can be read by someone familiar with the basic aspects of Lebesque integration and functional analysis.

For the reader who wishes to find out more about applications we recommend references [2], [13], [33], [35], and [50], of the Bibliography at the end of the book. For the reader who wishes to learn more about the mathematical aspects and about some topics not treated here, we recommend references [27], [28], [33], [48], [50], [59], and [61].

Theorems, displayed equations and formulas, displayed inequalities, definitions, etc. are numbered decimally as follows. Theorem III.7.2 is the second theorem in Section 7 of Chapter III.

References to this theorem that are made outside of Chapter III read "Theorem III.7.2". References within Chapter III simply read "Theorem 7.2". Similar remarks hold for displayed formulas, equations, inequalities, etc.

The author thanks Mrs. Nancy Eberle for typing various preliminary versions of the first four chapters for use as classroom notes at Purdue University. He thanks Dr. William Browning, who read the first four chapters, for his helpful suggestions and comments. Lastly, the author thanks Professor H. T. Banks for his critical reading and proof-reading of the complete final version of the book and for his comments and corrections. All final errors, however, are the author's sole responsibility.

Leonard D. Berkovitz

West Lafayette, Indiana
August 5, 1974

TABLE OF CONTENTS

CHAPTER I

EXAMPLES OF CONTROL PROBLEMS

1. Introduction

In recent years there has arisen in diverse areas a collection of important problems that have a common mathematical formulation. These are the so-called control problems. Despite their present day origins these problems, from a mathematical point of view, are variants of a class of problems that has been studied for several hundred years; namely, the problems of the calculus of variations.

In this chapter we shall present some examples of control problems drawn from different areas of application. The purpose of this list of problems is to illustrate the diversity of origins of control problems, to indicate their importance, and to motivate the mathematical formulation of the problems. It should not be construed that the list of examples is complete or that we have chosen the most significant problem in each area. In fact, we chose fairly simple problems in an effort to illustrate without excessive complication.

2. A Problem of Production Planning

The first problem, taken from economics, is a problem in production planning. Let T be a fixed time. Let $x(t)$ denote the stock of a commodity at time t, $0 \le t \le T$. Let $r(t) \ge 0$ denote the rate of demand for the commodity at time t; it is assumed here that r is a known function of time, defined and continuous for $0 \le t \le T$. Let $u(t)$ denote the rate of production at time t, $0 \le t \le T$. The function u is to be chosen by the production planner; it is the production plan, or <u>control</u>. We shall take u to be piecewise continuous on $0 \leq t \leq T$. We require that all demands are met. The stock inventory x is then determined by the differential equation

$$\frac{dx}{dt} = -r(t) + u(t) \qquad x(0) = x_0, \tag{2.1}$$

where x_0 is the initial stock level, and $x_0 > 0$. From the physical meaning of $x(t)$ it is clear that the production plan u must be chosen so that

$$x(t) \geq 0 \tag{2.2}$$

for all $0 \leq t \leq T$. Furthermore, since stock is not destroyed and the plant capacity places a limit on the ability to produce it is reasonable to require that the function u satisfy the constraints

$$0 \leq u(t) \leq A \tag{2.3}$$

for all $0 \leq t \leq T$. Here, $A > 0$ represents the maximum possible rate of production. A production plan u satisfying (2.3) and such that the corresponding solution of (2.1) exists and satisfies (2.2) for $0 \leq t \leq T$ will be called an <u>admissible</u> plan or an admissible control.

At this point the question arises as to whether any admissible plans exist. If A is sufficiently large, then there surely exist admissible plans. For example, if

$$M = \sup\ [r(t):\ 0 \leq t \leq T]$$

and $A > M$, then $u(t) = A$ is admissible. We shall henceforth suppose that admissible plans do exist.

Let us suppose that the cost of production per unit time is a function h of the rate of production. Thus, at time t the rate of production is $u(t)$ and the cost of production per unit time is $h(u(t))$. Let $b > 0$ be the cost per unit time of storing a unit of commodity. Then the cost per unit time at time t of operating the system is

$$f(t,x(t),u(t)) = h(u(t)) + bx(t). \tag{2.4}$$

The total cost is given by

$$C(u) = \int_0^T f(t,x(t),u(t))dt \tag{2.5}$$

where $x(t)$ is the solution of (2.1) corresponding to the admissible production plan u. We use the symbol $C(u)$ to designate the cost since the cost depends solely on the choice of the function u once the demand r and the initial stock x_0 are specified. We have here an example of a <u>functional</u>; that is an assignment of a real number to every function in a given class of functions.

The problem for the production planner is to choose an admissible control u such that $C(u)$, the total cost, is minimized.

3. Chemical Engineering

Let $x^1(t),\dots,x^n(t)$ denote the concentrations at time t of n substances in a reactor in which n simultaneous chemical reactions are taking place. Let the rates of the reactions be governed by a system of differential equations

$$\frac{dx^i}{dt} = G^i(x^1,\dots,x^n,\theta(t),p(t)) \qquad \begin{aligned} x^i(0) &= x_0^i \\ i &= 1,\dots,n. \end{aligned} \tag{3.1}$$

where $\theta(t)$ is the temperature in the reactor at time t and $p(t)$ is the pressure in the reactor at time t. We can control the temperature and pressure at each instant of time, subject to the constraints

$$\begin{aligned} \theta_b &\le \theta(t) \le \theta_a \\ p_b &\le p(t) \le p_a \end{aligned} \tag{3.2}$$

where θ_a, θ_b, p_a, and p_b are constants. These represent the minimum and maximum attainable temperature and pressure.

We let the reaction proceed for a time T. The concentrations at this time are $x^1(T),\dots,x^n(T)$. Associated with each product is an economic value, or price c^i, $i = 1,\dots,n$. The price may be negative, as in the case of hazardous wastes that must be disposed of at some expense. The value of the end product is

$$V(p,\theta) = \sum_{i=1}^{n} c^i x^i(T). \tag{3.3}$$

Given a set of initial concentrations x_0^i, the value of the end product is completely determined by the choice of functions p and θ if the functions G^i have certain nice properties. Hence the notation $V(p,\theta)$. This is another example of a functional; in this case we have an assignment of a real number to each pair of functions in a certain collection.

The problem here is to choose piecewise continuous functions p and θ on the interval $[0,T]$ so that (3.2) is satistied and so that $V(p,\theta)$ is maximized.

A variant of the preceding problem is the following. Instead of allowing the reaction to proceed for a fixed time T, we stop the reaction when one of the reactants, say x^1, reaches a preassigned concentration x_f^1. Now the final time t_f is not fixed beforehand, but is the smallest positive root of the equation $x^1(t) = x_f^1$. The problem now is to maximize

$$V(p,\theta) = \sum_{i=2}^{n} c^i x^i(t_f) - k^2 t_f.$$

The term $k^2 t_f$ represents the cost of running the reactor.

Still another variant of the problem is to stop the reaction when several of the reactants reach preassigned concentrations, say $x^1 = x_f^1$, $x^2 = x_f^2,\dots,x^j = x_f^j$. The value of the end product is now

$$\sum_{i=j+1}^{n} c^i x^i(t_f) - k^2 t_f.$$

We remark that in the last two variants of the problem there is another question that must be considered before one takes up the problem of maximization. Namely, can one achieve the desired final concentrations using pressure and temperature functions p and θ in the class of functions permitted?

4. Flight Mechanics

In this problem a rocket is taken to be a point of variable mass whose moments of inertia are neglected. The motion of the rocket is assumed to take place in a plane relative to a fixed frame. Let $y = (y^1, y^2)$ denote the position vector of the rocket and let $v = (v^1, v^2)$ denote the velocity vector of the rocket. Then

$$\frac{dy^i}{dt} = v^i \qquad y^i(0) = y_0^i \qquad i = 1,2, \tag{4.1}$$

where $y_0 = (y_0^1, y_0^2)$ denotes the initial position of the rocket.

Let $\beta(t)$ denote the rate at which the rocket burns fuel at time t and let $m(t)$ denote the mass of the rocket at time t. Thus

$$\frac{dm}{dt} = -\beta. \tag{4.2}$$

The mass of the rocket is equal to the mass of the fuel plus the mass $a > 0$ of the vehicle. Hence we have $m(t) \geq a$.

Let $\omega(t)$ denote the angle that the thrust vector makes with the positive y^1-axis at time t. The burning rate and the thrust angle will be at our disposal subject to the constraints

$$0 \leq \beta_0 \leq \beta(t) \leq \beta_1 \qquad \omega_0 \leq \omega(t) \leq \omega_1, \tag{4.3}$$

where β_0, β_1, ω_0, and ω_1 are fixed.

To complete the equations of motion of the rocket we analyze the momentum transfer in rectilinear rocket motion. At time t a rocket of mass m and velocity v has momentum mv. During an interval of time δt let the rocket burn an amount of fuel $\delta\mu > 0$. At time $t + \delta t$ let the ejected combustion products have velocity v'; their mass is cleary $\delta\mu$. At time $t + \delta t$ let the velocity of the rocket be $v + \delta v$; its mass is clearly $m - \delta\mu$. Let us consider the system which at time t consisted of the rocket of mass m and velocity v. At time $t + \delta t$ this system consists of the rocket and the ejected combustion products. The change in momentum of the system in the time interval δt is therefore

$$(\delta\mu)v' + (m - \delta\mu)(v + \delta v) - mv.$$

If we divide the last expression by $\delta t > 0$ and then let $\delta t \to 0$, we obtain the rate of change of momentum of the system, which must equal the sum of the external forces acting upon the system. Hence, if F is the resultant external force per unit mass acting upon the system we have

$$Fm - (v'-v)\frac{d\mu}{dt} = m\frac{dv}{dt}.$$

If we assume that $(v'-v)$, the velocity of the combustion products relative to the rocket is a constant c, and if we use $d\mu/dt = \beta$, we get

$$F - c\beta/m = dv/dt.$$

If we apply the preceding analysis to each component of the planar motion we get the following equations, which together with (4.1), (4.2) and (4.3) govern the planar rocket motion

$$\begin{aligned} \frac{dv^1}{dt} &= F^1 - \frac{c\beta}{m}\cos\omega \\ \frac{dv^2}{dt} &= F^2 - \frac{c\beta}{m}\sin\omega \qquad v^i(0) = v_0^i, \quad i = 1,2. \end{aligned} \tag{4.4}$$

Here, the components of the force F can be functions of y and v. This would be the case if the motion takes place in a non-constant gravitational field and if drag forces act on the rocket.

The control problems associated with the motion of the rocket are of the following type. The burning rate control β and the thrust direction control ω are to be chosen from the class of piecewise continuous functions (or some other appropriate class) in such a way that certain of the variables t, y, v, m attain specified terminal values. From among the controls that achieve these values the control that maximizes (or minimizes) a given function of the remaining terminal values is to be determined. In other problems an integral evaluated along the trajectory in the state space is to be extremized.

To be more specific, consider the "minimum fuel problem". It is required that the rocket go from a specified initial point y_0 to a specified terminal point y_f in such a way as to minimize the fuel consumed. This problem is important for the following reason. Since the total weight of rocket plus fuel plus payload that can be constructed and lifted is constrained by the state of the technology, it follows that the less fuel consumed, the larger the payload that can be carried by the rocket. From (4.2) we have

$$m_f = m_0 - \int_{t_0}^{t_f} \beta(t)dt,$$

where t_0 is the initial time, t_f is the terminal time (time at which y_f is reached), m_f is the final mass, and m_0 is the initial mass. The fuel consumed is therefore $m_0 - m_f$. Thus the problem of minimizing the fuel consumed is the problem of minimizing

$$P(\beta,\omega) = \int_{t_0}^{t_f} \beta(t)dt \tag{4.5}$$

subject to (4.1) to (4.4). This problem is equivalent to the problem of maximizing m_f. In the minimum fuel problem the terminal velocity vector v_f will be unspecified if a "hard landing" is permitted; it will be specified if a "soft landing" is required. The terminal time t_f may or may not be specified.

Another example is the problem of rendezvous with a moving object whose position vector at time t is $z(t) = (z^1(t), z^2(t))$ and whose velocity vector at time t is $w(t) = (w^1(t), w^2(t))$, where w^1 and w^2 are continuous functions. Let us suppose that there exist thrust programs β and ω satisfying (4.3) and such that a rendezvous can be effected. Mathematically this is expressed by the assumption that the solutions y, v of the equations of motion corresponding to the given choice of β and ω have the property that the equations

$$\begin{aligned} y(t) &= z(t) \\ v(t) &= w(t) \end{aligned} \tag{4.6}$$

have positive solutions. Such controls (β,ω) will be called admissible. Since for each admissible β and ω the corresponding solutions y and v are continuous, and since the functions z and w are continuous by hypothesis, it follows that for each admissible pair (β,ω) there is a smallest positive solution $t_f(\beta,\omega)$ for which (4.6) holds. The number $t_f(\beta,\omega)$ is the rendezvous time. Two problems are possible here. The first is to determine from among the admissible controls one that delivers the maximum payload; i.e. to maximize $m_f = m_f(t_f(\beta,\omega))$. The second is to minimize the rendezvous time $t_f(\beta,\omega)$.

5. Electrical Engineering

A control surface is to be kept at some arbitrary position by means of a servo-mechanism. Outside disturbances such as wind gusts occur infrequently and are short with respect to the time constant of the servo-mechanism. A direct-current electric motor is used to apply a torque to bring the control surface to its desired position. Only the armature voltage v into the motor can be controlled. For simplicity we take the desired position to be the zero angle and we measure deviations in the angle θ from this desired position. With a suitable normalization the differential equation for θ can be written as

$$\frac{d^2\theta}{dt^2} + a\frac{d\theta}{dt} + \omega^2\theta = u \qquad \theta(0) = \theta_0 \qquad \theta'(0) = \theta_0'. \tag{5.1}$$

Here u represents the restoring torque applied to the control surface and the term $a d\theta/dt$ represents the damping effect. If no damping occurs then $a = 0$. Since the source of voltage cannot deliver a voltage larger in absolute value than some v_0, the restoring torque must be bounded in absolute value. Hence it follows that we must have

$$|u(t)| \leq A, \tag{5.2}$$

where A is some positive constant.

If we set

$$x^1 = \theta \qquad x^2 = d\theta/dt$$

we can rewrite equation (5.1) as follows:

$$\begin{aligned} \frac{dx^1}{dt} &= x^2 & x^1(0) &= \theta_0 \\ \frac{dx^2}{dt} &= -ax^2 - \omega^2 x^1 + u & x^2(0) &= \theta_0'. \end{aligned} \tag{5.3}$$

The problem is the following. A short disturbance has resulted in a deviation $\theta = \theta_0$ and $d\theta/dt = \theta_0'$ from rest at the desired position. How should the voltage be applied over time so that the control surface is brought back to the set position $\theta = 0$, $d\theta/dt = 0$ in the shortest possible time. In terms of (5.3) the problem is to choose a function u from an appropriate class of functions, say piecewise continuous functions, such that u satisfies (5.2) at each instant of time and such that the solution (x^1, x^2) of (5.3) corresponding to u reaches the origin in (x^1, x^2)-space in minimum time.

6. The Brachistochrone Problem

We now present a problem from the calculus of variations; the brachistochrone problem, posed by John Bernoulli in 1696. This problem can be regarded as the starting point of the theory of the calculus of variations. Galileo also seems to have considered this problem in 1630 and 1638, but was not as explicit in his formulation.

Two points P_0 and P_1 are given in a vertical plane with P_0 higher than P_1. A particle, or point mass, acted upon solely by gravitational forces is to move along a curve C joining P_0 and P_1. Furthermore, the particle is to have an initial speed v_0 along the curve at P_0. The problem is to choose the curve C so that the time required for the particle to go from P_0 to P_1 is a minimum.

To formulate the problem analytically we set up a coordinate system in the plane as shown in Figure 1.

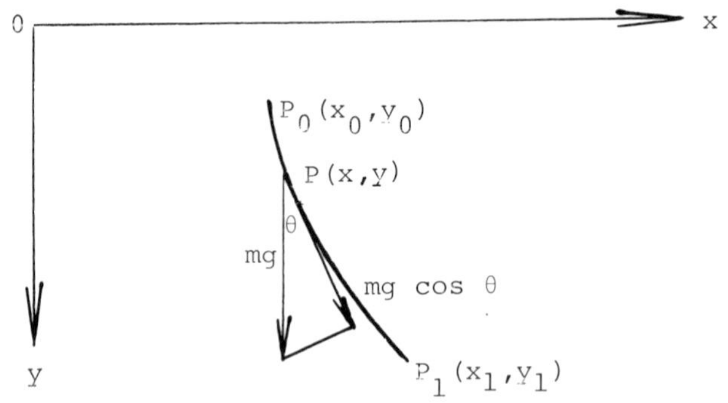

Figure 1

Let P_0 have coordinates (x_0, y_0), let P_1 have coordinates (x_1, y_1) and let C have $y = y(x)$ as its equation. Let s_1 denote the arc length of C between P_0 and P_1. We shall determine the time required to traverse C from P_0 to P_1.

Let P be a point on C with coordinates (x,y). At this point the component of the gravitational force acting along the curve is $mg\cos\theta$ where θ is the angle that the tangent makes with the downward (positive y) oriented vertical. Thus, if we let $s(t)$ denote the distance traversed along C from P_0 by the particle in t seconds, we have

$$m\frac{d^2s}{dt^2} = mg\cos\theta .$$

If we multiply both sides of this equation by $2m^{-1}(ds/dt)$ and use the relation $dy/ds = \cos\theta$, we get

$$\frac{d}{dt}\left(\frac{ds}{dt}\right)^2 = 2g\frac{dy}{dt} .$$

If we set $v = ds/dt$ and integrate this relationship between the initial point P_0 and a point P on the curve, we get

$$v^2 - v_0^2 = 2g(y-y_0).$$

This equation can be written in the form

$$\frac{ds}{dt} = [2g(y-\alpha)]^{1/2} \tag{6.1}$$

where $\alpha = y_0-(v_0^2/2g)$.

Hence, using the relation

$$ds = [1+(y')^2]^{1/2}dx$$

we get that the time τ to traverse C from P_0 to P_1 is

$$\tau = \int_0^\tau dt = \int_0^{s_1}\frac{ds}{[2g(y-\alpha)]^{1/2}} = (2g)^{-1/2}\int_{x_0}^{x_1}\left[\frac{1+(y')^2}{y-\alpha}\right]^{1/2}dx .$$

Thus, aside from the constant factor of $(2g)^{-1/2}$, the problem of finding a curve C that minimizes the time of transit between P_0 and P_1 is equivalent to the following problem. In the class of functions y that are differentiable on $[x_0,x_1]$ and satisy the conditions $y(x_0) = y_0$, $y(x_1) = y_1$ find a function that minimizes the integral

$$\int_{x_0}^{x_1} \left(\frac{1+(y')^2}{y-\alpha}\right)^{1/2} dx.$$

We can put this problem in a format similar to the previous ones as follows. Change the notation for the independent variable from x to t. Then set

$$y' = u \qquad y(t_0) = y_0. \tag{6.2}$$

A continuous function u will be called admissible if it is defined on $[t_0,t_1]$ and if the solution of (6.2) corresponding to u satisfies $y(t_1) = y_1$. Our problem is to determine the admissible function u that minimizes

$$J(u) = \int_{t_0}^{t_1} \left(\frac{1+u^2}{y-\alpha}\right)^{1/2} dt$$

in the class of all admissible u.

We point out that the brachistochrone problem can be formulated as a control problem in a different fashion. By (6.1), the speed of the particle along the curve C is given by $(2g(y-\alpha))^{1/2}$. Hence, if θ is as in Figure 1

$$\frac{dx}{dt} = (2g(y-\alpha))^{1/2} \sin\theta$$

$$\frac{dy}{dt} = (2g(y-\alpha))^{1/2} \cos\theta.$$

Let $u = \sin\theta$. Then the equations of motion become

$$\begin{aligned} \frac{dx}{dt} &= (2g(y-\alpha))^{1/2}u & x(t_0) &= x_0 \\ \frac{dy}{dt} &= (2g(y-\alpha))^{1/2}(1-u^2)^{1/2} & y(t_0) &= y_0. \end{aligned} \tag{6.3}$$

The problem is to choose a control u satisfying $|u| \le 1$ such that the point (x,y) which at initial time t_0 is at (x_0,y_0) reaches the prescribed point (x_1,y_1) in minimum time. If t_1 is the time at which P_1 is reached, then this is equivalent to minimizing t_1-t_0. This in turn is equivalent to minimizing

$$\int_{t_0}^{t_1} dt$$

subject to (6.3) and the constraint $|u(t)| \leq 1$.

The brachistochrone problem can be modified in the following fashion. One can replace the fixed point P_1 by a curve Γ_1 defined by $y = y_1(x)$ and seek the curve C joining P_0 to Γ_1 along which the mass particle must travel if it is to go from P_0 to Γ_1 in minimum time. We can also replace P_0 by a curve Γ_0 where Γ_0 is at positive distance from Γ_1 and ask for the curve C joining Γ_0 and Γ_1 along which the particle must travel in order to minimize the time of transit.

CHAPTER II

FORMULATION OF THE CONTROL PROBLEM

1. Introduction

In this chapter we first point out the common mathematical structure of the examples in the previous chapter. This permits a somewhat imprecise preliminary formulation of the mathematical problem of optimal control. It should, however, motivate the precise and more general formulation of the mathematical problem of optimal control which is given in Section 3. In Section 4 we discuss various equivalent formulations of the problem, and in Section 5 we show how some other control problems can be cast in the form given in Section 3. We conclude this chapter with a discussion of the relationship between problems in the calculus of variations and control problems. Sections 4 to 6 inclusive can be omitted at a first reading and can be read as the need arises.

2. Preliminary Formulation of the Control Problem

All the examples in the preceding chapter have the following form. The state of a system at time t is described by a point or vector

$$x(t) = (x^1(t),\ldots,x^n(t))$$

in n-dimensional euclidean space, $n \geq 1$. Initially, at time t_0, the state of the system is

$$x(t_0) = x_0 = (x_0^1,\ldots,x_0^n).$$

More generally, we can require that at the initial time t_0 the initial state x_0 is such that the point (t_0,x_0) belongs to some pre-assigned set $\mathcal{T}_0$ in (t,x)-space. The state of the system varies with time according to the system of differential equations

$$\frac{dx^i}{dt} = f^i(t,x,z) \qquad x^i(t_0) = x_0^i \quad i = 1,\dots,n, \tag{2.1}$$

where $z = (z^1,\dots,z^m)$ is a vector in real euclidean space E^m and the functions f^i are real valued continuous functions of the variables (t,x,z).

By the 'system varying according to (2.1)' we mean the following. A function u with values in m-dimensional euclidean space is chosen from some prescribed class of functions. In this section we shall take this class to be a subclass $\mathscr{L}$ of the class of piecewise continuous functions. When the substitution $z = u(t)$ is made in the right hand side of (2.1) we obtain a system of ordinary differential equations:

$$\frac{dx^i}{dt} = f^i(t,x,u(t)) = F_u^i(t,x) \qquad i = 1,\dots,n. \tag{2.2}$$

The subscript u on the F_u^i emphasizes that the right hand side of (2.2) depends on the choice of function u. For each u in $\mathscr{L}$ it is assumed that there exists a point (t_0,x_0) in $\mathscr{T}_0$ and a function $\phi = (\phi^1,\dots,\phi^n)$ defined on an interval $[t_0,t_2]$ with values in E^n such that (2.2) is satisfied. That is, we require that for every t in $[t_0,t_2]$

$$\phi'^i(t) = \frac{d\phi^i}{dt} = f^i(t,\phi(t),u(t)) \qquad \phi^i(t_0) = x_0^i$$
$$i = 1,\dots,n.$$

At points of discontinuity of u this equation is interpreted as holding for the one-sided limits. The function ϕ describes the evolution of the system with time and will sometimes be called a <u>trajectory</u>.

The function u is further required to be such that at some time t_1, where $t_0 < t_1$, the point $(t_1,\phi(t_1))$ belongs to a preassigned set $\mathcal{T}_1$ and for $t_0 \leq t < t_1$ the points $(t,\phi(t))$ do not belong to $\mathcal{T}_1$. The set $\mathcal{T}_1$ is called the terminal set for the problem. Examples of terminal sets, taken from Chapter I, are given in the next paragraph.

In the production planning problem $\mathcal{T}_1$ is the line $t = T$ in the (t,x) plane. In the first version of the chemical engineering problem the set $\mathcal{T}_1$ is the hyperplane $t = T$; that is those points in (t,x)-space with $x = (x^1,\ldots,x^n)$ free and t fixed at T. In the last version of the chemical engineering problem $\mathcal{T}_1$ is the set of points in (t,x)-space whose coordinates x^i are fixed at x_f^i for $i = 1,\ldots,j$ and whose remaining coordinates are free. In some problems it is required that the solution hit a moving target set $G(t)$. That is, at each time t of some interval $[\tau_0,\tau_1]$ there is a set $G(t)$ of points in x-space, and it is required that the solution ϕ hit $G(t)$ at some time t. Stated analytically, we require the existence of a point t_1 in $[\tau_0,\tau_1]$ such that $\phi(t_1)$ belongs to $G(t_1)$. An example of this type of problem is the rendezvous problem of Section 1.4. The set $\mathcal{T}_1$ in the moving target set problem is the set of all points (t,x) with $\tau_0 \leq t \leq \tau_1$ and $x \in G(t)$.

The discussion in the preceding paragraphs is sometimes summarized in less precise but somewhat more graphic language by the statement that the functions u are required to transfer the system from an initial state x_0 at time t_0 to a terminal state x_1 at time t_1, where $(t_0,x_0) \in \mathcal{T}_0$ and $(t_1,x_1) \in \mathcal{T}_1$. Note that to a given u in $\mathcal{C}$ there will generally correspond more than one trajectory ϕ. This results from different choices of initial points (t_0,x_0) in $\mathcal{T}_0$ or from non-uniqueness of solutions of (2.2) if no assumptions are made to guarantee the uniqueness of solutions of (2.2).

It is often further required that a function u in $\mathcal{C}$ and a corresponding solution ϕ must satisfy a system of inequality constraints

$$R^i(t,\phi(t),u(t)) \geq 0 \quad i = 1,2,\ldots,r, \qquad (2.3)$$

for all $t_0 \leq t \leq t_1$, where the functions $R^1,\ldots,R^r$ are given functions of (t,x,z). For example, in the production planning problem discussed in Section 1.2 the constraints can be written as $R^i \geq 0$, $i = 1,2,3$, where $R^1(t,x,z) = x$, $R^2(t,x,z) = z$, and $R^3(t,x,z) = A - z$. In the example of Section 1.5 the constraints can be written as $R^i \geq 0$, $i = 1,2$, where $R^1(t,x,z) = z + A$ and $R^2(t,x,z) = A - z$.

In the examples of Chapter I the control u is to be chosen so that certain functionals are minimized or maximized. These functionals have the following form. Let f^0 be a real valued continuous function of (t,x,z), let g_0 be a real valued function defined on $\mathscr{T}_0$, and let g_1 be a real valued function defined on $\mathscr{T}_1$. For each u in $\mathscr{C}$ and each corresponding solution ϕ of (2.2) define a cost or payoff as follows:

$$J(\phi,u) = g_0(t_0,\phi(t_0)) + g_1(t_1,\phi(t_1)) + \int_{t_0}^{t_1} f^0(s,\phi(s),u(s))ds.$$

If the functional J is to be minimized then a u^* in $\mathscr{C}$ and a corresponding solution ϕ^* of (2.2) are to be found such that $J(\phi^*,u^*) \leq J(\phi,u)$ for all u in $\mathscr{C}$ and corresponding ϕ. In other problems the functional J is to be maximized. Examples of J taken from Chapter I are given in the next paragraph.

In the examples of Chapter I the set $\mathscr{T}_0$ is always a point (t_0,x_0). The differential equations in the examples, except in Section 1.3, are such that the solutions are unique. In Section 1.3 let us assume that the functions G^i are such that the solutions are unique. Thus in these examples the choice of u completely determines the function ϕ. In the economics example $J(\phi,u)$ is the total cost $C(u)$ given by (2.5). The function f^0 is given by (2.4) and the functions g_0 and g_1 are identically zero. In the first chemical engineering example of Section 1.3, $J(\phi,u) = V(p,\theta)$, where $V(p,\theta)$ is given by (3.3). The functions f^0 and g_0 are identically zero. In the minimum fuel problem of Section 1.4, $J(\phi,u) = P(\beta,\omega)$, where P is given by (4.5). Here $f^0 = \beta$ and g_0 and g_1 are identically zero. An equivalent formulation is obtained if one takes $J(\phi,u) = -m_f$. Now $f^0 = 0$, $g_0 = 0$, and $g_1 = -m_f$.

We conclude this section with a discussion of two generalizations that will appear in the mathematical formulation to be given in the next section. The first deals with the initial and terminal data. The initial set $\mathscr{T}_0$ and the terminal set $\mathscr{T}_1$ determine a set $\mathscr{B}$ of points (t_0,x_0,t_1,x_1) in E^{2n+2} as follows:

$$\mathscr{B} = \{(t_0,x_0,t_1,x_1) : (t_0,x_0) \in \mathscr{T}_0, (t_1,x_1) \in \mathscr{T}_1\}. \quad (2.4)$$

Thus a simple generalization of the requirement that $(t_0,\phi(t_0)) \in \mathscr{T}_0$ and $(t_1,\phi(t_1)) \in \mathscr{T}_1$ is the following. Let there be given a set $\mathscr{B}$ of points in E^{2n+2}. It is required of a trajectory ϕ that $(t_0,\phi(t_0),t_1,\phi(t_1))$ belong to $\mathscr{B}$. That is, we now permit possible relationships between initial and terminal data. We shall show later that in some sense this situation is really no more general than the situation in which the initial and terminal data are assumed to be unrelated.

The second generalization deals with the description of the constraints on u. For each (t,x), a system of inequalities $R^i(t,x,z) \geq 0$, $i = 1,\ldots,r$ determines a set $U(t,x)$ in the m-dimensional z-space; namely

$$U(t,x) = \{z: R^i(t,x,z) \geq 0, \quad i = 1,\ldots,r\}.$$

The requirement that a function u and a corresponding trajectory satisfy constraints of the form (2.3) can therefore be written as follows:

$$u(t) \in U(t,\phi(t)) \qquad t_0 \le t \le t_1.$$

Thus the constraint (2.3) is a special case of the following more general constraint condition.

Let Ω be a function that assigns to each point (t,x) of some suitable subset of E^{n+1} a subset of the z-space E^m. Thus:

$$\Omega: \quad (t,x) \to \Omega(t,x),$$

where $\Omega(t,x)$ is a subset of E^m. The constraint (2.3) is replaced by the more general constraint

$$u(t) \in \Omega(t,\phi(t)).$$

3. Mathematical Formulation

The formulation will involve the Lebesgue integral. This is essential in the study of solutions to the problem. The reader who wishes to keep the formulation on a more elementary level can replace 'measurable controls' by 'piecewise continuous controls', replace 'absolutely continuous functions' by 'piecewise $C^{(1)}$ functions', and interpret the solution of equation (3.1) below as we interpreted the solution of equation (2.2) in Section 2 of this chapter.

We establish some notation and terminology. Let t denote a real number, which will sometimes be called time. Let x denote a vector in real euclidean space E^n, $n \ge 1$; thus $x = (x^1,\ldots,x^n)$. The vector x will be called the <u>state variable</u>. We shall use superscripts to denote components of vectors and we shall use subscripts to distinguish among vectors. Let z denote a vector in euclidean m-space E^m, $m \ge 1$; thus $z = (z^1,\ldots,z^m)$. The vector z will be called the <u>control variable</u>. Let $\mathscr{R}$ be a region of (t,x)-space and let $\mathscr{U}$ be a region of z-space, where by a region we mean an open connected set. Let $\mathscr{G} = \mathscr{R} \times \mathscr{U}$, the cartesian product of $\mathscr{R}$ and $\mathscr{U}$. Let $f^0, f^1, \ldots, f^n$ be real valued functions defined on $\mathscr{G}$. We shall write

$$f = (f^1, \ldots, f^n) \qquad \hat{f} = (f^0, f^1, \ldots, f^n).$$

Let $\mathscr{B}$ be a set of points

$$(t_0, x_0, t_1, x_1) = (t_0, x_0^1, \ldots, x_0^n, \; t_1, x^1, \ldots, x_1^n)$$

in E^{2n+2} with $t_1 \geq t_0 + \delta$, for some fixed $\delta > 0$. The set $\mathscr{B}$ will be said to define the <u>end conditions</u> for the problem.

Let Ω be a mapping that assigns to each point (t,x) in $\mathscr{R}$ a subset $\Omega(t,x)$ of the region $\mathscr{U}$ in z-space. The mapping Ω will be said to define the <u>control constraints</u>. If $\Omega(t,x) = \mathscr{U}$ for all (t,x) in $\mathscr{R}$, then we say that there are no control constraints.

Henceforth we shall usually use vector-matrix notation. The system of differential equations (2.2) will be written simply as

$$\frac{dx}{dt} = f(t,x,u(t)),$$

where we follow the usual convention in the theory of differential equations and take dx/dt and $f(t,x,u(t))$ to be column vectors. We shall not distinguish between a vector and its transpose if it is clear whether a vector is a row vector or a column vector or if it is immaterial whether the vector is a row vector or a column vector. The inner product of two vectors u and v will be written as $\langle u,v \rangle$. We shall use the symbol $|x|$ to denote the ordinary euclidean norm of a vector. Thus

$$|x| = \left(\sum_{i=1}^{n} |x^i|^2\right)^{1/2} = \langle x,x \rangle^{1/2}.$$

If A and B are matrices, then we write their product as AB.

If $f = (f^1, \ldots, f^n)$ is a vector valued function from a set Δ

in some euclidean space to the euclidean space E^n such that each of the real valued functions $f^1,\ldots,f^n$ is continuous (or $C^{(k)}$, or measurable, etc.) then we shall say that f is continuous (or $C^{(k)}$, or measurable, etc.) on the set Δ. Similarly if a matrix A has entries that are continuous functions (or $C^{(k)}$, or measurable functions, etc.) defined on a set Δ in some euclidean space then we shall say that A is continuous (or $C^{(k)}$, or measurable, etc.) on Δ.

DEFINITION 3.1. A measurable function u defined on an interval $[t_0,t_1]$ with range in $\mathscr{U}$ is said to be a control on $[t_0,t_1]$ if there exists an absolutely continuous function ϕ defined on $[t_0,t_1]$ with range in E^n such that:

(i) $(t,\phi(t)) \in \mathscr{R}$ for all t in $[t_0,t_1]$

(ii) ϕ is a solution of the system of differential equations

$$\frac{dx}{dt} = f(t,x,u(t)); \tag{3.1}$$

that is,

$$\phi'(t) = f(t,\phi(t),u(t)) \quad \text{a.e. on} \quad [t_0,t_1].$$

The function ϕ is called a trajectory corresponding to u. The point $(t_0,\phi(t_0))$ will be called the initial point of the trajectory and the point $(t_1,\phi(t_1))$ will be called the terminal point of the trajectory. The point $(t_0,\phi(t_0),t_1,\phi(t_1))$ will be called the end point of the trajectory.

Note that since ϕ is absolutely continuous, it is the integral of its derivative. Hence (ii) contains the statement that the function $t \to f(t,\phi(t),u(t))$ is Lebesgue integrable on $[t_0,t_1]$.

The system of differential equations (3.1) will be called the state equations.

We emphasize the following point about our notation. We are using the letter z to denote a point of $\mathscr{U}$; we are using the letter u to denote a function with range in $\mathscr{U}$.

DEFINITION 3.2. A control u is said to be an admissible control if there exists a trajectory ϕ corresponding to u such that

(i) $t \to f^0(t,\phi(t),u(t))$ is in $L_1[t_0,t_1]$

(ii) $u(t) \in \Omega(t,\phi(t))$ a.e. on $[t_0,t_1]$

(iii) $(t_0,\phi(t_0),t_1,\phi(t_1)) \in \mathscr{B}$.

A trajectory corresponding to an admissible control as in Definition 3.2 will be called an admissible trajectory.

DEFINITION 3.3. A pair of functions (ϕ,u) such that u is an admissible control and ϕ is an admissible trajectory corresponding to u will be called an admissible pair.

Note that to a given admissible control there may correspond more than one admissible trajectory as a result of different choices of permissible end points. Also, even if we fix the endpoint, there may be several trajectories corresponding to a given control because we do not require uniqueness of solutions of (3.1) for given initial conditions.

We now state the control problem.

PROBLEM 1. Let $\mathscr{A}$ denote the set of all admissible pairs (ϕ,u) and let $\mathscr{A}$ be non-empty. Let

$$J(\phi,u) = g(t_0,\phi(t_0),t_1,\phi(t_1)) + \int_{t_0}^{t_1} f^0(t,\phi(t),u(t))dt, \qquad (3.2)$$

where (ϕ,u) is an admissible pair and g is a given real valued function defined on $\mathscr{B}$. Let $\mathscr{A}_1$ be a non-empty subset of $\mathscr{A}$. Find a pair (ϕ^*,u^*) in $\mathscr{A}_1$ that minimizes (3.2) in the class $\mathscr{A}_1$. That is, find an element (ϕ^*,u^*) in $\mathscr{A}_1$ such that

$$J(\phi^*,u^*) \le J(\phi,u) \quad \text{for all} \quad (\phi,u) \quad \text{in } \mathscr{A}_1.$$

The precise formulation of Problem 1 is rather lengthy. Therefore the following statement, which gives the essential data of the problem, is often used to mean that we are considering Problem 1. Minimize (3.2) subject to the state equations (3.1), the end conditions $\mathscr{B}$, and the control constraints Ω.

We have stated Problem 1 as a minimization problem. In some applications it is required that the functional J be maximized. There is, however, no need to consider maximum problems separately since the problem of maximizing J is equivalent to the problem of minimizing $-J$. Hence we shall confine our attention to minimum problems.

DEFINITION 3.4. A pair (ϕ^*,u^*) that solves Problem 1 is called an optimal pair. The trajectory ϕ^* is called an optimal trajectory and the control u^* is called an optimal control.

The first term on the right in (3.2) is the function g evaluated at the end points of an admissible trajectory. Thus it assigns a real number to every admissible trajectory and is thus a functional G_1 defined on the admissible trajectories. The functional G_1 is defined by the formula

$$G_1(\phi) = g(t_0,\phi(t_0),t_1,\phi(t_1)).$$

Other examples of functionals defined on admissible trajectories are

$$G_2(\phi) = \max\{|\phi(t)|: \quad t_0 \leq t \leq t_1\}$$

and

$$G_3(\phi) = \max\{|\phi(t)-h(t)|: \quad t_0 \leq t \leq t_1\},$$

where h is a given continuous function defined on an interval I containing all the intervals $[t_0,t_1]$ of definition of admissible trajectories. The functionals G_2 and G_3 arise in problems in which in addition to minimizing (3.2) it is also desired to keep the state of the system close to some preassigned state.

The preceding discussion justifies the consideration of the following generalization of Problem 1.

PROBLEM 2. Let everything be as in Problem 1, except that (3.2) is replaced by

$$\hat{J}(\phi,u) = G(\phi) + \int_{t_0}^{t_1} f^0(t,\phi(t),u(t))dt, \qquad (3.3)$$

where G is a functional defined on the admissible trajectories. Find a pair (ϕ^*,u^*) in $\mathscr{A}_1$ that minimizes (3.3) in the class $\mathscr{A}_1$.

4. Equivalent Formulations

Certain special cases of Problem 1 are actually equivalent to Problem 1 in the sense that Problem 1 can be formally transformed into the special case in question. This information is useful in certain investigations where it is more convenient to study one of the special cases than to study Problem 1. The reader is warned that in making the transformation to the special case some of the properties of the original problem, such as linearity, continuity, convexity, etc., may be altered. In any particular investigation one must check that the pertinent hypotheses made for the original problem are valid for the transformed problem.

Special cases of Problem 1 are obtained by taking $f^0 = 0$ or $g = 0$. In keeping with the terminology for related problems in the calculus of variations we shall call a problem in which $f^0 = 0$ a Mayer problem and we shall call a problem in which $g = 0$ a Lagrange problem. Problem 1 of Section 3 is sometimes called a Bolza problem, also as in the calculus of variations. We shall show that the Mayer formulation and the Lagrange formulation are as general as the Bolza formulation by showing that Problem 1 can be written either as a

Mayer problem or as a Lagrange problem.

We formulate Problem 1 as a Mayer problem in a higher dimensional euclidean space. Let $\hat{x} = (x^0, x) = (x^0, x^1, \ldots, x^n)$. Let $\hat{\mathscr{R}} = E^1 \times \mathscr{R}$ and let $\hat{\mathscr{G}} = \hat{\mathscr{R}} \times \mathscr{U}$. The functions $f^0, f^1, \ldots, f^n$ are defined on $\hat{\mathscr{G}}$ since they are defined on $\mathscr{G}$ and they are independent of x^0. Let the mapping $\hat{\Omega}$ be defined on $\hat{\mathscr{R}}$ by the equation $\hat{\Omega}(t, \hat{x}) = \Omega(t, x)$. Let

$$\hat{\mathscr{B}} = \{(t_0, \hat{x}_0, t_1, \hat{x}_1) : (t_0, x_0, t_1, x_1) \in \mathscr{B},\ x_0^0 = 0\}.$$

Let (ϕ, u) be an admissible pair for Problem 1. Let $\hat{\phi} = (\phi^0, \phi)$, where ϕ^0 is an absolutely continuous function such that

$$\phi^{0\prime}(t) = f^0(t, \phi(t), u(t)) \qquad \phi^0(t_0) = 0$$

for almost every t in $[t_0, t_1]$. By virtue of (i) of Definition 3.2 such a function ϕ^0 exists and is given by

$$\phi^0(t) = \int_{t_0}^{t} f^0(s, \phi(s), u(s))\,ds.$$

Then $(\hat{\phi}, u)$ is an admissible pair for a problem in which $\mathscr{R}, \mathscr{G}, \Omega, \mathscr{B}$, are replaced by $\hat{\mathscr{R}}, \hat{\mathscr{G}}, \hat{\Omega}, \hat{\mathscr{B}}$, respectively and in which the system of state equations (3.1) is replaced by

$$\begin{aligned} \frac{dx^0}{dt} &= f^0(t, x, u(t)) \\ \frac{dx}{dt} &= f(t, x, u(t)). \end{aligned} \tag{4.1}$$

Equations (4.1) can be written as

$$\frac{d\hat{x}}{dt} = \hat{f}(t, x, u(t))$$

if we set $\hat{f} = (f^0, f)$. Conversely, to every admissible pair $(\hat{\phi}, u)$ for a problem involving $\hat{\mathscr{R}}, \hat{\mathscr{G}}, \hat{\Omega}, \hat{\mathscr{B}}$ and (4.1) there corresponds the admissible pair (ϕ, u) for Problem 1, where ϕ consists of the last n-components of $\hat{\phi}$. Let

$$\hat{g}(t_0,\hat{x}_0,t_1,\hat{x}_1) = g(t_0,x_0,t_1,x_1) + x_1^0$$

and let

$$\hat{J}(\hat{\phi},u) = \hat{g}(t_0,\hat{\phi}(t_0),t_1,\hat{\phi}(t_1)).$$

Then $\hat{J}(\hat{\phi},u) = J(\phi,u)$, where $\hat{\phi} = (\phi^0,\phi)$. Hence the Mayer problem of minimizing $\hat{J}$ subject to state equations (4.1), control constraints $\hat{\Omega}$ and end conditions $\hat{\mathscr{B}}$ is equivalent to Problem 1.

We now show that Problem 1 can be formulated as a Lagrange problem. Let $\hat{x},\hat{\mathscr{R}},\hat{\mathscr{G}},\hat{\Omega}$ be as in the previous paragraph. Let

$$\hat{\mathscr{B}} = \{(t_0,\hat{x}_0,t_1,\hat{x}_1): (t_0,x_0,t_1,x_1) \in \mathscr{B}, \quad x_0^0 = g(t_0,x_0,t_1,x_1)/(t_1 - t_0)\}. \tag{4.2}$$

(Recall that for all points in $\mathscr{B}$ we have $t_1 > t_0$.) Let (ϕ,u) be an admissible pair for Problem 1 and let $\hat{\phi} = (\phi^0,\phi)$ where $\phi^0(t) \equiv g(t_0,x_0,t_1,x_1)/(t_1 - t_0)$. Then $(\hat{\phi},u)$ is an admissible pair for a problem in which $\mathscr{R},\mathscr{G},\Omega,\mathscr{B}$ are replaced by roofed quantities with $\hat{\mathscr{B}}$ as in (4.2) and with state equations

$$\frac{dx^0}{dt} = 0 \qquad \frac{dx}{dt} = f(t,x,u(t)). \tag{4.3}$$

Conversely, to every admissible pair $(\hat{\phi},u)$ for the problem with roofed quantities there corresponds the admissible pair (ϕ,u) for Problem 1, where ϕ consists of the last n components of $\hat{\phi}$. If we replace f^0 of Problem 1 by $f^0 + x^0$ and let

$$\hat{J}(\hat{\phi},u) = \int_{t_0}^{t_1} (f^0(t,\phi(t),u(t)) + \phi^0(t))dt \tag{4.4}$$

then $\hat{J}(\hat{\phi},u) = J(\phi,u)$. Hence the Lagrange problem of minimizing (4.4) subject to state equations (4.3), control constraints $\hat{\Omega}$, and end conditions $\hat{\mathscr{B}}$ is equivalent to Problem 1.

In Problem 1 the initial time t_0 and the terminal time t_1 need not be fixed. We now show that Problem 1 can be written as a problem with fixed initial time and fixed terminal time. The device for reducing the variable initial and terminal time problem to a fixed initial and terminal time problem is the change in time parameter

$$t = t_0 + s(t_1 - t_0) \qquad 0 \leq s \leq 1$$

and the introduction of new state variables as follows.

Let w be a scalar and consider the problem with state variables (t,x,w), where x is an n-vector and t is a scalar. Let s denote the time variable. Let the state equations be

$$\begin{aligned} &\frac{dt}{ds} = w \qquad \frac{dw}{ds} = 0 \\ &\frac{dx}{ds} = f(t,x,\bar{u}(s))w \end{aligned} \tag{4.5}$$

where $\bar{u}$ is the control and f is as in Problem 1. Let

$$\begin{aligned} \bar{\mathscr{B}} = \{ &(s_0,t_0,x_0,w_0,s_1,t_1,x_1,w_1): s_0 = 0, \quad s_1 = 1, \\ &(t_0,x_0,t_1,x_1) \in \mathscr{B},\ w_0 = t_1 - t_0 \}. \end{aligned} \tag{4.6}$$

Note that the initial and terminal times are now fixed. Let $\bar{\Omega}(s,t,x,w) = \Omega(t,x)$. Let $\bar{\phi} = (\tau,\xi,\omega)$ be a solution of (4.5) corresponding to a control $\bar{u}$, where the Greek-Latin correspondence between (τ,ξ,ω) and (t,x,w) indicates the correspondence between components of $\bar{\phi}$ and the system (4.5). Let

$$\bar{J}(\bar{\phi},\bar{u}) = g(\tau(0),\xi(0),\tau(1),\xi(1)) + \int_0^1 f^0(\tau(s),\xi(s),\bar{u}(s))\omega(s)ds. \tag{4.7}$$

Consider the fixed end-time problem of minimizing (4.7) subject to the state equations (4.5), the control constraints $\bar{\Omega}$ and the end conditions $\bar{\mathscr{B}}$.

Since $t_1 - t_0 > 0$ it follows that for any solution of (4.5) satisfying (4.6) we have $\omega(s) = t_1 - t_0$, a positive constant, for $0 \le s \le 1$. Let (ϕ, u) be an admissible pair for Problem 1. It is readily verified that if

$$\tau(s) = t_0 + s(t_1 - t_0) \qquad \xi(s) = \phi(t_0 + s(t_1 - t_0))$$
$$\bar{u}(s) = u(t_0 + s(t_1 - t_0)) \qquad \omega(s) = t_1 - t_0,$$

then $(\bar{\phi}, \bar{u}) = (\tau, \xi, \omega, \bar{u})$ is an admissible pair for the fixed end-time problem and $\bar{J}(\bar{\phi}, \bar{u}) = J(\phi, u)$. Conversely, let $(\bar{\phi}, \bar{u})$ be an admissible pair for the fixed end-time problem. If we set

$$\phi(t) = \xi((t-t_0)/(t_1-t_0)) \qquad u(t) = \bar{u}((t-t_0)/(t_1-t_0)),$$
$$t_0 \le t \le t_1,$$

then since $\tau(s) = t_0 + s(t_1 - t_0)$, we have $t = \tau(s)$ for $0 \le s \le 1$. It is readily verified that (ϕ, u) is admissible for Problem 1 and that $J(\phi, u) = \bar{J}(\bar{\phi}, \bar{u})$. Hence Problem 1 is equivalent to a fixed end-time problem.

The following observation will be useful in the sequel. Since for any admissible solution of the fixed time problem we have $\omega(s) = t_1 - t_0 > 0$ we can take the set $\bar{\mathscr{R}}$ for the fixed end-time problem to be $[0,1] \times \mathscr{R} \times E^+$, where $E^+ = \{w: w > 0\}$.

A special case of the end conditions occurs if the initial and terminal data are separated. In this event a set $\mathscr{T}_0$ of points (t_0, x_0) in E^{n+1} and a set $\mathscr{T}_1$ of points (t_1, x_1) in E^{n+1} are given and an admissible trajectory is required to satisfy the conditions

$$(t_i, \phi(t_i)) \in \mathscr{T}_i \qquad i = 0,1. \tag{4.8}$$

The set $\mathscr{B}$ in this case is given by (2.4). We shall show that the apparently more general requirement (iii) of Definition 3.2 can be reduced to the form (4.8) by embedding the problem in a space of higher dimension as follows.

Let $y = (y^1, \dots, y^n)$ and let y^0 be a scalar. Let $\hat{y} = (y^0, y)$. Let the sets $\mathscr{R}$ and $\mathscr{G}$ of Problem 1 be replaced by sets $\tilde{\mathscr{R}} = \mathscr{R} \times E^{n+1}$ and $\tilde{\mathscr{G}} = \tilde{\mathscr{R}} \times \mathscr{U}$. Then the vector function $\hat{f} = (f^0, f)$ is defined on $\tilde{\mathscr{G}}$ since it is independent of $\hat{y}$. Let $\tilde{\Omega}(t,x,\hat{y}) = \Omega(t,x)$. Let the state equations be

$$\frac{dx}{dt} = f(t,x,u(t))$$
$$\frac{d\hat{y}}{dt} = 0. \tag{4.9}$$

Let

$$\tilde{\mathscr{T}}_0 = \{(t_0, x_0, y_0^0, y_0) : (t_0, x_0, y_0^0, y_0) \in \mathscr{B}\}$$

$$\tilde{\mathscr{T}}_1 = \{(t_1, x_1, y_1^0, y_1) : y_1^0 = t_1, \ y_1^i = x_1^i, \quad i = 1, \dots, n\}$$

Replace condition (iii) of Definition 3.2 by the condition

$$(t_i, \tilde{\phi}(t_i)) \in \tilde{\mathscr{T}}_i \qquad i = 0,1, \tag{4.10}$$

where $\tilde{\phi}$ is a solution of (4.9). Then it is easily seen that a function u is an admissible control for Problem 1 if and only if it is an admissible control for the system (4.9) subject to control constraints $\tilde{\Omega}$ and end-condition (4.10). Moreover, the admissible trajectories $\tilde{\phi}$ are of the form $\tilde{\phi} = (\phi, t_1, x_1)$. Hence if we take the cost functional to be $\tilde{J}$, where

$$\tilde{J}(\tilde{\phi}, u) = J(\phi, u),$$

then Problem 1 is equivalent to a problem with end conditions of the form (4.8).

5. Isoperimetric Problems and Parameter Optimization

In some control problems, in addition to the usual constraints there exists constraints of the form

$$\int_{t_0}^{t_1} h^i(t,\phi(t),u(t))dt \leq c^i \qquad i = 1,\ldots,q$$
$$\int_{t_0}^{t_1} h^i(t,\phi(t),u(t))dt = c^i \qquad i = q+1,\ldots,p, \tag{5.1}$$

where the functions h^i are defined on $\mathscr{G}$ and the constants c^i are prescribed. Constraints of the form (5.1) are called isoperimetric constraints. A problem with isoperimetric constraints can be reduced to a problem without isoperimetric constraints as follows.

Introduce additional state variables $x^{n+1},\ldots,x^{n+p}$ and let $\tilde{x}$ denote a vector in E^{n+p}. Thus $\tilde{x} = (x,\bar{x})$, where $\bar{x} = (x^{n+1},\ldots,x^{n+p})$. Let the state equations be

$$\frac{dx^i}{dt} = f^i(t,x,u(t)) \qquad i = 1,\ldots,n$$
$$\frac{dx^{n+i}}{dt} = h^i(t,x,u(t)) \qquad i = 1,\ldots,p \tag{5.2}$$

or

$$\frac{d\tilde{x}}{dt} = \tilde{f}(t,x,u(t)),$$

where $\tilde{f} = (f,h)$. Let the control constraints be given by the mapping $\tilde{\Omega}$ defined by the equation $\tilde{\Omega}(t,\tilde{x}) = \Omega(t,x)$. Let the end conditions be given by the set $\tilde{\mathscr{B}}$ consisting of all points $(t_0,\tilde{x}_0,t_1,\tilde{x}_1)$ such that: (i) $(t_0,x_0,t_1,x_1) \in \mathscr{B}$; (ii) $x_0{}^i = 0$, $i = n+1,\ldots,n+p$; (iii) $x_1{}^i \leq c^i$, $i = n+1,\ldots,n+q$; and (iv) $x_1{}^i = c^i$, $i = n+q+1,\ldots,n+p$. For the system with state variable $\tilde{x}$, let $\mathscr{R}$ be replaced by $\tilde{\mathscr{R}} = \mathscr{R} \times E^p$ and let $\mathscr{G}$ be replaced by $\tilde{\mathscr{G}} = \tilde{\mathscr{R}} \times \mathscr{U}$.

Let (ϕ,u) be an admissible pair for Problem 1 such that the constraints (5.1) are satisfied. Let $\tilde{\phi} = (\phi,\bar{\phi})$, where

$$\bar{\phi}(t) = \int_0^t h(s,\phi(s),u(s))ds \qquad \bar{\phi}(0) = 0.$$

Then $(\tilde{\phi},u)$ is an admissible pair for the system with state variable

$\tilde{x}$. Conversely, if $(\tilde{\phi},u)$ is admissible for the $\tilde{x}$ system then (ϕ,u), where ϕ consists of the first n components of $\tilde{\phi}$, is admissible for Problem 1 and satisfies the isoperimetric constraints. Hence by taking the cost functional for the problem in $\tilde{x}$-space to be $\tilde{J}$, where $\tilde{J}(\tilde{\phi},u) = J(\phi,u)$, we can write the problem with constraints (5.1) as an equivalent problem in the format of Problem 1.

In Problem 1 the functions $f^0,f^1,\ldots,f^n$ defining the cost functional and the system of differential equations (3.1) are regarded as being fixed. In some applications these functions are dependent upon a parameter vector $w = (w^1,\ldots,w^k)$, which is at our disposal. For example, in the rocket problem of Section 1.4 we may be able to vary the effective exhaust velocity over some range $c_0 \leq c \leq c_1$ by proper design changes. The system of differential equations (3.1) will now read

$$\frac{dx}{dt} = f(t,x,w,u(t)) \qquad w \in W ,$$

where W is some preassigned set in E^k. For a given choice of control u a corresponding trajectory ϕ will in general now depend on the choice of parameter value w. Hence, so will the value $J(\phi,u,w)$ of the cost functional. The problem now is to choose a parameter value w^* in W for which there exists an admissible pair (ϕ^*,u^*) such that $J(\phi^*,u^*,w^*) \leq J(\phi,u,w)$ for all w in W and corresponding admissible pairs (ϕ,u).

The problem just posed can be reformulated in the format of Problem 1 in $(n+k+1)$-dimensional space as follows. Introduce new state variables $w = (w^1,\ldots,w^k)$ and consider the system

$$\begin{aligned} \frac{dx^i}{dt} &= f^i(t,x,w,u(t)) \qquad & i = 1,\ldots,n \\ \frac{dw^i}{dt} &= 0 \qquad & i = 1,\ldots,k . \end{aligned} \tag{5.3}$$

Let $\tilde{x} = (x,w)$, let $\tilde{\mathscr{R}} = \mathscr{R} \times E^k$, let $\tilde{\mathscr{G}} = \tilde{\mathscr{R}} \times \mathscr{U}$, and let $\tilde{\Omega}(t,x,w) = \Omega(t,x)$. Let the end conditions be given by

$$\tilde{\mathscr{B}} = \{(t_0,x_0,w_0,t_1,x_1,w_1): \quad (t_0,x_0,t_1,x_1) \in \mathscr{B},\ w_0 \in W\}$$

Let $\tilde{J}(\tilde{\phi},u) = J(\phi,w,u)$. It is readily verified that the problem of minimizing J subject to (5.3), the control constraints $\tilde{\Omega}$ and end conditions $\tilde{\mathscr{B}}$ is equivalent to the problem involving the optimization of parameters.

6. Relationship with the Calculus of Variations

The brachistochrone problem formulated in Section 1.6 is an example of the simple problem in the calculus of variations, which can be stated as follows. Let t be a scalar, let x be a vector in E^n and let x' be a vector in E^n. Let $\mathscr{G}$ be a region in (t,x,x')-space. Let f^0 be a real valued function defined on $\mathscr{G}$. Let $\mathscr{B}$ be a given set of points (t_0,x_0,t_1,x_1) in E^{2n+2} and let g be a real valued function defined on $\mathscr{B}$. An admissible trajectory is defined to be an absolutely continuous function ϕ defined on an interval $[t_0,t_1]$ such that:

$$\begin{aligned}
&\text{(i)} \quad (t,\phi(t),\phi'(t)) \in \mathscr{G} \quad \text{for almost all } t \text{ in } [t_0,t_1] \\
&\text{(ii)} \quad (t_0,\phi(t_0),t_1,\phi(t_1)) \in \mathscr{B} \qquad\qquad (6.1)\\
&\text{(iii)} \quad t \to f^0(t,\phi(t),\phi'(t)) \text{ is integrable on } [t_0,t_1].
\end{aligned}$$

The problem is to find an admissible arc that minimizes

$$g(t_0,\phi(t_0),t_1,\phi(t_1)) + \int_{t_0}^{t_1} f^0(t,\phi(t),\phi'(t))dt.$$

As with the brachistochrone problem, the general simple problem in the calculus of variations can be written as a control problem by relabelling x' as z; i.e. we set $u = \phi'$. (Recall that z denotes the control variable and u denotes the control function.) The simple problem in the calculus of variations becomes the following control problem. Minimize

$$g(t_0,\phi(t_0),t_1,\phi(t_1)) + \int_{t_0}^{t_1} f^0(t,\phi(t),u(t))dt$$

subject to the state equations

$$\frac{dx^i}{dt} = u^i(t) \qquad i = 1,\ldots,n,$$

end conditions (ii) of (6.1) and control constraints Ω, where

$$\Omega(t,x) = \{z: (t,x,z) \in \mathscr{G}\}.$$

The problem of Bolza in the calculus of variations differs from the simple problem in that in addition to (6.1) an admissible arc is required to satisfy a system of differential equations

$$F^i(t,\phi(t),\phi'(t)) = 0 \qquad i = 1,\ldots,\mu. \qquad (6.2)$$

The functions $F^1,\ldots,F^\mu$ are defined and continuous on $\mathscr{G}$ and $\mu < n$.

In the development of the necessary conditions in the theory of the problem of Bolza the assumption is usually made that the functions f^0 and $F = (F^1,\ldots,F^\mu)$ are of class $C^{(1)}$ on the region $\mathscr{G}$ of (t,x,x')-space and the matrix of partial derivatives $F_{x'} = (\partial F^i/\partial x'^j)$, $i = 1,\ldots,\mu$, $j = 1,\ldots,n$, has rank μ everywhere on $\mathscr{G}$. Hence in the neighborhood of any point (t_2,x_2,x_2') at which

$$F^i(t_2,x_2,x_2') = 0 \qquad i = 1,\ldots,\mu \qquad (6.3)$$

holds we can solve for μ components of x' in terms of t,x and the remaining $n-\mu$ components of x'. Moreover these μ components of x' will be $C^{(1)}$ functions of their arguments. Let us now suppose that we can solve (6.3) globally in this fashion. Since we can relabel components we can assume that we solve the first μ components in terms of the remaining $n-\mu$, and get

$$x'^i = G^i(t,x,\tilde{x}') \qquad i = 1,\ldots,\mu,$$

where $\tilde{x}' = (x'^{\mu+1},\ldots,x'^n)$. Thus equation (6.2) is equivalent to

$$\phi'^i(t) = G^i(t,\phi(t),\tilde{\phi}'(t)) \qquad i = 1,\ldots,\mu,$$

where $\tilde{\phi}' = (d\phi^{\mu+1}/dt,\ldots,d\phi^n/dt)$. Let $m = n-\mu$ and let $z = (z^1,\ldots,z^m) = (x'^{\mu+1},\ldots,x'^n) = \tilde{x}'$. It then follows that under the assumptions made here that the Bolza problem is equivalent to the following control problem with control variable $z = (z^1,\ldots,z^m)$.

The functional to be minimized is defined by the equation

$$\bar{J}(\phi,u) = g(t_0,\phi(t_0),t_1,\phi(t_1)) + \int_{t_0}^{t_1} \bar{f}^0(t,\phi(t),u(t))dt$$

where

$$\bar{f}^0(t,x,z) = f^0(t,x,G^1(t,x,z),\ldots,G^\mu(t,x,z),z^1,\ldots,z^m).$$

The system equations are

$$\frac{dx^i}{dt} = G^i(t,x,u(t)) \qquad i = 1,\ldots,\mu$$

$$\frac{dx^{\mu+i}}{dt} = u^i(t) \qquad i = 1,\ldots,m.$$

The end conditions are defined by the set $\mathscr{B}$ of the Bolza problem and the control constraints Ω are defined as follows:

$$\Omega(t,x) = \{z:(t,x,G^1(t,x,z),\ldots,G^\mu(t,x,z),z^1,\ldots,z^m) \in \mathscr{G}\}.$$

It is, of course, also required of an admissible (ϕ,u) that the mapping $t \to \bar{f}^0(t,\phi(t),u(t))$ be integrable.

Conversely, under certain conditions the control problem can be written as a problem of Bolza in the calculus of variations. Let us first suppose that $\Omega(t,x) = E^m$ for all (t,x). That is, there are no constraints on the control. We introduce new coordinates $y^1,\dots,y^m$ and let

$$\frac{dy^i}{dt} = u^i(t) \qquad i = 1,\dots,m.$$

Then equation (3.1) can be written as

$$\frac{dx^i}{dt} - f^i(t,x,\frac{dy}{dt}) = 0 \qquad i = 1,\dots,n.$$

If we set

$$F^i(t,x,y,x',y') = x'^i - f^i(t,x,y') \qquad i = 1,\dots,n, \tag{6.4}$$

then the control problem can be written as the following problem of Bolza in $(n+m+1)$-dimensional (t,x,y)-space. The class of admissible arcs is the set of absolutely continuous functions $\hat{\phi} = (\phi,\eta) = (\phi^1,\dots,\phi^n, \eta^1,\dots,\eta^m)$ defined on intervals $[t_0,t_1]$ such that: (i) $(t,\phi(t),\eta'(t))$ is in the domain of definition of the function $\tilde{f} = (f^0,f)$; (ii) $(t_0,\phi(t_0),t_1,\phi(t_1))$ is in $\mathscr{B}$ and $\eta(t_0) = 0$; (iii) the function $t \to f^0(t,\phi(t),\eta'(t))$ is integrable; and (iv)

$$F(t,\hat{\phi}(t),\hat{\phi}'(t)) = \phi'(t) - f(t,\phi(t),\eta'(t)) = 0 \qquad \text{a.e. on } [t_0,t_1], \tag{6.5}$$

where $F = (F^1,\dots,F^n)$. The problem is to minimize the functional

$$g(t_0,\phi(t_0),t_1,\phi(t_1)) + \int_{t_0}^{t_1} f^0(t,\phi(t),\eta'(t))dt$$

in the class of admissible arcs.

It is clear from (6.4) that a typical row of the $n \times (n+m)$ matrix $(F_{x'}\ F_{y'})$ for the Bolza problem obtained from the control problem has the form

$$(0\dots0\ 1\ 0\dots0\ \ \partial f^i/\partial y'^1 \dots \partial f^i/\partial y'^m)$$

where the entry 1 occurs in the i-th column and all other entries in the first n columns are zero. Thus the $n \times (n+m)$ matrix $(F_{x'}, F_{y'})$ has rank n as usually required in the theory of the necessary conditions for the Bolza problem.

Let us now suppose that control constraints Ω are present and that the sets $\Omega(t,x)$ are defined by a system of inequalities. We suppose that there are r functions $R^1,\ldots,R^r$ of class $C^{(1)}$ on $\mathscr{G}$. The set $\Omega(t,x)$ is defined as follows:

$$\Omega(t,x) = \{z\colon R^i(t,x,z) \geq 0, \quad i = 1,\ldots,r\}.$$

We impose a further restriction, which we call the <u>constraint qualification</u>.

(i) If m, the number of components of z, is less than or equal to r then at any point (t,x,z) of $\mathscr{G}$ at most m of the functions $R^1,\ldots,R^r$ can vanish at that point.

(ii) At the point (t,x,z) let $i_1,\ldots,i_\rho$ denote the set of indices such that $R^i(t,x,z) = 0$. Let $R_{z,\rho}(t,x,z)$ denote the matrix formed by taking the rows $i_1,\ldots,i_\rho$ of the matrix $R_z(t,x,z) = \left(\frac{\partial R^i(t,x,z)}{\partial z^j}\right)$ $\begin{matrix} i = 1,\ldots r \\ j = 1,\ldots m. \end{matrix}$ Then $R_{z,\rho}(t,x,z)$ has rank ρ. (6.6)

To formulate the control problem as a Bolza problem we proceed as before and let $y' = z$. The constraints take the form $R^i(t,\phi(t),\eta'(t)) \geq 0$. This restriction is not present in the classical Bolza formulation. We can, however, write the variational problem with constraints as a Bolza problem by introducing a new variable $w = (w^1,\ldots,w^r)$ and r additional state equations

$$R^i(t,x,y') - (w'^i)^2 = 0 \quad i = 1,\ldots,r. \tag{6.7}$$

The Bolza problem now is to minimize

$$g(t_0,\phi(t_0),t_1,\phi(t_1)) + \int_{t_0}^{t_1} f^0(t,\phi(t),\eta'(t))dt$$

subject to the differential equations (6.5) and (6.7) and the end conditions

$$(t_0,\phi(t_0),t_1,\phi(t_1)) \varepsilon \mathscr{B} \qquad \eta(t_0) = 0 \qquad \omega(t_0) = 0,$$

where the function ω is the component of the admissible arc corresponding to the variable w.

Let

$$\begin{aligned} F^i(t,x,y,w,x',y',w') &= x'^i - f^i(t,x,y') \quad i = 1,\ldots,n, \\ F^{n+i}(t,x,y,w,x',y',w') &= R^i(t,x,y') - (w'^i)^2 \quad i = 1,\ldots,r. \end{aligned} \tag{6.8}$$

We shall show that the $(n+r) \times (n+m+r)$ matrix

$$M = \left(\frac{\partial F^q}{\partial x'^j} \frac{\partial F^q}{\partial y'^k} \frac{\partial F^q}{\partial w'^s}\right) = (F_{x'} \; F_{y'} \; F_{w'})$$

has rank $n+r$ as usually required in the theory of the Bolza problem. It is a straightforward calculation using (6.4) and (6.8) to see that

$$M = \begin{pmatrix} I & -f_{y'} & 0_1 \\ 0_2 & R_{y'} & -W \end{pmatrix}$$

where I is the $n \times n$ identity matrix, $f_{y'}$ is the $n \times m$ matrix with typical entry $\partial f^i/\partial y'^k$, 0_1 is an $n \times r$ zero matrix, 0_2 is an $r \times n$ zero matrix, $R_{y'}$ is the $r \times m$ matrix with typical entry $\partial R^i/\partial y'^k$ and W is an $r \times r$ diagonal matrix with diagonal entries $2w'^i$.

From the form of the matrix M it is clear that to prove that it has rank $n+r$ it suffices to show that the $r \times (m+r)$ matrix $(R_{y'} - W)$ has rank r. To do this let us suppose that the indexing is such that the indices $i_1,\ldots,i_\rho$ for which $R^{i_j}(t,x,y') = 0$ are the indices $1,\ldots,\rho$. Let $(R_{y'})_\rho$ denote the submatrix of $R_{y'}$ consisting of the first ρ rows of $R_{y'}$ and let $(R_{y'})_{r-\rho}$ denote the submatrix of $R_{y'}$ consisting of the remaining rows. Thus if $i > \rho$, then $R^i(t,x,y') > 0$; if $i \leq \rho$, then $R^i(t,x,y') = 0$. Hence since

$$(w'^i)^2 = R^i(t,x,y') \qquad i = 1,\ldots,r,$$

it follows that

$$2w'^i = 0 \quad \text{if} \quad i \leq \rho \qquad 2w'^i \neq 0 \quad \text{if} \quad i > \rho.$$

Hence

$$(R_{y'} - W) = \begin{pmatrix} (R_{y'})_\rho & 0_3 & 0_4 \\ (R_{y'})_{r-\rho} & 0_5 & D \end{pmatrix},$$

where D is a diagonal matrix of dimension $(r-\rho) \times (r-\rho)$ with non zero entries $2w'^i$, $i > \rho$, and where 0_3, 0_4 and 0_5 are zero matrices. By the constraint qualification (6.6) the matrix $(R_{y'})_\rho$ has rank ρ. Since D has rank $r-\rho$ it follows that $(R_{y'} - W)$ has rank r, as required.

CHAPTER III

EXISTENCE THEOREMS WITH CONVEXITY ASSUMPTIONS

1. Introduction

In this chapter we develop the basic existence theorems for problems in which a certain convexity condition is present. The key theorem on which the entire development is based is Theorem 4.1, which guarantees compactness of certain sets of trajectories together with a property related to lower-semicontinuity. Theorem 4.1 will also be used in Chapters 4 and 6.

Theorem 4.1 and the existence theorems based on it permit the constraint sets to depend on the time and the state and require that a certain condition introduced by Cesari be satisfied. Although this condition is not usually verifiable by inspection in a given example, it will be shown to hold in a wide class of problems that are of interest in applications. For problems in which the constraint sets depend on time but not on the state, existence theorems with hypotheses that are relatively easy to verify are given in Section 8. For these theorems the Cesari condition is replaced either by a generalized Lipschitz condition or by the requirement that the controls in a minimizing sequence lie in a fixed ball of some L_p space, $1 \leq p \leq \infty$. These conditions are also usually fulfilled in problems of interest in applications. The theorems of Sections 4 and 8 have a large area of overlap, but neither set contains the other.

Another very important theorem in this chapter is Theorem 7.1, which is an extension of Filippov's lemma. Theorem 6.2 is a classical existence theorem for ordinary problems.

The reader who is primarily interested in applications can at first reading confine his attention to Section 2, Section 3 up to Definition 3.2, Section 5 up to Lemma 5.2 and Exercises 5.1, 6.3, 6.4,

6.5 and 6.6.

The mathematical tools that we use to establish the existence of optimal controls are such that we can only guarantee the existence of an optimal control that is a measurable function. In a practical problem a greater degree of regularity in the behavior of the optimal control would be desirable. In Chapter 5 we shall obtain theorems that describe an optimal control. Hopefully, in problems of practical interest this additional information will enable us to conclude that an optimal control is more than just measurable and is one that can be implemented.

2. Non-Existence and Non-Uniqueness of Optimal Controls

In the statement of Problem 1 it was assumed that the set of admissible pairs was not void. Given a system of state equations together with end conditions and control constraints there is no guarantee that the set of admissible pairs is not void. The following simple example is introduced to emphasize this point.

EXAMPLE 2.1. Let x be one-dimensional. Let the state equation be

$$\frac{dx}{dt} = u(t) \tag{2.1}$$

Let $\mathscr{B}$ consist of the single point $(t_0, x_0, t_1, x_1) = (0,0,1,2)$ and let

$$\Omega(t,x) = \{z: \ |z| \leq 1\}.$$

Thus, the set of controls is the set of real valued integrable functions u defined on $[0,1]$. An admissible control satisfies the inequality $|u(t)| \leq 1$ for almost all t in $[0,1]$ and transfers the system from $x_0 = 0$ at time $t_0 = 0$ to the state $x_1 = 2$ at time $t_1 = 1$. From (2.1) it is clear that to each control u there corresponds a unique trajectory ϕ such that $\phi(0) = 0$, namely the trajectory given by $\phi(t) = \int_0^t u(s)ds$. These pairs are not admissible since

$$|\phi(1)| \leq \int_0^1 |u(t)|dt \leq \int_0^1 dt = 1$$

and we require $\phi(1) = 2$ for admissibility.

The problem of the existence of admissible pairs is related to the problem of controllability, which will not be taken up in these notes.

If the class of admissible controls is not void, it does not necessarily follow that an optimal control exists. The following examples illustrate the non-existence of optimal controls.

EXAMPLE 2.2. Let x be one dimensional. Let the state equation be $dx/dt = u(t)$. Let $\mathscr{B}$ consist of the single point $(t_0,x_0,t_1,x_1) = (0,1,1,0)$ and let $\Omega(t,x) = E^1$. Let

$$J(\phi,u) = \int_0^1 t^2u^2(t)dt. \tag{2.2}$$

The set of controls is the set of functions in $L_1[0,1]$. To each control u there corresponds a unique trajectory ϕ satisfying $\phi(0)=1$, namely the trajectory given by

$$\phi(t) = 1 + \int_0^t u(s)ds.$$

For each $0 < \varepsilon < 1$ define a control u_ε as follows:

$$u_\varepsilon(t) = \begin{cases} 0 & \text{if} \quad \varepsilon \leq t \leq 1 \\ -\varepsilon^{-1} & \text{if} \quad 0 \leq t < \varepsilon. \end{cases}$$

Let ϕ_ε denote the unique trajectory corresponding to u_ε and satisfying $\phi_\varepsilon(0) = 1$. Since $\phi_\varepsilon(1) = 0$ and tu_ε is in $L_2[0,1]$ it follows that $(\phi_\varepsilon,u_\varepsilon)$ is an admissible pair. Hence the class $\mathscr{A}$ of admissible pairs is not void. Moreover,

$$J(\phi_\varepsilon, u_\varepsilon) = \int_0^\varepsilon t^2 \varepsilon^{-2} dt = \varepsilon/3.$$

Since $J(\phi,u) \geq 0$ for all admissible pairs (ϕ,u), it follows that $0 = \inf \{J(\phi,u): (\phi,u) \in A\}$. From (2.2) it is clear that $J(\phi,u) = 0$ if and only if $u(t) = 0$ a.e. on $[0,1]$. But $u^* = 0$ is not admissible since if ϕ^* is the unique trajectory corresponding to $u^* = 0$ and satisfying $\phi^*(0) = 1$, then $\phi^*(1) = 1$.

EXAMPLE 2.2 (a). Let everything be as in Example 2.2 except that the control constraints are given as follows:

$$\Omega(t,x) = \{z: |z| \leq 1/t\} \quad \text{if} \quad 0 < t \leq 1$$
$$\Omega(0,x) = E^1.$$

The arguments of Example 2 are still valid and an optimal control fails to exist.

EXAMPLE 2.2 (b). Let everything be as in Example 2.2 except that the control constraints are given as follows:

$$\Omega(t,x) = \{z: |z| \leq 1/t\} \quad \text{if} \quad 0 < t \leq 1$$
$$\Omega(0,x) = 0.$$

If we now define

$$u_\varepsilon(t) = \begin{cases} 0 & \text{if} \quad \varepsilon \leq t \leq 1 \\ -\varepsilon^{-1} & \text{if} \quad 0 < t < \varepsilon \\ 0 & \text{if} \quad t = 0 \end{cases}$$

and proceed as in Example 2.2, we again find that an optimal control fails to exist.

EXAMPLE 2.3. The state equations are

$$dx^1/dt = u^1(t)$$
$$dx^2/dt = u^2(t)$$
$$dx^3/dt = 1.$$

Let the end conditions be

$$\mathcal{T}_0: \quad (t_0, x_0^1, x_0^2, x_0^3) = (0,0,0,0)$$

$$\mathcal{T}_1: \quad (t_1, x_1^1, x_1^2, x_1^3) = (1,0,0,1).$$

Let $\Omega(t,x)$ be defined by the condition

$$\Omega(t,x) \equiv \{z = (z^1, z^2): (z^1)^2 + (z^2)^2 = 1\}.$$

Let

$$J(\phi,u) = \int_0^1 [(\phi^1)^2 + (\phi^2)^2]dt.$$

To see that the set of admissible controls is not void we first note that any control u_k defined by

$$u_k(t) = (u_k^1(t), u_k^2(t)) = (\sin 2\pi kt, \cos 2\pi kt) \quad k = \pm 1, \pm 2, \ldots,$$

satisfies the conditions $(u_k^1(t))^2 + (u_k^2(t))^2 = 1$. The trajectory ϕ_k with components

$$\phi_k^1(t) = (1-\cos 2\pi kt)/2\pi k$$
$$\phi_k^2(t) = (\sin 2\pi kt)/2\pi k$$

$$\phi_k^3(t) = t$$

is the unique trajectory corresponding to u_k and satisfying the end conditions.

Since $J(\phi,u) \geq 0$ for all admissible u and since $J(\phi_k, u_k) \leq (\pi k)^{-2}$, it follows that $0 = \inf\{J(\phi,u): (\phi,u) \in \mathscr{A}\}$. Therefore, if there exists an optimal control u^* we must have $J(\phi^*, u^*) = 0$. But if $J(\phi^*, u^*) = 0$, then we must have $\phi^{*1}(t) = \phi^{*2}(t) = 0$ a.e. This, however, can only happen if $u^{*1}(t) = u^{*2}(t) = 0$

a.e. Such a control does not satisfy $(u^1)^2 + (u^2)^2 = 1$. Hence an optimal control fails to exist in this example.

EXAMPLE 2.4. Let x be a scalar and let the state equation be

$$\frac{dx}{dt} = 2x^2(1-t)-1+u(t). \tag{2.3}$$

Let $\Omega(t,x) = \{z: |z| \le 1\}$. Let the end conditions be given by $\mathcal{T}_0 = \{(0,x_0): 0 \le x_0 \le 1\}$ and $\mathcal{T}_1 = \{(a,x_1): 0 \le x_1 \le (1-a)^{-2}\}$, where a is a fixed number > 1. Let $J(\phi,u) = -\phi(a)$. Hence if u is an admissible control and ϕ is a corresponding admissible trajectory it is required to maximize $\phi(a)$ over all admissible pairs (ϕ,u).

The set of admissible controls for this problem is a subset of the measurable functions u on $[0,a]$ such that $|u(t)| \le 1$ a.e. If we substitute $u(t) = 1$ into the right hand side of (2.3) we get

$$\frac{dx}{dt} = 2x^2(1-t). \tag{2.4}$$

The solution of this differential equation satisfying the initial condition $\phi(0) = x_0$, $x_0 \neq 0$, is

$$\phi(t) = [(1-t)^2+c]^{-1}, \tag{2.5}$$

where $c = (1-x_0)/x_0$. The solution of (2.4) satisfying the initial condition $\phi(0) = 0$ is $\phi(t) \equiv 0$. The field of trajectories corresponding to $u = 1$ is indicated in Figure 1. Values of $c \ge 0$ correspond to initial points x_0 in the interval $0 < x_0 \le 1$. Note that if $x_0 = 1$, then $c = 0$ and $u = 1$ is not an admissible control.

Let $\mathcal{F}$ denote the field of trajectories corresponding to $u(t) = 1$ and initial conditions $0 \le x_0 < 1$. Note that $\mathcal{F}$ does not include the trajectory starting from $x_0 = 1$ at $t_0 = 0$. It is clear from (2.3) and from the properties of the field of trajectories $\mathcal{F}$

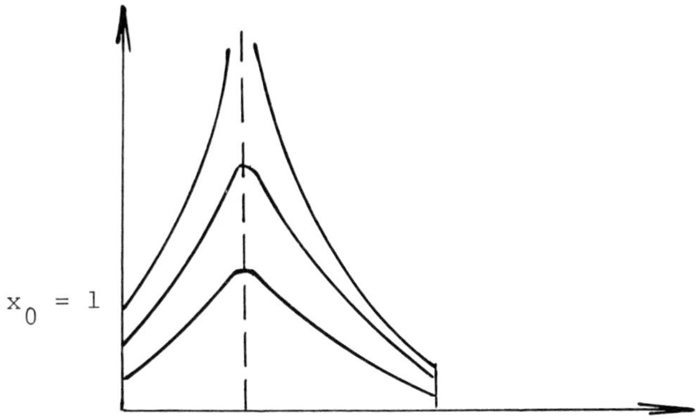

Figure 1

that if an optimal pair (ϕ^*,u^*) exists and if $\phi^*(0) = x_0 < 1$, then we must have $u^*(t) = 1$ a.e. It then follows from (2.5) (See Fig. 1) that $u^*(t) = 1$ and $0 \leq x_0 < 1$ cannot be optimal. For if we take a new initial state x_0', where $x_0 < x_0' < 1$, then the solution ϕ of (2.4) corresponding to x_0' will give $\phi(a) > \phi^*(a)$. On the other hand an optimal trajectory cannot have $x_0 = 1$ as initial point. For if $x_0 = 1$, then $u(t) \equiv 1$ is not admissible. Moreover, once we take $u(t) < 1$ on a set of positive measure the trajectory goes into the interior of $\mathscr{F}$. It is then possible to modify the control so as to increase the value $\phi(a)$. We leave the rigorous formulation of this argument to the reader.

We conclude this section with an example showing that there may be more than one optimal control.

EXAMPLE 2.5. Let x be one-dimensional. Let the state equation be $dx/dt = u(t)$. Let $\mathscr{B}$ consist of the single point $(t_0,x_0,t_1,x_1) = (0,0,1,0)$. Let $\Omega(t,x) = \{z: |z| \leq 1\}$, and let

$$J(\phi,u) = \int_0^1 (1-u^2(t))dt.$$

Clearly, $J(\phi,u) \geq 0$. Define a control u_1^* as follows: $u_1^*(t) = 1$

if $0 \leq t < \frac{1}{2}$ and $u_1^*(t) = -1$ if $\frac{1}{2} \leq t \leq 1$. Then u_1^* is admissible and $J(\phi_1^*, u_1^*) = 0$, where ϕ_1^* is the unique trajectory corresponding to u_1^*. Hence u_1^* is optimal. We now show that there are infinitely many optimal controls. For each integer $n = 1,2,3,\ldots$, define a control u_n^* as follows:

$$u_n^*(t) = (-1)^k \quad \text{if} \quad \frac{k}{2^n} \leq t < \frac{k+1}{2^n}, \quad k = 0,1,2,\ldots,2^n-1.$$

Then for each integer n, u_n^* is admissible and $J(\phi_n^*, u_n^*) = 0$, where ϕ_n^* is the trajectory corresponding to u_n^*. Hence each u_n^* is optimal.

3. Convexity Conditions, Regularity Conditions, and Conditions for Weak L_1 Convergence

In this section we introduce certain convexity and regularity conditions that are required in the discussion of the existence of optimal controls and trajectories.

Let (t_0, x_0) be a point in $\mathscr{R}$, and let $\delta > 0$. Let $N_\delta(t_0, x_0)$ denote the relatively closed δ-neighborhood of (t_0, x_0); thus

$$N_\delta(t_0,x_0) = \{(t,x): (t,x) \in \mathscr{R}, \text{dist}((t,x),(t_0,x_0)) \leq \delta\},$$

where by dist (a,b) we mean the euclidean distance between the points a and b.

Let Λ be a mapping that assigns subsets $\Lambda(t,x)$ of E^k to points (t,x) in $\mathscr{R}$. Then by $\Lambda(N_\delta(t_0,x_0))$ we shall mean the following:

$$\Lambda(N_\delta(t_0,x_0)) = \cup[\Lambda(t,x): (t,x) \in N_\delta(t_0,x_0)].$$

DEFINITION 3.1. The mapping Λ is said to be upper semi-continuous at a point (t_0, x_0) in $\mathscr{R}$ if

$$\bigcap_{\delta>0} \text{cl}\, \Lambda(N_\delta(t_0,x_0)) \subseteq \Lambda(t_0,x_0), \tag{3.1}$$

where cl denotes closure.

Note that since the inclusion opposite to that in (3.1) always holds, an equivalent definition is obtained if we replace $\subseteq$ by equality in (3.1). Hence, if Λ is to be upper semicontinuous at (t_0,x_0) then $\Lambda(t_0,x_0)$ must be a closed set.

The mapping Λ is said to be upper semicontinuous on $\mathscr{R}$ if it is upper semicontinuous at every point of $\mathscr{R}$.

An example of an upper semicontinuous mapping is the one defined by $\Omega(t,x) = U$, where U is a fixed closed set. Although this example appears to be trivial, it is important because in many applications the constraint sets $\Omega(t,x)$ do not depend on (t,x), but are fixed. Another example of an upper semicontinuous mapping is the mapping Ω of Example 2.2(a). At $(t_0,x_0) = (0,0)$, for every $\delta > 0$ we have $\Omega(N_\delta(0,0)) = E^1$. Since $\Omega(0,0) = E^1$, (3.1) follows immediately. We leave the verification that Ω is upper semicontinuous at other points to the reader. An example of a mapping that fails to be upper semicontinuous at a point is the mapping Ω of Example 2.2(b) at $(0,0)$. We again have $\Omega(N_\delta(0,0)) = E^1$ for every $\delta > 0$, but now $\Omega(0,0) = 0$, so (3.1) fails.

The following equivalent characterization of an upper semicontinuous mapping on a closed set $\mathscr{R}$ will be used in the proof of the existence theorems.

LEMMA 3.1. Let $\mathscr{R}$ be closed. A necessary and sufficient condition that the mapping Λ be upper semicontinuous on $\mathscr{R}$ is that the set $\Delta = \{(t,x,\lambda): \lambda \in \Lambda(t,x), (t,x) \in \mathscr{R}\}$ be closed.

We first suppose that (3.1) holds. Let $\{(t_n,x_n,\lambda_n)\}$ be a sequence of points in Δ converging to a point (t_0,x_0,λ_0). Thus, $\lambda_n \in \Lambda(t_n,x_n)$ and $(t_n,x_n) \to (t_0,x_0)$. Since $\mathscr{R}$ is closed, $(t_0,x_0) \in \mathscr{R}$. Moreover, for every $\delta > 0$ there is an integer $n(\delta)$

such that if $n > n(\delta)$ then $(t_n, x_n) \in N_\delta(t_0, x_0)$. Thus $\lambda_0 \in$ cl $\Lambda(N_\delta(t_0, x_0))$. But δ is arbitrary, so $\lambda_0 \in \bigcap_{\delta>0}$ cl $\Lambda(N_\delta(t_0, x_0))$.

Hence by (3.1), $\lambda_0 \in \Lambda(t_0, x_0)$ and so Δ is closed.

Conversely, let Δ be closed, and let $\lambda_0 \in$ cl $\Lambda(N_\delta(t_0, x_0))$ for every $\delta > 0$. Then there exists a sequence $\{(t_n, x_n)\}$ of points in $\mathscr{R}$, a sequence of positive numbers $\{\delta_n\}$, and a sequence of points $\{\lambda_n\}$ such that the following hold: (i) $\delta_n \to 0$; (ii) $(t_n, x_n) \in N_{\delta_n}(t_0, x_0)$; (iii) $\lambda_n \in \Lambda(t_n, x_n)$; and (iv) $\lambda_n \to \lambda_0$. Thus, $(t_n, x_n, \lambda_n) \to (t_0, x_0, \lambda_0)$. Since $\mathscr{R}$ and Δ are closed, $\lambda_0 \in \Lambda(t_0, x_0)$ and therefore (3.1) holds.

Consider

$$\hat{f}(t,x,z) = (f^0(t,x,z), \ldots, f^n(t,x,z)),$$

where the functions $f^0, f^1, \ldots, f^n$ are real valued and defined on $\mathscr{G} = \mathscr{R} \times \mathscr{U}$. For fixed (t,x) as z ranges over the set $\Omega(t,x)$ the vectors $\hat{f}(t,x,z)$ will trace out a set in E^{n+1}. We denote this set by $\mathscr{Q}(t,x)$. Thus:

$$\mathscr{Q}(t,x) = \{\hat{y} = (y^0, y) : y^0 = f^0(t,x,z),\ y = f(t,x,z),\ z \in \Omega(t,x)\}. \tag{3.2}$$

We shall also write $\mathscr{Q}(t,x) = \hat{f}(t,x,\Omega(t,x))$. We shall also need to consider the following set, which is related to $\mathscr{Q}(t,x)$.

$$\begin{aligned}\mathscr{Q}^+(t,x) &= \{\hat{y} = (y^0, y) : y^0 \geq f^0(t,x,z),\ y = f(t,x,z),\ z \in \Omega(t,x)\} \\ &= \{\hat{y} = (y^0, y) : y^0 \geq f^0(t,x,\Omega(t,x)),\ y = f(t,x,\Omega(t,x))\}.\end{aligned} \tag{3.3}$$

We illustrate these ideas by means of Examples 2.2 and 2.3. In Example 2.2

$$\mathscr{Q}(t,x) = \{(y^0,y): y^0 = t^2z^2,\ y = z,\ z \in E^1\}$$
$$\mathscr{Q}^+(t,x) = \{(y^0,y): y^0 \geq t^2z^2,\ y = z,\ z \in E^1\}.$$

In Figure 2, for fixed (t,x), $t \neq 0$, $\mathscr{Q}(t,x)$ is the parabola $y^0 = t^2y^2$ bounding the shaded region while $\mathscr{Q}^+(t,x)$ is the parabola plus the shaded region. The set $\mathscr{Q}(t,x)$ is not convex, while the set $\mathscr{Q}^+(t,x)$ is convex. If $t = 0$, then $\mathscr{Q}(t,x)$ is the y-axis, while $\mathscr{Q}^+(t,x)$ is the upper half plane $y^0 \geq 0$.

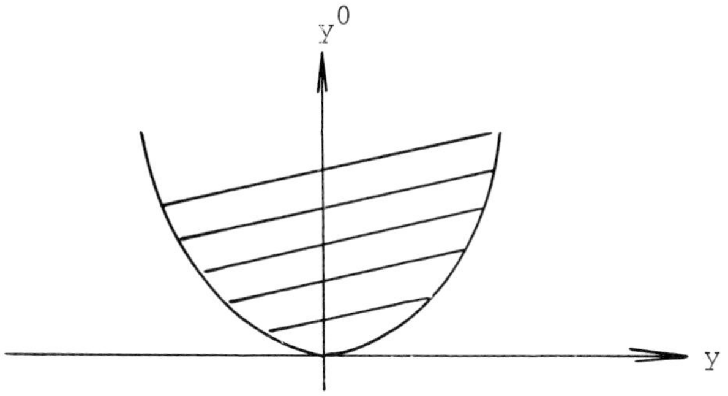

Figure 2

In Example 2.3,

$$\mathscr{Q}(t,x) = \{(y^0,y): y^0 = (x^1)^2+(x^2)^2,\ y^1 = z^1,\ y^2 = z^2,$$
$$y^3 = 1;\ (z^1)^2 + (z^2)^2 = 1\}$$
$$\mathscr{Q}^+(t,x) = \{(y^0,y): y^0 \geq (x^1)^2 + (x^2)^2,\ y^1 = z^1,\ y^2 = z^2,$$
$$y^3 = 1;\ (z^1)^2 + (z^2)^2 = 1\}.$$

If we fix (t,x) and take the intersection of $\mathscr{Q}(t,x)$ with the hyperplane $y^3 = 1$ we obtain a circle C of radius one with center at the origin in the plane $y^0 = (x^1)^2 + (x^2)^2$. (See Fig. 3). The intersection of $\mathscr{Q}^+(t,x)$ with the hyperplane $y^3 = 1$ is the surface of a right circular cylinder erected above the circle C. Thus neither $\mathscr{Q}(t,x)$ nor $\mathscr{Q}^+(t,x)$ is convex.

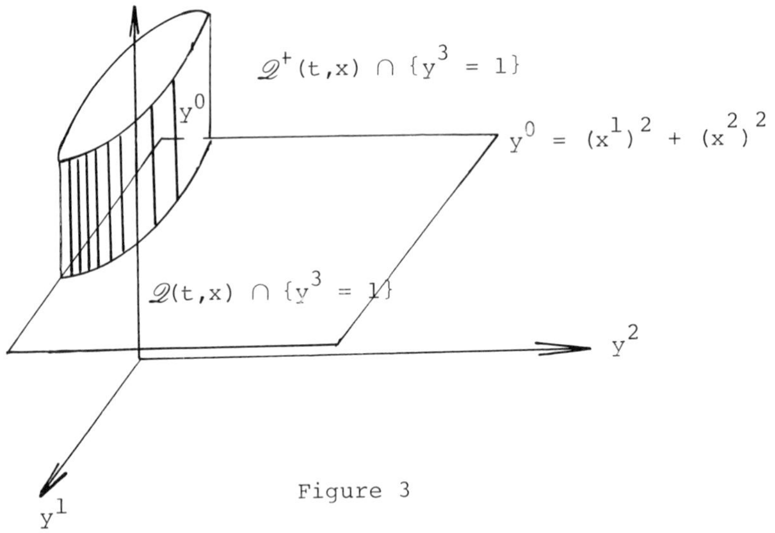

Figure 3

DEFINITION 3.2. A mapping Λ from $\mathscr{R}$ to subsets of E^k is said to have the Cesari property at a point (t_0,x_0) in $\mathscr{R}$ if

$$\bigcap_{\delta>0} \text{cl co } \Lambda(N_\delta(t_0,x_0)) \subseteq \Lambda(t_0,x_0),$$

where by cl co A we mean the closure of the convex hull of A. We say that Λ has the Cesari property on $\mathscr{R}$ if it has the Cesari property at every point of $\mathscr{R}$.

Note that the inclusion opposite to that in (3.4) always occurs. Hence (3.4) is equivalent to a statement in which the inclusion is replaced by equality. It therefore follows that if Λ is to satisfy the Cesari property at (t_0,x_0) then $\Lambda(t_0,x_0)$ must be a closed convex set. Thus in Example 2.3 the mapping $\mathscr{Q}^+$ defined by $(t,x) \to \mathscr{Q}^+(t,x)$ cannot satisfy the Cesari property since the sets $\mathscr{Q}^+(t,x)$ are not convex. In Examples 2.2 and 2.2(a) the mapping $\mathscr{Q}^+$ satisfies the Cesari property. On the other hand, in Example 2.2(b) the mapping $\mathscr{Q}^+$ does not satisfy the Cesari property even though all sets $\mathscr{Q}^+(t,x)$ are convex. To see this note that for each $\delta > 0$ we have $\mathscr{Q}^+(N_\delta(0,0)) = \{(y^0,y): y^0 > 0\} \cup \{(0,0)\} =$

co $\mathcal{Q}^+(N_\delta(0,0))$. Thus the intersection of the sets cl co $\mathcal{Q}^+(N_\delta(0,0))$ is the closed upper half plane while $\mathcal{Q}^+(0,0) = \{(y^0,y): y^0 \geq 0, y = 0\}$.

Sufficient conditions for the Cesari property to hold in terms of the functions $\hat{f}$ and Ω will be given in connection with the existence theorems of Sections 5 and 6.

Let (t_0,x_0) be a point in $\mathcal{R}$ and let $N_{x\delta}(t_0,x_0)$ denote the set of points (t_0,x) in $\mathcal{R}$ such that $|x-x_0| \leq \delta$. Let

$$\Lambda(N_{x\delta}(t_0,x_0)) = \{\Lambda(t_0,x): (t_0,x) \in N_{x\delta}(t_0,x_0)\}$$

DEFINITION 3.3. The mapping Λ is said to satisfy the weak Cesari property at (t_0,x_0) if

$$\bigcap_{\delta>0} \text{cl co } \Lambda(N_{x\delta}(t_0,x_0)) \subseteq \Lambda(t_0,x_0). \tag{3.5}$$

Again, since we always have that the right hand side is contained in the left hand side, (3.5) is equivalent to a statement in which the inclusion is replaced by equality. We let the reader check that if Λ satisfies the Cesari condition at a point (t_0,x_0), then Λ satisfies the weak Cesari condition at (t_0,x_0).

Any mapping $\mathcal{Q}^+$ that is independent of x will satisfy the weak Cesari property, provided the sets $\mathcal{Q}^+(t,x)$ are closed and convex. Thus, the mapping $\mathcal{Q}^+$ corresponding to Example 2.2(b) has the weak Cesari property at $(0,0)$ even though it does not have the Cesari property.

We shall require a measure of nearness of trajectories. Since the initial and terminal times are not fixed, the following metric is introduced. Let $\mathcal{X}$ denote the class of continuous functions from arbitrary compact intervals in E^1 to E^n. Let x be a function in $\mathcal{X}$ defined on $[a,b]$ and let y be a function in $\mathcal{X}$ defined on $[c,d]$. We extend the domain of definition of x to $(-\infty,\infty)$ by setting $x(t) = x(a)$ for $t \leq a$ and by setting $x(t) = x(b)$ for $t \geq b$. We extend the domain of definition of y in similar fashion. We now define

$$\rho(x,y) = |a-c| + |b-d| + \max\{|x(t)-y(t)|: -\infty < t < +\infty\}. \qquad (3.6)$$

We assert that ρ is a metric and that $\mathscr{X}$ is a complete metric space under this metric. We shall henceforth denote the corresponding metric space by $\mathscr{X}_\rho$.

EXERCISE 3.1. Prove the assertion just made.

We conclude this section with a listing of some well known facts and definitions that are sometimes omitted from introductory courses in integration and functional analysis. We shall use these facts and definitions in our discussion of existence theorems. We shall refer the reader to standard references for the proofs of many of the major results.

<u>DEFINITION 3.3.</u> A set $\mathscr{F}$ of functions f in $L_1[a,b]$, where $[a,b] = \{t: a \leq t \leq b\}$, is said to have equi-absolutely continuous integrals if given an $\varepsilon > 0$ there is a $\delta > 0$ such that for all Lebesgue measurable sets $E \subset [a,b]$ with meas $(E) < \delta$ and all f in $\mathscr{F}$,

$$\left|\int_E f\,dt\right| < \varepsilon.$$

Note that since $[a,b]$ is a finite interval and we are dealing with Lebesgue measure, it follows that if the functions f in $\mathscr{F}$ have equi-absolutely continuous integrals, then there is a constant $K > 0$ such that for all f in $\mathscr{F}$

$$\int_a^b |f|\,dt < K. \qquad (3.7)$$

That is, the set $\mathscr{F}$ is bounded in $L_1[a,b]$.

DEFINITION 3.4. A set $\mathscr{F}$ of absolutely continuous functions f defined on $[a,b]$ is said to be equi-absolutely continuous if given an $\varepsilon > 0$ there is a $\delta > 0$ such that for any finite collection of non overlapping intervals $[\alpha_i, \beta_i]$ contained in $[a,b]$, with $\Sigma_i |\beta_i - \alpha_i| < \delta$, the inequality $\Sigma_i |f(\beta_i) - f(\alpha_i)| < \varepsilon$ holds for all f in $\mathscr{F}$.

We leave it to the reader to verify that a set of absolutely continuous functions is equi-absolutely continuous if and only if the derivatives f' have equi-absolutely continuous integrals.

For us, the importance of the notion of equi-absolute continuity stems from the following theorem.

THEOREM 3.1. Let $[a,b]$ be a finite interval and let $\{f_n\}$ be a sequence of functions in $L_1[a,b]$. The sequence of functions $\{f_n\}$ converges weakly to a function f in $L_1[a,b]$ if and only if the following conditions are satisfied: (i) the functions f_n have equi-absolutely continuous integrals and (ii) for every t in $[a,b]$

$$\lim_{n\to\infty} \int_a^t f_n(s)ds = \int_a^t f(s)ds.$$

We shall sketch a proof of the theorem, referring the reader to standard texts for some of the arguments and leaving other parts to the reader.

We first consider the necessity of conditions (i) and (ii). Weak convergence of f_n to f means that for every bounded measurable function g defined on $[a,b]$

$$\int_a^b g\ f_n dt \to \int_a^b gf\ dt. \tag{3.8}$$

Hence by taking g to be the characteristic function of $[a,t]$ we obtain (ii). By taking g to be the characteristic function of a measurable set E we get that (3.8) holds when the integrals are taken over any measurable set E. Condition (i) then follows from Corollary 1 of Theorem 3, page 156, in Natanson [46].

Now suppose that (i) and (ii) hold. We have already remarked that (i) implies that (3.7) holds with f replaced by f_n; i.e. the sequence $\{f_n\}$ is bounded in $L_1[a,b]$. Condition (ii) implies that condition (ii) holds when the interval of integration is taken to be $[t',t'']$, where $[t',t'']$ is any interval contained in $[a,b]$. From this statement and (i) it follows that (ii) holds when the integrals are taken over any measurable set E in $[a,b]$. It then follows that (3.8) holds for any step function g. If g is an arbitrary measurable function then g is the almost everywhere limit of a sequence of step functions $\{\sigma_k\}$. By Egorov's theorem, for every $\delta > 0$ there is a set E of measure $< \delta$ such that on the complement of E relative to $[a,b]$, $\sigma_k \to g$ uniformly. From the last observation, from (3.8) with g replaced by a step function, the uniform L_1 bound for the functions f_n, and the equi-absolute continuity of the $\{f_n\}$ there follows the validity of (3.8) for arbitrary bounded measurable g.

Another fact that we shall use is the following.

THEOREM 3.2. Let $\mathscr{Y}$ be a Banach space and let $\{y_n\}$ be a sequence of elements in $\mathscr{Y}$ converging weakly to an element $y_0 \in \mathscr{Y}$. Then given an integer n and an $\varepsilon > 0$ there exists a finite set of real numbers $\{\alpha_i\}$, with $\alpha_i \geq 0$, $\Sigma_i\alpha_i = 1$ such that $||y_0-\Sigma_i\alpha_i y_{n+i}|| < \varepsilon$, where $||\cdots||$ denotes the norm in $\mathscr{Y}$.

This theorem will be deduced from the following fact, which in turn is a consequence of the Hahn-Banach theorem. Every strongly closed convex subset of $\mathscr{Y}$ is also weakly closed. For the latter statement the reader is referred to Hille-Phillips [26], Theorem 2.9.3 or to Dunford-Schwartz, [20], Thoerem V 3.13.

Let $\mathscr{M} = \text{cl co } \{y_n\}$, where the closure is taken in the norm topology of $\mathscr{Y}$. Then $\mathscr{M}$ is a strongly closed convex set in $\mathscr{Y}$. Hence it is weakly closed. Therefore since y_0 is a weak limit point it is in $\mathscr{M}$ and hence can be approximated in norm to any degree of accuracy by points in co $\{y_n\}$.

4. A General Existence Theorem

In this section we state and prove a general existence theorem, Theorem 4.2, for Problem 2 of Chapter 2. Two other general existence theorems, Theorem 8.3 and Theorem 8.6, will be given in Section 8. These theorems do not subsume Theorem 4.2 nor are they subsumed by it.

While some of the hypotheses of this theorem can be checked directly in a given example, the hypotheses are not in a convenient form. The role of Theorem 4.2 is that of a parent theorem from which we can easily derive existence theorems whose hypotheses are easily verified for entire classes of problems. This will be done in Sections 5 and 6. In Section 6 the classical existence theorems for the ordinary problems in the calculus of variations will be obtained from Theorem 4.2. The reader primarily interested in applications may wish to read most of Section 5 before proceeding with the remainder of this section, referring back to Assumption 4.1 below when necessary.

In order to understand matters better, we review the proof of the well known result that a real valued lower semicontinuous function f defined on a compact metric space $\mathscr{Y}$ attains its minimum on $\mathscr{Y}$; i.e. there is a point y_0 in $\mathscr{Y}$ such that $f(y_0) \leq f(y)$ for all y in $\mathscr{Y}$. If μ denotes the infimum of f on $\mathscr{Y}$, then $\mu < +\infty$ and there is a sequence $\{y_n\}$ such that $f(y_n) \to \mu$. The sequence $\{y_n\}$ is called a minimizing sequence. Since $\mathscr{Y}$ is compact, there is a subsequence, which we again denote by $\{y_n\}$, and a point y_0 in $\mathscr{Y}$ such that $y_n \to y_0$. From the lower semicontinuity of f we obtain

$\liminf f(y_n) \geq f(y_0)$. But by the definition of μ, $f(y_0) \geq \mu$. Also $\liminf f(y_n) = \mu$. Hence $\mu > -\infty$ and $f(y_0) = \mu$. We remark that only the conditional compactness of the minimizing sequence was used.

Using the preceding argument as a guide one might attack the existence question for Problem 2 as follows. First place a topology on the set $\mathscr{A}$ of admissible pairs (ϕ,u) so that the subset $\mathscr{A}_1$ is compact and then show that J is lower semi-continuous on $\mathscr{A}_1$ in this topology. Although this can be done in some special cases, in general, the non-linearity of the functions f in the state equations necessitates a modification of this procedure. This is done in Theorem 4.1 and also in Theorems 8.2 and 8.4. It turns out that a fruitful way to proceed is the following. Let $\mathscr{A}_T$ denote the set of all admissible trajectories. We consider $\mathscr{A}_T$ as a set in $\mathscr{X}_\rho$. Let $\mathscr{A}_{1T}$ denote the set of trajectories corresponding to pairs (ϕ,u) in $\mathscr{A}_1$. Theorem 4.1 gives criteria for the set $\mathscr{A}_{1T}$ to be conditionally compact in $\mathscr{A}_T$. That is we give criteria for the following to be true. If $\{\phi_k\}$ is a sequence in $\mathscr{A}_{1T}$, then there is a subsequence $\{\phi_{k_j}\}$ and an admissible pair (ϕ,u) such that $\phi_{k_j} \to \phi$ in $\mathscr{X}_\rho$. Note that we require more than the conditional compactness of $\mathscr{A}_{1T}$ in $\mathscr{X}_\rho$. Theorem 4.1 also furnishes a generalization of the notion of lower semicontinuity that is appropriate for our problem.

Theorem 4.1 is a fundamental result. The existence theorem of this section will be based on it as will the existence theorem involving generalized controls in Chapter 4. This theorem will also be used to obtain properties of the "attainable set" in Chapter IV.

The following set of assumptions will be made in Theorem 4.1 and in several of our existence theorems.

ASSUMPTION 4.1. (i) There exists a compact set $\mathscr{R}_0 \subset \mathscr{R}$ such that for all admissible trajectories ϕ, the points $(t,\phi(t)) \in \mathscr{R}_0$ for all t in $[t_0,t_1]$. (ii) The set $\mathscr{B}$ is closed. (iii) The mapping Ω is upper semicontinuous on $\mathscr{R}_0$. (iv) For each (t,x) in $\mathscr{R}_0$ the set $\mathscr{Q}^+(t,x)$ is closed and convex. (v) There exists a constant $K \geq 0$ such that $f^0(t,x,z) \geq -K$ for all (t,x,z) in $\mathscr{D}$, where

$$\mathscr{D} = \{(t,x,z):\ (t,x) \in \mathscr{R}_0,\ z \in \Omega(t,x)\}. \tag{4.1}$$

(vi) The function f^0 is lower semicontinuous on $\mathscr{G} = \mathscr{R} \times \mathscr{U}$ and f is continuous on $\mathscr{G}$.

We shall say that a trajectory ϕ <u>lies in a set</u> $\mathscr{R}'$ of (t,x) space if its graph does; i.e. if $(t,\phi(t)) \in \mathscr{R}'$ for all t in $[t_0,t_1]$. Thus, in Assumption 4.1-(i) we are supposing that all admissible trajectories lie in $\mathscr{R}_0$.

THEOREM 4.1. Let Assumption 4.1 hold and let the mapping $\mathscr{Q}^+$ satisfy the weak Cesari property at every point of $\mathscr{R}_0$. Let

$$I(\phi,u) = \int_{t_0}^{t_1} f^0(t,\phi(t),u(t))dt \qquad (\phi,u) \in \mathscr{A}. \tag{4.2}$$

Let $\mathscr{A}_0$ be a set of admissible pairs (ϕ,u) such that the trajectories ϕ are equi-absolutely continuous and such that $\sup\{I(\phi,u):\ (\phi,u) \in \mathscr{A}_0\} < +\infty$. Then there exists a sequence $\{(\phi_k,u_k)\}$ in $\mathscr{A}_0$ and an admissible pair (ϕ^*,u^*) in $\mathscr{A}$ with the following properties. (i) $\phi_k \to \phi^*$ in $\mathscr{X}_\rho$. (ii)

$$\lim_{k\to\infty} I(\phi_k,u_k) \geq I(\phi^*,u^*). \tag{4.3}$$

We emphasize that <u>no assertion is made</u> concerning the relationship between the controls u_k and the control u^*. Furthermore, it is not true in general that (4.3) holds for <u>all</u> u such that (ϕ^*,u) is admissible.

EXERCISE 4.1. Prove the last statement by means of an example. One such example was given by Cesari in [18], p. 528.

A theorem equivalent to Theorem 4.1 in which the emphasis is slightly different is given in Corollary 4.1 below. The state of affairs described in Corollary 4.1 is sometimes summarized by saying that the lower closure property holds on $\mathscr{A}_0$.

COROLLARY 4.1. Let $\{(\phi_k,u_k)\}$ be a sequence in $\mathscr{A}_0$ and let ϕ be an element of $\mathscr{X}_\rho$ such that $\phi_k \to \phi$ in $\mathscr{X}_\rho$. Then there exists a control u such that (ϕ,u) is admissible and $\liminf I(\phi_k,u_k) \geq I(\phi,u)$.

EXERCISE 4.2. Prove that Corollary 4.1 and Theorem 4.1 are equivalent.

The proof of Theorem 4.1 will be given in Section 7. The intervening material is not needed for the proof so that the reader may pass directly to the proof should he so desire.

We now state and prove an existence theorem for Problem 2 of II.3.

THEOREM 4.2. Let Assumption 4.1 hold and let the mapping $\mathscr{Q}^+$ satisfy the weak Cesari property at every point of $\mathscr{R}_0$. Let the functional G be defined, lower semicontinuous, and bounded from below on $\mathscr{A}_T$, the set of admissible trajectories viewed as a subset of $\mathscr{X}_\rho$. Let $\mu = \inf\{\hat{J}(\phi,u): (\phi,u) \in \mathscr{A}\}$. Let there exist a minimizing sequence $\{(\phi_k,u_k)\}$ such that the trajectories ϕ_k are equi-absolutely continuous. Then μ is finite and there exists a (ϕ^*,u^*) in $\mathscr{A}$ such that $\hat{J}(\phi^*,u^*) = \mu$.

Proof: Since for admissible (ϕ,u) the function $t \to f^0(t,\phi(t),u(t))$ is in $L_1[t_0,t_1]$, it follows that $\mu < +\infty$. Since $\hat{J}(\phi,u) = G(\phi) + I(\phi,u)$, since G is bounded below on $\mathscr{A}_T$, and since $\mu < +\infty$, it follows that the sequence $\{I(\phi_k,u_k)\}$ is bounded above. Thus we may take the minimizing sequence whose existence is postulated in the hypotheses of the present theorem to be the set $\mathscr{A}_0$ of Theorem 4.1 and obtain the following conclusion. There exists a subsequence, which we again label as $\{(\phi_k,u_k)\}$, and a (ϕ^*,u^*) in $\mathscr{A}$ defined on an interval $[t_0^*,t_1^*]$ such that $\phi_k \to \phi^*$ in $\mathscr{X}_\rho$ and such that $\lim I(\phi_k,u_k) \geq I(\phi^*,u^*)$. From the lower semicontinuity of G we get $\liminf G(\phi_k) \geq G(\phi^*)$. Since

$$\begin{aligned}\mu &= \lim \hat{J}(\phi_k,u_k) = \lim (G(\phi_k) + I(\phi_k,u_k)) \\ &\geq \liminf G(\phi_k) + \lim I(\phi_k,u_k) \\ &\geq G(\phi^*) + I(\phi^*,u^*) = \hat{J}(\phi^*,u^*) \geq \mu.\end{aligned}$$

Hence μ is finite and $\hat{J}(\phi^*,u^*) = \mu$.

Note that in Theorem 4.2 $\mathscr{A}_1 = \mathscr{A}$. If $\mathscr{A}_1 \subset \mathscr{A}$, then to obtain the existence theorem one must show that the pair (ϕ^*,u^*) obtained from Theorem 4.1 possess the properties that define $\mathscr{A}_1$ as a proper subset of $\mathscr{A}$.

We may formulate Theorem 4.1 and Corollary 4.1 in a way that appears to be excessively sophisticated in the present context, but it is the formulation that carries over to distributed parameter systems. We sketch this formulation briefly. The reader who is not familiar with the elementary facts about Sobolev spaces that we will use can omit this material without impairing his understanding of Theorem 4.1 and Corollary 4.1.

We suppose that Assumption 4.1 holds. Then there exists a compact interval $\mathscr{I}$ such that all admissible trajectories are defined on subintervals of $\mathscr{I}$. To simplify the exposition we shall suppose that the problem is one with fixed initial time t_0 and fixed terminal time t_1, and that $\mathscr{I} = [t_0,t_1]$. An admissible trajectory is an absolutely continuous function ϕ whose derivative is in

$L_1[\mathscr{I}]$. Thus every admissible trajectory is an element of the Sobolev space $H_1^1(\mathscr{I})$, where we are identifying the equivalence class which constitutes an element of $H_1^1(\mathscr{I})$ with one of its representatives. Since $\mathscr{I}$ is a one dimensional interval, every element of $H_1^1(\mathscr{I})$ has an absolutely continuous representative. We shall identify each element of $H_1^1(\mathscr{I})$ with its absolutely continuous representative.

Let $\mathscr{A}_0$ be a set of admissible pairs (ϕ,u) such that the trajectories ϕ in $\mathscr{A}_{0T}$ are equi-absolutely continuous. Therefore, since the trajectories are uniformly bounded, it follows from Ascoli's theorem and Theorem 3.1 that the set $\mathscr{A}_{0T}$ of trajectories is a weakly compact set in $H_1^1(\mathscr{I})$. Conversely, if the trajectories form a weakly compact set in $H_1^1(\mathscr{I})$, then they are equi-absolutely continuous. Thus, in the statement of Theorem 4.1 the phrase, "the trajectories ϕ are equi-absolutely continuous" can be replaced by the phrase, "the trajectories ϕ are weakly compact in $H_1^1(\mathscr{I})$". Thus Theorem 4.1 gives conditions ensuring, among other things, that a set of trajectories $\mathscr{A}_{0T}$ that is weakly compact in $H_1^1(\mathscr{I})$ is also conditionally compact in $\mathscr{A}_T$, where $\mathscr{A}_{0T}$ and $\mathscr{A}_T$ are considered subsets of $C[\mathscr{I}]$.

5. An Existence Theorem in the Presence of Compact Constraints

In this section we shall discuss a theorem that guarantees the existence of optimal controls when the sets $\Omega(t,x)$ are compact. The theorem covers important classes of problems that arise in applications. It requires a greater degree of regularity in the dependence of $\Omega(t,x)$ upon (t,x) than is afforded by upper semicontinuity.

Let U be a set in E^m and let $\varepsilon > 0$. Let $[U]_\varepsilon$ denote the closed ε-neighborhood of U; i.e.

$$[U]_\varepsilon = \{z\colon z \in E^m,\ \text{dist}\,(z,U) \le \varepsilon\},$$

where by dist (z,U) we mean $\inf\{\text{dist}(z,\eta): \eta \in U\}$.

<u>DEFINITION 5.1</u>. A mapping Λ that assigns subsets $\Lambda(t,x)$ of E^m to points (t,x) in $\mathscr{R}$ is said to be <u>upper semi-continuous with respect to inclusion</u> or u.s.c.i. at a point (t_0,x_0) in $\mathscr{R}$ if for every $\varepsilon > 0$ there exists a $\delta > 0$ such that

$$\Lambda(t,x) \subseteq [\Lambda(t_0,x_0)]_\varepsilon \tag{5.1}$$

for every (t,x) in $N_\delta(t_0,x_0)$. The mapping is said to be <u>u.s.c.i.</u> on $\mathscr{R}$ if it is <u>u.s.c.i.</u> at every point of $\mathscr{R}$.

Clearly, if $\Omega(t,x) = U$, a fixed set, for all (t,x) in $\mathscr{R}$ then the mapping is <u>u.s.c.i.</u> on $\mathscr{R}$.

If Λ is u.s.c.i. at a point (t_0,x_0) and $\Lambda(t_0,x_0)$ is closed then Λ is upper semicontinuous at (t_0,x_0), but the converse need not hold. If Λ is u.s.c.i. at (t_0,x_0) then for every $\varepsilon > 0$ there is a $\delta(\varepsilon) > 0$ such that if $(t,x) \in N_{\delta(\varepsilon)}(t_0,x_0)$ then $\Lambda(t,x) \subseteq [\Lambda(t_0,x_0)]_\varepsilon$. Hence $\Lambda(N_{\delta(\varepsilon)}(t_0,x_0)) \subseteq [\Lambda(t_0,x_0)]_\varepsilon$ and the same is true of $\text{cl}\,\Lambda(N_{\delta(\varepsilon)}(t_0,x_0))$. Also, $\bigcap_{\delta>0}\text{cl}\,\Lambda(N_\delta(t_0,x_0)) \subseteq \bigcap_{\delta(\varepsilon)}\text{cl}\,\Lambda(N_{\delta(\varepsilon)}(t_0,x_0)) \subseteq \bigcap_{\varepsilon>0}[\Lambda(t_0,x_0)]_\varepsilon$. Since $\Lambda(t_0,x_0)$ is closed the last intersection is equal to $\Lambda(t_0,x_0)$, and therefore Λ is upper semicontinuous at (t_0,x_0). An example of a mapping that is upper semicontinuous but is not u.s.c.i. is the following. Let

$$\Lambda(t,x) = \{z \in E^1: 0 \le z \le 1\} \cup \{z = 1/t\} \qquad t \ne 0$$
$$\Lambda(0,x) = \{z \in E^1: 0 \le z \le 1\}.$$

Then Λ is upper semicontinuous at any point $(0,x_0)$ but is not u.s.c.i. at $(0,x_0)$. Note that each set $\Lambda(t,x)$ is compact.

THEOREM 5.1. Let the class $\mathscr{A}$ of admissible pairs be non-empty and let the following hypotheses hold. (i) There exists a compact set $\mathscr{R}_0 \subset \mathscr{R}$ such that for all admissible trajectories ϕ, we have $(t,\phi(t)) \varepsilon \mathscr{R}_0$ for all t in $[t_0,t_1]$. (ii) The set $\mathscr{B}$ is closed. (iii) The mapping Ω is u.s.c.i. on $\mathscr{R}_0$. (iv) For each (t,x) in $\mathscr{R}_0$ the set $\Omega(t,x)$ is compact. (v) For each (t,x) in $\mathscr{R}_0$ the set $\mathscr{Q}^+(t,x)$ is convex. (vi) The function f^0 is lower semicontinuous on $\mathscr{G} = \mathscr{R} \times \mathscr{U}$ and the function f is continuous on $\mathscr{G}$. Let J be as in II (3.2) and let g be lower semicontinuous on $\mathscr{B}$. Then there exists a (ϕ^*,u^*) in $\mathscr{A}$ such that $J(\phi^*,u^*) \leq J(\phi,u)$ for all (ϕ,u) in $\mathscr{A}$.

Theorem 5.1 will be proved by showing that the hypotheses of Theorem 5.1 imply that the hypotheses of Theorem 4.2 are fulfilled. This will be done at the end of this section after we have discussed the theorem.

REMARK 5.1. Theorem 5.1 holds if we replace J by the functional $\hat{J}$ defined in II(3.3), where G is lower semicontinuous and bounded below on $\mathscr{A}_T$. As noted in Section II.3 in the discussion preceding the statement of Problem 2 this covers functionals such as $\max |\phi(t)|$ or $\max |\phi(t)-h(t)|$.

REMARK 5.2. In the proof of Theorem 5.1 we shall show that (iii) and (iv) of the hypotheses imply that the set $\mathscr{D}$ defined in (4.1) is compact and that the compactness is what is actually utilized. Therefore, (iii) and (iv) can be replaced by the hypothesis that $\mathscr{D}$ is compact. Conditions (iii) and (iv), however, are easier to verify in applications since they involve the data of the problem directly.

We now point out some important classes of control problems to which Theorem 5.1 is applicable. We suppose that the hypotheses of Theorem 5.1 concerning $\mathscr{R}_0$, Ω, g and $\mathscr{B}$ hold. We make the additional hypothesis that <u>for each</u> (t,x) <u>in</u> $\mathscr{R}_0$ <u>the set</u> $\Omega(t,x)$ <u>is convex</u>. We shall consider certain special classes of functions $f^0, f^1, \ldots, f^n$ that guarantee that $\mathscr{Q}^+(t,x)$ is a convex subset of

E^{n+1} whenever $\Omega(t,x)$ is convex.

First we consider problems in which the functions f^i are <u>linear</u> in x and z. Thus,

$$f^i(t,x,z) = \sum_{j=1}^{n} a_i^j(t)x^j + \sum_{j=1}^{m} b_i^j(t)z^j + h_i(t) \qquad i = 0,1,\ldots,n.$$

Hence the state equations are

$$\frac{dx^i}{dt} = \sum_{j=1}^{n} a_i^j(t)x^j + \sum_{j=1}^{m} b_i^j(t)u^j(t) + h_i(t) \qquad i = 1,\ldots,n$$

and the functional J is given by

$$J(\phi,u) = g(t_0,\phi(t_0),t_1,\phi(t_1)) + \int_{t_0}^{t_1}\{\sum_{j=1}^{n} a_0^j(t)\phi^j(t)$$
$$+ \sum_{j=1}^{m} b_0^j(t)u^j(t) + h_0(t)\}dt.$$

In vector matrix notation the state equations become

$$\frac{dx}{dt} = A(t)x + B(t)u(t) + h(t) \tag{5.2}$$

and the cost functional J is written as

$$J(\phi,u) = g(t_0,\phi(t_0),t_1,\phi(t_1)) + \int_{t_0}^{t_1}\{\langle a_0(t),\phi(t)\rangle$$
$$+ \langle b_0(t),u(t)\rangle + h_0(t)\}dt,$$

where $A(t)$, $B(t)$, $a_0(t)$, $b_0(t)$ and $h_0(t)$ have the obvious meaning. Let the real valued functions h_i, a_i^j, and b_i^k, $i = 0,1,\ldots,n$; $j = 1,\ldots,n$; $k = 1,\ldots,m$ be defined and continuous on some fixed interval $[T_0,T_1]$. Then the set $\mathscr{R}$ is the slab $T_0 \le t \le T_1$, $-\infty < x^i < \infty$, $i = 1,\ldots,n$ in (t,x)-space. The set $\mathscr{U}$ is all of E^m.

We leave it as an exercise for the reader to show directly that in the linear problem the sets $\mathscr{Q}^+(t,x)$ are convex if the sets $\Omega(t,x)$ are convex.

REMARK 5.3. In Chapter 6, Theorem 6.2, we shall show that to obtain the existence of an optimal pair in linear systems we can dispense with the requirement that the sets $\Omega(t,x)$ be convex, provided Ω depends only on t and not on x.

An important problem in the class of problems discussed in the next to the last paragraph is the "time optimal problem with linear plant". In this problem the state equations are of the form (5.2) and it is required to bring the system from a given initial position x_0 at a given initial time t_0 to a given terminal position x_1 in such a way as to minimize the time to carry this out. The regulator problem of Sec. 5, Chapter I is an example of such a problem. If t_1 denotes the time at which the trajectory reaches x_1, then we wish to minimize $t_1 - t_0$, and the cost functional becomes $J(\phi,u) = t_1 - t_0$. Thus we can consider J as being obtained either by setting $g(t_0,x_0,t_1,x_1) = t_1 - t_0$ and $f^0 \equiv 0$ or by setting $g \equiv 0$ and $f^0 = 1$.

Another class of problems to which Theorem 5.1 can be applied is the so called class of problems with "linear plant and convex integral cost criterion". In these problems the state equations are given by (5.2) and the cost functional is given by

$$J(\phi,u) = g(t_0,\phi(t_0),t_1,\phi(t_1)) + \int_{t_0}^{t_1} f^0(t,\phi(t),u(t))dt,$$

where for each (t,x) in $\mathscr{R}$ f^0 is a convex function of z on the set $\Omega(t,x)$. Recall that a real valued function ψ defined on a convex set S in E^m, $m \geq 2$, is said to be convex if for all x, y in S and all real $\alpha \geq 0$, $\beta \geq 0$, such that $\alpha + \beta = 1$,

$$\psi(\alpha x + \beta y) \leq \alpha\psi(x) + \beta\psi(y).$$

An important problem in the class of linear problems with convex integral cost criterion is the minimum fuel problem for linear systems. In this problem a linear system is to be brought from a given initial state x_0 to any state x_1 in a specified set of terminal states in such a way as to minimize the fuel consumed during the transfer. The terminal time can either be fixed or free. The control u is required to satisfy constraints $|u^i(t)| \le 1$, $i = 1,\dots,m$. The rate of fuel flow at time t, which we denote by $\beta(t)$, is assumed to be proportional to the magnitude of the control vector as follows

$$\beta(t) = \sum_{i=1}^{n} c^i |u^i(t)| \qquad c^i > 0, \text{ constant.}$$

Thus, the fuel consumed in transferring the system from x_0 to x_1 is

$$J(\phi,u) = \int_{t_0}^{t_1} \Big(\sum_{i=1}^{n} c^i |u^i(t)| \Big) dt.$$

The functional J is to be minimized. Here

$$f^0(t,x,z) = \sum_{i=1}^{n} c^i |z^i|$$

and f^0 is convex in z.

Another important problem in the class of linear problems with convex integral cost criterion is the "quadratic criterion" problem which arises in the following way. An absolutely continuous function ξ is specified on a fixed interval $[t_0,t_1]$. This is usually a desired trajectory for the system. It is required to choose an admissible control u so that the mean square error over $[t_0,t_1]$ between the trajectory ϕ and the given trajectory ξ be minimized and that this be accomplished with minimum energy consumption. If one takes the integral $\int_{t_0}^{t_1} |u|^2 dt$ to be a measure of the energy consumption one is led to consider the cost functional

$$J(\phi,u) = |\phi(t_1)-\xi(t_1)|^2 + \int_{t_0}^{t_1} |\phi(t)-\xi(t)|^2 dt + \int_{t_0}^{t_1} |u(t)|^2 dt.$$

If we set $\overline{\phi}(t) = \phi(t)-\xi(t)$, then since ϕ is a solution of (5.2), $\overline{\phi}$ will also be a solution of a linear system of the form (5.2).

Hence we can suppose that the functional J has the form

$$J(\phi,u) = |\phi(t_1)|^2 + \int_{t_0}^{t_1} |\phi(t)|^2 dt + \int_{t_0}^{t_1} |u(t)|^2 dt.$$

If one assigns non-negative weights to the coordinates of the trajectory and to the components of u, the functional becomes

$$J(\phi,u) = \langle \phi(t_1), R\phi(t_1)\rangle + \int_{t_0}^{t_1} \langle \phi(t), X(t)\phi(t)\rangle dt$$
$$+ \int_{t_0}^{t_1} \langle u(t), U(t)u(t)\rangle dt, \tag{5.3}$$

where X and U are continuous diagonal matrices with non-negative diagonal entries and R is a constant diagonal matrix with non-negative diagonal entries.

More generally, we can take X and U to be continuous positive semi-definite symmetric matrices on $[t_0,t_1]$. Later, when we consider non-compact constraint sets, the matrix U will be required to be positive definite. The generality in assuming that U is not necessarily diagonal is somewhat spurious, as the following discussion shows. There exists a real orthogonal matrix P such that $U = P'DP$, where D is diagonal and the prime denotes "transpose". Under the change of variable $v = Pu$ the quadratic form $\langle u,Uu\rangle$ becomes $\langle v,Dv\rangle$ with D diagonal. The state equations (5.2) become

$$\frac{dx}{dt} = A(t)x + C(t)v(t) + h(t),$$

where $C(t) = B(t)P^{-1}(t)$. If X is a constant matrix then there is a change of variable $y = Qx$, where Q is orthogonal and constant, such that the quadratic form $\langle x,Xx\rangle$ is replaced by $\langle y,Yy\rangle$, with Y diagonal, and the state equations are transformed in equations that are linear in y and v.

The linear problems and the linear problems with convex integral cost criteria are special cases of the following problem, in which the existence of an optimal control and trajectory is a corollary of Theorem 5.1.

COROLLARY 5.1. Let all the hypotheses of Theorem 5.1 hold with the exception of (v) and (vi) and let Ω have the further property that for each (t,x) in $\mathscr{R}_0$ the set $\Omega(t,x)$ is a convex set in E^m. Let h be a continuous function from $\mathscr{R}$ to E^n, let B be an $n \times m$ matrix continuous on $\mathscr{R}$ and let f^0 be a real valued lower semicontinuous function on $\mathscr{G}$ such that for each (t,x) in $\mathscr{R}_0$ f^0 is a convex function of z on $\Omega(t,x)$. Let the state equations be

$$\frac{dx}{dt} = h(t,x) + B(t,x)u(t),$$

and let the cost functional be

$$J(\phi,u) = g(t_0,\phi(t_0),t_1,\phi(t_1)) + \int_{t_0}^{t_1} f^0(t,\phi(t),u(t))dt.$$

Then for each (t,x) in $\mathscr{R}_0$ the set $\mathscr{Q}^+(t,x)$ is convex and the functional J attains its minimum in $\mathscr{A}$.

REMARK 5.4. In an important class of problems with state equations linear in x, the trajectories will always lie in a compact set, provided the initial points lie in a compact set. In these problems hypothesis (i) of Theorem 5.1 is always fulfilled. This will be explored in Exercises 5.1 and 5.2 below.

To establish the corollary we need only show that for each (t,x) in $\mathscr{R}_0$ the set $\mathscr{Q}^+(t,x)$ is convex, for then all of the hypotheses of Theorem 5.1 will hold. (The validity of (vi) follows from the continuity properties of f^0, h and B).

Let $\hat{y} = (y^0,y)$. In the present problem

$$\mathscr{Q}^+(t,x) = \{\hat{y} = (y^0,y): y^0 \geq f^0(t,x,z),\ y = h(t,x) + B(t,x)z,\ z \in \Omega(t,x)\}.$$

Let $\hat{y}_1$ and $\hat{y}_2$ be any two points of $\mathscr{Q}^+(t,x)$. Then there exists points z_1 and z_2 in $\Omega(t,x)$ such that

$$y_1^0 \geq f^0(t,x,z_1) \qquad y_1 = h(t,x) + B(t,x)z_1$$
$$y_2^0 \geq f^0(t,x,z_2) \qquad y_2 = h(t,x) + B(t,x)z_2.$$

Let $\alpha \geq 0$, $\beta \geq 0$, $\alpha + \beta = 1$ and let $\hat{y}_3 = \alpha\hat{y}_1 + \beta\hat{y}_2$. To show that $\mathscr{Q}^+(t,x)$ is convex we must show that $\hat{y}_3 \in \mathscr{Q}^+(t,x)$. We have

$$\begin{aligned} y_3 = \alpha y_1 + \beta y_2 &= h(t,x) + \alpha[B(t,x)z_1] + \beta[B(t,x)z_2] \\ &= h(t,x) + B(t,x)(\alpha z_1 + \beta z_2). \end{aligned}$$

Since $\Omega(t,x)$ is convex, there exists a z_3 in $\Omega(t,x)$ such that $z_3 = \alpha z_1 + \beta z_2$. Hence

$$y_3 = h(t,x) + B(t,x)z_3 \qquad z_3 \in \Omega(t,x).$$

From the convexity of f^0 and the definition of z_3 we have

$$\begin{aligned} y_3^0 = \alpha y_1^0 + \beta y_2^0 &\geq \alpha f^0(t,x,z_1) + \beta f^0(t,x,z_2) \\ &\geq f^0(t,x,\alpha z_1 + \beta z_2) = f^0(t,x,z_3). \end{aligned}$$

Hence $\hat{y}_3 \in \mathscr{Q}^+(t,x)$.

We next discuss the relevance of the various hypotheses.

Example 2.2 illustrates the need for the assumption that the sets $\Omega(t,x)$ are compact. In this example the mapping Ω is u.s.c.i. since $\Omega(t,x) = E^1$ for all (t,x). The sets $\mathscr{Q}^+(t,x)$ were already seen to be convex. Moreover, if we consider any subclass $\mathscr{A}_0$ of $\mathscr{A}$ defined by requiring the trajectories to lie in a compact set $\mathscr{R}_0$ of the form

$$\mathscr{R}_0 = \{(t,x)\colon 0 \leq t \leq 1,\ a \leq x \leq b\} \qquad a \leq 0,\ b \geq 1,$$

then the argument in Example 2.2 shows that the minimum does not exist in $\mathscr{A}_0$. Hence, since $\mathscr{B}$ is a point and $g \equiv 0$, all of the hypotheses of Theorem 5.1 are satisfied, except the hypothesis that $\Omega(t,x)$ is compact for all (t,x) in $\mathscr{R}_0$.

Example 2.2(a) shows that if the compactness of the sets $\Omega(t,x)$ fails to hold at a single point then the conclusion of the theorem may fail. Example 2.2(b) shows that if the upper semicontinuity with respect to inclusion fails, even at one point, then the conclusion of Theorem 5.1 may fail. Note that in Examples 2.2(a) and 2.2(b) the set $\mathscr{D}$ defined in (4.1) is not compact. In Example 2.2(b) $\mathscr{D}$ even fails to be closed. Note that Ω is not upper semicontinuous at points of the form $(0,x)$.

In Example 2.3 we can again restrict our attention to a compact set $\mathscr{R}_0$ and fail to get existence. Now all of the hypotheses are fulfilled except the requirement that $\mathscr{Q}^+(t,x)$ be convex.

Example 2.4 illustrates the need for the compactness of $\mathscr{R}_0$. If we take $\mathscr{R}_0$ to be the closure of the region covered by $\mathscr{F}$, then $\mathscr{R}_0$ is not compact. Since $f^0 \equiv 0$ in this example it is clear from (2.3) and the definition of Ω that for each (t,x) the set $\mathscr{Q}^+(t,x)$ is convex. Clearly $\Omega(t,x)$ is compact for all (t,x), and is u.s.c.i. Since $\mathscr{T}_0$ and $\mathscr{T}_1$ are compact, the set $\mathscr{B}$ is compact. Since $g(t_0,x_0,t_1,x_1) = x_1$, the function g is continuous. Thus all of the conditions of Theorem 5.1 except the compactness of $\mathscr{R}_0$ are fulfilled.

We next consider conditions that guarantee the existence of a compact set $\mathscr{R}_0$ such that the graphs of all admissible trajectories lie in $\mathscr{R}_0$ as required in (i) of the hypotheses. These conditions are not included in the statement of the theorem because they are too restrictive. Trajectories can lie in compact sets even if these conditions fail. Furthermore, in specific problems one can often establish directly that the trajectories lie in a compact set.

LEMMA 5.1. Let $\mathscr{R}$ be contained in the slab $a \leq t \leq b$,

$-\infty < x^i < \infty$, $i = 1,\dots,n$. Let $\Delta = \{(t,x,z): (t,x) \in \mathscr{R},\ z \in \Omega(t,x)\}$. Let the function $f = (f^1,\dots,f^n)$ satisfy

$$|\langle x, f(t,x,z)\rangle| \leq K(t)(|x|^2 + 1) \tag{5.4}$$

for all (t,x,z) in $\mathscr{D}$, where K is a function that is integrable on $[a,b]$. Let each admissible trajectory contain at least one point $(t_2,\phi(t_2))$ that belongs to a given compact set C in $\mathscr{R}$. Then there exists a compact set $\mathscr{R}_0$ contained in $\mathscr{R}$ such that all trajectories in $\mathscr{A}$ lie in $\mathscr{R}_0$. If we require that all initial points of trajectories in $\mathscr{A}$ lie in C then we can omit the absolute value in the left hand side of (5.4).

Proof. For any trajectory ϕ let $\Phi(t) = |\phi(t)|^2+1 = \langle\phi(t),\phi(t)\rangle+1$. Then $\Phi'(t) = 2\langle\phi(t), f(t,\phi(t),u(t))\rangle$, and by virtue of (5.4)

$$|\Phi'(t)| \leq 2K(t)(|\phi(t)|^2+1) = 2K(t)\Phi(t).$$

Hence

$$-2K(t)\Phi(t) \leq \Phi'(t) \leq 2K(t)\Phi(t). \tag{5.5}$$

If $(t_2,\phi(t_2))$ is a point of the trajectory that belongs to C, then upon integrating (5.5) we get

$$\Phi(t) \leq \Phi(t_2)\exp\left(2\left|\int_{t_2}^{t} K(s)ds\right|\right) \leq \Phi(t_2)\exp\left(2\int_a^b K(s)ds\right)$$

for all points of the trajectory. Since C is compact there exists a constant D such that if (t,x) is in C then $|x| \leq D$. Hence

$$\Phi(t) \leq (D^2+1)\exp\left(2\int_a^b K(s)ds\right).$$

Since the right hand side of this inequality is a constant and is independent of the trajectory ϕ, it follows that all trajectories lie in some compact set $\mathscr{R}_0$.

If the initial points (t_0,x_0) all lie in a compact set, we need only utilize the rightmost inequality in (5.5) to obtain a bound on $\Phi(t)$ that is independent of ϕ. We may therefore omit the absolute value in the left hand side of (5.4) in this case.

In some problems it is possible to show that there is a compact set $\mathscr{R}_0$ such that those trajectories that do not lie in $\mathscr{R}_0$ give larger values to J or $\hat{J}$ than do those that lie in $\mathscr{R}_0$. In that event one can ignore the trajectories that do not lie in $\mathscr{R}_0$. One simply redefines $\mathscr{R}$ to be $\mathscr{R}_0$ and redefines the set of admissible pairs to be those pairs whose trajectories lie in $\mathscr{R}_0 \equiv \mathscr{R}$. An example of this will be given in Section 6 in connection with the linear quadratic problem with non-compact constraints and in Exercise 5.

We now prove Theorem 5.1. We begin with the following lemma.

LEMMA 5.2. Under the hypotheses of Theorem 5.1 the set $\mathscr{D}$ defined in (4.1) is compact and for each (t,x) in $\mathscr{R}_0$ the set $\mathscr{Q}^+(t,x)$ is closed and convex.

In the paragraph preceding the statement of Theorem 5.1 we showed that hypotheses (iii) and (iv) of Theorem 5.1 imply that Ω is upper semicontinuous on $\mathscr{R}_0$. Hence, by Lemma 3.1 $\mathscr{D}$ is closed. To show that $\mathscr{D}$ is compact we must therefore show that it is bounded. If $\mathscr{D}$ were unbounded, there would exist a sequence $\{(t_n,x_n,z_n)\}$ and a point (t_0,x_0) in $\mathscr{R}_0$ such that $(t_n,x_n) \to (t_0,x_0)$ and $|z_n| \to \infty$. Since Ω is u.s.c.i. there exists an n_0 such that for $n > n_0$, $\Omega(t_n,x_n) \subseteq [\Omega(t_0,x_0)]_1$. Since $\Omega(t_0,x_0)$ is compact, so is $[\Omega(t_0,x_0)]_1$. Since $z_n \in \Omega(t_n,x_n)$ it follows that $|z_n| \to \infty$ is not possible.

By (v) of Theorem 5.1 each set $\mathscr{Q}^+(t,x)$ is convex. To show that each $\mathscr{Q}^+(t,x)$ is closed let (t,x) be fixed and let $\{\eta_n\} = (y_n^0,y_n)$ be a sequence of points in $\mathscr{Q}^+(t,x)$ converging to a point

$\eta_0 = (y_0^0, y_0)$. By the definition of $\mathscr{Q}^+(t,x)$ there exists a sequence of points $z_n \in \Omega(t,x)$ such that $y_n = f(t,x,z_n)$ and $y_n^0 \geq f^0(t,x,z_n)$. Since $\Omega(t,x)$ is compact, there exists a subsequence, again called $\{z_n\}$, and a point z_0 in $\Omega(t,x)$ such that $z_n \to z_0$. From the lower semicontinuity of f^0 and the continuity of f we get

$$y^0 = \lim y_n^0 \geq \lim\inf f^0(t,x,z_n) \geq f^0(t,x,z_0)$$
$$y = \lim y_n = \lim f(t,x,z_n) = f(t,x,z_0).$$

Hence (y_0^0, y_0) is in $\mathscr{Q}^+(t,x)$ and the lemma is proved.

We now show that Assumption 4.1 of Theorem 4.2 holds. Assumptions (i), (ii) and (vi) of Theorem 5.1 are identical with (i), (ii) and (vi) of Assumption 4.1. In the course of proving Lemma 5.2 we showed that Ω is upper semicontinuous on $\mathscr{R}_0$ so that (iii) of Assumption 4.1 holds. Since f^0 is lower semicontinuous and $\mathscr{D}$ is compact, f^0 is bounded from below on $\mathscr{D}$. Thus (v) of Assumption 4.1 holds. The last assertion of Lemma 5.2 is (iv) of Assumption 4.1. Thus all of Assumption 4.1 holds.

We next show that under the hypotheses of Theorem 5.1, the mapping $\mathscr{Q}^+$ has the Cesari property at each (t,x) in $\mathscr{R}_0$. This, of course, implies that the weak Cesari property holds. In the demonstration we shall need a certain theorem of Caratheodory about convex sets. Since this theorem is used in several important places in the theory of optimal control, we state it here for future reference. We refer the reader to any text on convexity for the proof. See e.g. Eggleston [21].

THEOREM 5.2. Let $\mathscr{X}$ be a subset of E^n and let y be a point of co $\mathscr{X}$, the convex hull of $\mathscr{X}$. Then y can be written as a convex combination of at most $n+1$ points in $\mathscr{X}$. If, in addition, $\mathscr{X}$ has at most n components then y can be written as convex combination of at most n points in $\mathscr{X}$.

Let (t,x) be fixed, let $y = (y^1,\ldots,y^n)$ and let

$$\hat{y} = (y^0,y) \in \bigcap_{\delta>0} \text{cl co } \mathscr{Q}^+(N_\delta(t,x)).$$

Then there exists a sequence of positive numbers $\{\delta_k\}$ and a sequence of points $\{\hat{y}_k\} = \{(y_k^0,y_k)\}$ in E^{n+1} such that $\delta_k \to 0$ and

$$\hat{y}_k = (y_k^0,y_k) \in \text{co } \mathscr{Q}^+(N_{\delta_k}(t,x))$$

$$\hat{y}_k = (y_k^0,y_k) \to (y^0,y) = \hat{y}.$$

Hence for each integer k there exists an integer $j = j(k)$, real numbers $\alpha_{k1},\ldots,\alpha_{kj}$ with

$$\alpha_{ki} \geq 0 \qquad \sum_{i=1}^{j} \alpha_{ki} = 1, \tag{5.6}$$

points $(t_{k1},x_{k1}),\ldots,(t_{kj},x_{kj})$ in $\mathscr{R}_0$ and points $\hat{y}_{k1},\ldots,\hat{y}_{kj}$ such that

$$\begin{aligned} &|(t_{ki},x_{ki})-(t,x)| < \delta_k \\ &\hat{y}_{ki} \in \mathscr{Q}^+(t_{ki},x_{ki}) \qquad i = 1,\ldots,j \end{aligned} \tag{5.7}$$

and

$$\hat{y}_k = \sum_{i=1}^{j} \alpha_{ki}\hat{y}_{ki}. \tag{5.8}$$

From the second relation in (5.7) it follows that there exist points $z_{k1},\ldots,z_{kj}$ with $z_{ki} \in \Omega(t_{ki},x_{ki})$ such that

$$\begin{aligned} &y_{ki}^0 \geq f^0(t_{ki},x_{ki},z_{ki}) \\ &y_{ki} = f(t_{ki},x_{ki},z_{ki}) \qquad i = 1,\ldots,j. \end{aligned} \tag{5.9}$$

The set $\text{co } \mathscr{Q}^+(N_{\delta_k}(t,x))$ is a set in E^{n+1}. Hence by Theorem 5.2 every point in $\text{co } \mathscr{Q}^+(N_{\delta_k}(t,x))$ can be written as a convex combination of $n+2$ points in $\mathscr{Q}^+(N_{\delta_k}(t,x))$. Therefore, we suppose that $j = j(k) = n+2$, for every integer k.

From (5.6) it follows that the sequence $\{(\alpha_{k1},\ldots,\alpha_{k,n+2})\}_{k=1}^{\infty}$ is bounded in E^{n+2}. Hence there exists a subsequence, again denoted as $\{(\alpha_{k1},\ldots,\alpha_{k,n+2})\}$ and a point $(\alpha_1,\ldots,\alpha_{n+2})$ such that

$$(\alpha_{k1},\ldots,\alpha_{k,n+2}) \to (\alpha_1,\ldots,\alpha_{n+2}). \tag{5.10}$$

Moreover, $\alpha_i \geq 0$ and $\Sigma\alpha_i = 1$. For each $i = 1,\ldots,n+2$ let $\{(t_{ki},x_{ki},z_{ki})\}_{k=1}^{\infty}$ denote the subsequence corresponding to the convergent sequence $\{(\alpha_{k1},\ldots,\alpha_{k,n+2})\}$. Since $\mathscr{D}$ is compact there exists a subsequence and points $z_1,\ldots,z_{n+2}$ in E^m such that for all $i = 1,2,\ldots,n+2$

$$(t_{ki},x_{ki},z_{ki}) \to (t,x,z_i), \tag{5.11}$$

where $(t,x,z_i) \in \mathscr{D}$; i.e., $z_i \in \Omega(t,x)$.

From (5.8) and (5.9) we have

$$y_k^0 \geq \sum_{i=1}^{n+2} \alpha_{ki} f^0(t_{ki},x_{ki},z_{ki})$$

$$y_k = \sum_{i=1}^{n+2} \alpha_{ki} f(t_{ki},x_{ki},z_{ki}).$$

From the convergence of $\hat{y}_k$ to $\hat{y}$, (5.10), (5.11), the continuity of f and the lower semicontinuity of f^0 we get

$$y = \sum_{i=1}^{n+2} \alpha_i f(t,x,z_i)$$

and

$$\begin{aligned} y^0 = \lim y_k^0 &\geq \lim \Big(\sum_{i=1}^{n+2} \alpha_{ki} f^0(t_{ki},x_{ki},z_{ki})\Big) \\ &\geq \sum_{i=1}^{n+2} \liminf \alpha_{ki} f^0(t_{ki},x_{ki},z_{ki}) \\ &\geq \sum_{i=1}^{n+2} \alpha_i f^0(t,x,z_i). \end{aligned}$$

Hence $\hat{y} = (y^0,y)$ is in co $\mathscr{Q}^+(t,x)$. But $\mathscr{Q}^+(t,x)$ is convex, so that co $\mathscr{Q}^+(t,x) = \mathscr{Q}^+(t,x)$. Hence $\hat{y} \in \mathscr{Q}^+(t,x)$ and the Cesari property holds.

Since all trajectories lie in a compact set $\mathscr{R}_0$ and since $\mathscr{B}$ is closed, the set of end points $\{(t_0,\phi(t_0),t_1,\phi(t_1))\}$ of admissible trajectories will be contained in a compact subset of $\mathscr{B}$. Hence we may assume that $\mathscr{B}$ itself is compact.

Since g is lower semicontinuous on $\mathscr{B}$ and $\mathscr{B}$ is compact, g is bounded below on $\mathscr{B}$. Hence the mapping G, which in this case is defined by $G(\phi) = g(t_0,\phi(t_0),t_1,\phi(t_1))$ is lower semicontinuous on $\mathscr{A}_T$ and is bounded below.

We complete the proof of Theorem 5.1 by showing that the admissible trajectories ϕ in $\mathscr{A}_T$ are equi-absolutely continuous. It will therefore be true that the trajectories in any minimizing sequence are equi-absolutely continuous, and all of the hypotheses of Theorem 4.2 will be verified.

Since $\mathscr{D}$ is compact and f is continuous there exists a constant $C > 0$ such that $|f(t,x,z)| \leq C$ for all (t,x,z) in $\mathscr{D}$. Let $\varepsilon > 0$ be given. Then for any ϕ in $\mathscr{A}_T$ defined on $[t_0,t_1]$ and any measurable set $E \subset [t_0,t_1]$ with meas $(E) < \varepsilon/C$,

$$\left|\int_E \phi'(t)dt\right| \leq \int_E |\phi'(t)|dt = \int_E |f(t,\phi(t),u(t)|dt < \varepsilon.$$

Hence the functions ϕ' have equi-absolutely continuous integrals. As noted after Definition 3.4 this implies that the functions ϕ in $\mathscr{A}_T$ are equi-absolutely continuous.

EXERCISE 5.1. Consider a system whose state equations are (5.2) where A, B, and h are integrable on an interval $[a,b]$. Let the mapping Ω depend on t alone and Ω be defined on $[a,b]$. Thus, $\Omega\colon t \to \Omega(t)$ for all t in $[a,b]$. Furthermore, let the mapping Ω be u.s.c.i. on $[a,b]$ and let each set $\Omega(t)$ be compact. Show that (5.4) holds, and hence that all admissible trajectories lie in a compact set $\mathscr{R}_0$, provided each trajectory has at least one point in a given compact set C. Prove the last statement without using Lemma 5.1. (Hint: Use the variation of parameters formula). Show that any measurable function u defined on $[a,b]$ such that $u(t) \in \Omega(t)$ a.e. is admissible.

EXERCISE 5.2. Consider a system whose state equations are $dx/dt = A(t)x + B(t,u(t))$, where A is integrable on $[a,b]$ and B is continuous on $[a,b] \times E^m$. Let all the other conditions be as in Exercise 5.1. Carry out the demonstrations required in Exercise 5.1 in the present case.

EXERCISE 5.3. Show that Lemma 5.1 holds if we replace (5.4) by the following hypothesis. There exists a positive function V defined and $C^{(1)}$ on $\mathscr{R}$ and a positive constant K such that (i)

$$|\langle V_x(t,x), f(t,x,z)\rangle + V_t(t,x)| \le KV(t,x)$$

for all (t,x,z) in $\mathscr{D}$ and (ii) the set $\{(t,x): V(t,x) \le \alpha;$ $(t,x) \in \mathscr{R}\}$ is compact for every α.

EXERCISE 5.4. Show that in Theorem 5.1 the condition, "$\mathscr{Q}^+(t,x)$ is convex" cannot be replaced by the following assumption: The sets $\mathscr{Q}(t,x)$ are convex and f^0 is a convex function of z for each fixed (t,x) in $\mathscr{R}$. (Cesari [17], p. 399-400).

6. Non Compact Constraints

In this section we consider problems in which the constraint sets $\Omega(t,x)$ need not be compact. The principal contribution of Theorem 6.1 and its corollaries is the statement of conditions guaranteeing the equi-absolute continuity of the trajectories in a minimizing sequence. Theorem 6.2 is the classical Nagumo-Tonelli existence theorem for ordinary problems in the calculus of variations. In the exercises at the end of this section we shall take up the important special class of problems with "linear plant, convex integral cost-criterion and unbounded controls".

Note that if Assumption 4.1 holds then there exists a compact interval $\mathscr{I} = [a,b]$ such that the interval of definition $[t_0,t_1]$ of any admissible trajectory is contained in $\mathscr{I}$.

THEOREM 6.1. Let Assumption 4.1 hold and let the mapping $\mathscr{Q}^+$ satisfy the weak Cesari property at every point of $\mathscr{R}_0$. Let G be lower semicontinuous and bounded below on $\mathscr{A}_T$. Let $\{(\phi_k,u_k)\}$ be a minimizing sequence. For each $i = 1,\ldots,n$ let there exist a non-negative function H^i defined on $\mathscr{D}$ and a constant A^i such that for all (ϕ_k,u_k) in the minimizing sequence

$$\int_{t_{0k}}^{t_{1k}} H^i(t,\phi_k(t),u_k(t))dt \leq A^i, \tag{6.1}$$

where $[t_{0k},t_{1k}]$ is the interval of definition of (ϕ_k,u_k). For each $i = 1,\ldots,n$ and for every $\varepsilon > 0$ let there exist a non-negative function F^i_ε in $L_1[\mathscr{I}]$ such that for all (ϕ_k,u_k) in the minimizing sequence

$$|f^i(t,\phi_k(t),u_k(t))| \leq F^i_\varepsilon(t)+\varepsilon H^i(t,\phi_k(t),u_k(t)) \quad \text{a.e.} \tag{6.2}$$

Then there exists a (ϕ^*,u^*) in $\mathscr{A}$ such that $\hat{J}(\phi^*,u^*) \leq \hat{J}(\phi,u)$.

All of the hypotheses of Theorem 4.2, except the equi-absolute continuity of the functions ϕ_k, are either restatements or immediate consequences of the hypotheses of Theorem 6.1. Before showing that (6.1) and (6.2) or (6.3) below imply the desired equi-absolute continuity we point out the following.

REMARK 6.1. Since G is bounded below, the sequence $\{I(\phi_k,u_k)\}$ is bounded above. Hence if we take $H^i = f^0 + K$, where K is as in (v) of Assumption 4.1, then (6.1) holds.

REMARK 6.2. If (6.1) and (6.2) hold with $H^i = f^0 + K$ and $A^i = A$, then it can be shown that the optimal pair satisfies (6.1) and the corresponding components of $\phi^{*\prime}$ satisfy (6.2) with the argument of H^i now being $(t,\phi^*(t),u^*(t))$. This will be done in Section 7 as a corollary to the proof of Theorem 4.1.

We now show that (6.1) and (6.2) imply the equi-absolute continuity of the ϕ_k. Let $\eta > 0$ be given and let ε in (6.2) be equal to $\eta/2A^i$.

There exists a $\delta > 0$ such that if E is a measurable set in $\mathscr{I}$ with meas $(E) < \delta$, then $\int_E F_\varepsilon^i dt < \eta/2$. The number δ depends on F_ε^i as well as on η, but since $\varepsilon = \eta/2A^i$, the number δ ultimately depends on η alone. If we set $f^i(t,\phi_k(t),u_k(t)) = 0$ if $t \notin [t_{0k},t_{1k}]$ then from (6.1), (6.2) and the choice of ε we get

$$\int_E |f^i(t,\phi_k(t),u_k(t))|dt \leq \int_E F_\varepsilon^i dt+\varepsilon \int_E H^i(t,\phi_k(t),u_k(t))dt$$
$$< \eta/2 + (\eta/2A^i)A^i = \eta,$$

for all (ϕ_k,u_k) and measurable E in $\mathscr{I}$ with meas $(E) < \delta$. Since $\phi_k'(t) = f(t,\phi_k(t),u_k(t))$, the equi-absolute continuity of the functions ϕ_k^i follows.

COROLLARY 6.1. Let Φ be a positive non decreasing function defined on $[0,\infty)$ such that $\Phi(\xi) \to \infty$ as $\xi \to \infty$. Let $f_k^i(t) = f^i(t,\phi_k(t),u_k(t))$ and let there exist a constant C^i such that

$$\int_{t_{0k}}^{t_{1k}} |f_k^i(t)|\Phi(|f_k^i(t)|)dt \leq C^i . \tag{6.3}$$

Then the functions ϕ_k^i are equi-absolutely continuous. Thus conditions (6.1) and (6.2) can be replaced by (6.3) for any $i = 1,\ldots,n$.

We first show that the composite functions $\Phi(|f_k^i|)$ are measurable. Let α be given. Since Φ is non decreasing

$\{\xi: \Phi(\xi) \le \alpha\}$ is a set of the form $[0,\xi_0)$ or $[0,\xi_0]$. Since $|f_k^i|$ is measurable $\{t: |f_k^i(t)| \le \xi_0\}$ or $\{t: |f_k^i(t)| < \xi_0\}$ is measurable. Hence $\{t: \Phi(|f_k^i(t)|) \le \alpha\}$ is measurable.

Corollary 6.1 is obtained by showing that (6.3) implies that (6.1) and (6.2) hold. Let $H^i(t,x,z) = |f^i(t,x,z)|\Phi(|f^i(t,x,z)|)$. Let $\varepsilon > 0$ be given. Then there exists an $M_\varepsilon > 0$ such that if $\xi \ge M_\varepsilon$, then $\Phi(\xi) > 1/\varepsilon$. For each (ϕ_k, u_k) let

$$E_{1k} = \{t: |f^i(t,\phi_k(t),u_k(t))| \le M_\varepsilon\}$$
$$E_{2k} = \{t: |f^i(t,\phi_k(t),u_k(t))| > M_\varepsilon\}.$$

For t in E_{2k}

$$|f_k^i(t)| = \frac{|f_k^i(t)|\Phi(|f_k^i(t)|)}{\Phi(|f_k^i(t)|)} \le \frac{|f_k^i(t)|\Phi(|f_k^i(t)|)}{\Phi(M_\varepsilon)}$$

$$< \varepsilon H^i(t,\phi_k(t),u_k(t)).$$

Since $[t_{0k}, t_{1k}] = E_{1k} \cup E_{2k} \cup Z$, where Z is a set of measure zero,

$$|f_k^i(t)| \le M_\varepsilon + \varepsilon H^i(t,\phi_k(t),u_k(t)), \quad \text{a.e.}$$

which is (6.2). The relation (6.1) follows from (6.3) and the definition of H^i.

COROLLARY 6.2. Let there exist a constant $C^i > 0$ and a $p_i > 1$ such that for all elements of the minimizing sequence

$$\int_{t_{0k}}^{t_{1k}} |\phi_k^{i\prime}(t)|^{p_i} dt \le C^i.$$

Then the functions ϕ_k^i are equi-absolutely continuous.

Corollary 6.2 is obtained from Corollary 6.1 by taking $\Phi(\xi) = \xi^{p_i - 1}$. It is also an immediate consequence of Hölder's inequality.

In II.6 we formulated the simple problem in the calculus of

variations and showed how it can be written as a control problem. We now further assume that the region $\mathscr{G}$ in II.6 is of the form $\mathscr{R} \times E^n$, where $\mathscr{R}$ is a region in (t,x)-space. Thus, in the control formulation of the variational problem the set $\Omega(t,x)$ is E^n for all (t,x). We now state and prove one of the standard existence theorems in the calculus of variations.

THEOREM 6.2. Let f^0 be lower semicontinuous on $\mathscr{G} = \mathscr{R} \times E^n$ and let $f^0(t,x,z) \geq 0$ for all (t,x,z) in $\mathscr{G}$. For each (t,x) in $\mathscr{R}$ let f^0 be a convex function of z. Let the set of admissible trajectories be non empty and let there exist a compact subset $\mathscr{R}_0$ of $\mathscr{R}$ such that the graphs of all trajectories lie in $\mathscr{R}_0$. Let $\mathscr{B}$ be closed and let g be lower semicontinuous on $\mathscr{B}$. Let there exist a non-negative function Φ defined on $[0,\infty)$ such that $\Phi(\xi)/\xi \to \infty$ as $\xi \to \infty$ and such that for all (t,x,z) in $\mathscr{R}_0 \times E^n$, $f^0(t,x,z) \geq \Phi(|z|)$. Then there exists an absolutely continuous function ϕ^* that is admissible and that minimizes

$$J(\phi) = g(t_0,\phi(t_0),t_1,\phi(t_1)) + \int_{t_0}^{t_1} f^0(t,\phi(t),\phi'(t))dt.$$

It is immediately verified that if the variational problem is written as a control problem, then all of the statements of Assumption 4.1 hold under the hypotheses of Theorem 6.2. Since the trajectories all lie in a compact set $\mathscr{R}_0$ and since $\mathscr{B}$ is closed, it follows that all of the end points of trajectories lie in a compact subset of $\mathscr{B}$. Therefore, since g is lower semicontinuous on $\mathscr{B}$ it is bounded below on the set of end points. Hence the functional G, defined by $G(\phi) = g(t_0,\phi(t_0),t_1,\phi(t_1))$, is lower semicontinuous and bounded below on $\mathscr{A}_T$. From this and from the inequality $f^0(t,x,z) \geq 0$, it follows that if $\mu = \inf\{J(\phi) : \phi \in \mathscr{A}_T\}$, then μ is finite. Moreover, if $\{\phi_k\}$ is a minimizing sequence, there exists a positive constant A such that

$$\int_{t_{0k}}^{t_{1k}} f^0(t,\phi_k(t),u_k(t))dt \le A,$$

where $u_k(t) = \phi_k'(t)$. Note that u_k is finite almost everywhere.

Let $\varepsilon > 0$ be given. Then there exists a positive number M_ε such that if $|u_k(t)| \ge M_\varepsilon$, then

$$\frac{f^0(t,\phi_k(t),u_k(t))}{|u_k(t)|} \ge \frac{\Phi(|u_k(t)|)}{|u_k(t)|} \ge \frac{1}{\varepsilon} .$$

Hence for almost all t in $[t_{0k},t_{1k}]$

$$|u_k(t)| \le M_\varepsilon + \varepsilon f^0(t,\phi_k(t),u_k(t)).$$

Since $f(t,x,z) = z$ in the control formulation of the simple problem in the calculus of variations it follows that (6.1) and (6.2) hold with $H^i = f^0$ for all elements $(\phi_k,u_k) = (\phi_k,\phi_k')$ of a minimizing sequence and all $i = 1,\ldots,n$.

Thus far we have verified that under the hypotheses of Theorem 6.2 all of the hypotheses of Theorem 6.1 hold, with the exception of the assumption that $\mathscr{Q}^+$ satisfies the weak Cesari property. In Lemma 6.1 below we shall show that $\mathscr{Q}^+$ satisfies the Cesari property, and thereby show that all of the conditions of Theorem 6.1 are satisfied. This will complete the proof of Theorem 6.2.

LEMMA 6.1. Under the hypotheses of Theorem 6.2 the mapping $\mathscr{Q}^+$ satisfies the Cesari property.

<u>Proof</u>. Let (t,x) be a point in $\mathscr{R}$, let $y = (y^1,\ldots,y^n)$ and let

$$\hat{y} = (y^0,y) \in \bigcap_{\delta>0} \text{cl co } \mathscr{Q}^+(N_\delta(t,x)).$$

We must show that $\hat{y} \in \mathscr{Q}^+(t,x)$.

Let $\{\delta_k\}$, $\{\hat{y}_k\}$, $\{\hat{y}_{ki}\}$, $\{z_{ki}\}$, $\{\alpha_{ki}\}$, and $\{(t_{ki},x_{ki})\}$,

$i = 1,2,\ldots,n+2$ be as in the proof of Theorem 5.1 where it is shown that the mapping $\mathscr{Q}^+$ of Theorem 5.1 has the Cesari property. Since now $\mathscr{D} = \mathscr{G}$ and is not compact, we must proceed differently. Since now $f(t,x,z) = z$, we have

$$\{y_{ki}\} = \{z_{ki}\} \qquad i = 1,\ldots,n+2. \tag{6.4}$$

From (5.7) and (5.8) we have

$$\begin{aligned} y_k^0 &= \sum_{i=1}^{n+2} \alpha_{ki} y_{ki}^0 \\ y_{ki}^0 &\geq f^0(t_{ki},x_{ki},z_{ki}) \geq 0. \end{aligned} \tag{6.5}$$

Since $\{y_k^0\}$ is convergent it is bounded. Since for each i, $\alpha_{ki} \geq 0$ and $y_{ki}^0 \geq 0$ it follows that for each i the sequence $\{\alpha_{ki}, y_{ki}^0\}$ is also bounded. Also, for each i the sequence $\{\alpha_{ki}\}$ is bounded. Hence there exist non-negative numbers $\alpha_1,\ldots,\alpha_{n+2}$ with $\Sigma\alpha_i = 1$, real numbers $\eta_1,\ldots,\eta_{n+2}$, and a subsequence of the index set $\{k\}$ which we again label as $\{k\}$ such that for every i

$$\alpha_{ki} \to \alpha_i \qquad \alpha_{ki} y_{ki}^0 \to \eta_i. \tag{6.6}$$

Since $\Sigma\alpha_i = 1$ and $\alpha_i \geq 0$ it follows that, after a relabeling of components if necessary, there is a positive integer $s \leq n+2$ such that $\alpha_i > 0$ for $i = 1,\ldots,s$ and $\alpha_i = 0$ for $i = s+1,\ldots,n+2$ if $s < n+2$. Hence for $i = 1,\ldots,s$

$$y_{ki}^0 \to \eta_i/\alpha_i. \tag{6.7}$$

Let $y_i^0 = \eta_i/\alpha_i$. By hypothesis we have

$$\frac{f^0(t,x,z)}{|z|} \geq \frac{\Phi(|z|)}{|z|} \to \infty \qquad \text{as} \quad |z| \to \infty. \tag{6.8}$$

From the inequality in (6.5), and from (6.7) and (6.8) it follows that the sequence $\{(z_{k1},\ldots,z_{ks})\}$ is bounded. Hence there exist a subsequence which we relabel as $\{(z_{k1},\ldots,z_{ks})\}$ and points $z_1,\ldots,z_s$ such that

$$z_{ki} \to z_i. \tag{6.9}$$

Now let $i > s$. If the sequence $\{z_{ki}\}$ is bounded, then since $\alpha_{ki} \to 0$, it follows that $\alpha_{ki}z_{ki} \to 0$. If the sequence $\{z_{ki}\}$ is unbounded, there exists a subsequence $\{z_{ki}\}$ such that $|z_{ki}| \to \infty$. Then by virtue of (6.5) and (6.8)

$$\alpha_{ki}y^0_{ki} \geq \alpha_{ki}f^0(t_{ki},x_{ki},z_{ki}) \geq \alpha_{ki}|z_{ki}|A_{ki},$$

where $A_{ki} \to +\infty$ as $k \to \infty$. But from (6.6), $\alpha_{ki}y^0_{ki} \to \eta_i$, so that we must have $\alpha_{ki}z_{ki} \to 0$ if $i > s$. Combining this result with (6.6) and (6.9) gives

$$\sum_{i=1}^{n+2} \alpha_{ki}z_{ki} \to \sum_{i=1}^{s} \alpha_i z_i, \tag{6.10}$$

where $\alpha_i \geq 0$ and $\Sigma\alpha_i = 1$.

Hence, since f^0 is convex in z for each fixed (t,x) and is lower semicontinuous,

$$\begin{aligned}
y^0 = \lim_k y^0_k &= \lim_k \sum_{i=1}^{n+2} \alpha_{ki}y^0_{ki} \\
&\geq \liminf_k \sum_{i=1}^{n+2} \alpha_{ki}f^0(t_{ki},x_{ki},z_{ki}) \\
&\geq \liminf_k f^0(t_{ki},x_{ki},\sum_{i=1}^{n+2} \alpha_{ki}z_{ki}) \\
&\geq f^0(t,x,\sum_{i=1}^{s} \alpha_i z_i).
\end{aligned}$$

From (6.4) we have

$$y_k = \sum_{i=1}^{n+2} \alpha_{ki}z_{ki}.$$

Since $y_k \to y$ it follows from (6.10) that $y = \sum_{i=1}^{s} \alpha_i z_i$. Hence $y^0 \geq f^0(t,x,y)$. But this says that $(y^0,y) \in \mathscr{Q}^+(t,x)$, and the lemma is proved.

EXERCISE 6.1. Let Φ be a non-decreasing non-negative function on $[0,\infty)$ such that $\Phi(\xi)/\xi \to \infty$ as $\xi \to \infty$. Let f^0 be a real valued lower semicontinuous function on $\mathscr{G} = \mathscr{R} \times E^m$ such that f^0 is a convex function of z for each (t,x) in $\mathscr{R}$ and $f^0(t,x,z) \geq \Phi(|z|)$ for all (t,x,z) in $\mathscr{G}$. Let $f(t,x,z) = h(t,x) + B(t,x)z$, where h and B are continuous functions of x for fixed t. Let

$$\mathscr{Q}^+(t,x) = \{(\eta,\xi): \eta \geq f^0(t,x,z), \quad \xi = f(t,x,z), \quad z \in E^m\}.$$

Show that $\mathscr{Q}^+$ has the weak Cesari property in $\mathscr{R}$.

EXERCISE 6.2. Let $\mathscr{R}_0$ be a compact subset of (t,x)-space. Let Ω, $\mathscr{D}$, f^0 and f be as in Assumption 4.1, except that $K = 0$. The function $f = (f^1,\dots,f^n)$ is said to be of slower growth than f^0, uniformly on $\mathscr{R}_0$ whenever the following holds. For every $\varepsilon > 0$ there is a $K(\varepsilon)$ such that for all (t,x,z) in $\mathscr{D}$ with $|z| > K(\varepsilon)$ the inequality $|f(t,x,z)| \leq \varepsilon f^0(t,x,z)$ holds. Let the function that is identically equal to one and the function f both be of slower growth than f^0 uniformly in $\mathscr{R}_0$ and let $\mathscr{Q}^+(t,x)$ be convex for every $(t,x) \in \mathscr{R}_0$. Show that $\mathscr{Q}^+(t,x)$ has the Cesari property. (See Cesari [18], p. 539-540.)

EXERCISE 6.3. In this exercise we shall develop an existence theorem for the problem with "linear plant, convex integral cost criterion and unbounded controls". Other methods of proof will be taken up in Exercise 8.6 of this chapter. A generalization of the result proved in this exercise will be taken up in Exercise 7.4.

Let Φ be a non-decreasing non-negative function on $[0,\infty)$

such that $\Phi(\xi)/\xi \to \infty$ as $\xi \to \infty$. Let $\mathscr{R}$ be the slab $[a,b] \times R^n$ in R^{n+1}. Let f^0 be a real valued lower semicontinuous function on $\mathscr{G} = \mathscr{R} \times E^m$ such that f^0 is a convex function of z for each (t,x) in $\mathscr{R}$ and $f^0(t,x,z) \geq \Phi(|z|)$ for all (t,x,z) in $\mathscr{G}$. Let

$$f(t,x,z) = A(t)x + B(t)z + h(t), \tag{6.11}$$

where A, B and h are continuous on $[a,b]$. Let $t_0 = a$, $t_1 = b$, and $x_0 = \xi_0$, where ξ_0 is fixed. Let $\mathscr{X}$ be a closed set in R^n and let $\mathscr{B} = \{(a,\xi_0,b,x_1): x_1 \in \mathscr{X}\}$. Let $\Omega(t,x) = E^m$. Let the set of admissible pairs be non-empty and let G be lower semicontinuous and bounded below on $\mathscr{A}_T$.

Show that under the preceding hypotheses there exists an admissible pair that minimizes the functional $\hat{J}$ defined in II (3.3).

Hint: The argument of Theorem 6.2 shows that the controls u_k in a minimizing sequence have equi-absolutely continuous integrals. This observation combined with the variation of parameters formula and the hypothesis that the initial points of all admissible trajectories are fixed yields the fact that all trajectories in a minimizing sequence lie in a compact set $\mathscr{R}_0$. From Exercise 6.1 we see that the weak Cesari property holds.

EXERCISE 6.4. In this exercise we study several functionals G that occur in applications. Let $\mathscr{I}$ and $\mathscr{X}$ be as in Exercise 6.3 and let g be lower semicontinuous on $\mathscr{X}$. Show that the functional G defined by the relation $G(\phi) = g(\phi(b))$ satisfies the hypothesis of Exercise 6.3. Do the same for the functional G defined by $G(\phi) = \max\{|\phi(t)-\xi(t)|: t \in [a,b]\}$, where ξ is a given continuous function defined on $\mathscr{I} = [a,b]$.

EXERCISE 6.5. In this exercise we obtain an existence theorem for the "linear plant quadratic integral cost criterion problem".

Other proofs of this theorem will be taken up in Exercises 8.3 and 8.6.

Let $\mathscr{R}$, $\mathscr{G}$, $\mathscr{X}$, Ω, G and f be as in Exercise 6.3. For all (t,x,z) in $\mathscr{G}$, let

$$f^0(t,x,z) = \langle x, X(t)x\rangle + \langle z, R(t)z\rangle,$$

where X and R are matrix functions defined and continuous on [a,b] with X(t) positive semidefinite for all t in [a,b] and R(t) positive definite for all t in [a,b].

(a) Show that $\int_a^b \langle u(t), R(t)u(t)\rangle dt < \infty$ if and only if $u \in L_2[a,b]$.

Hint: For any positive semi-definite symmetric quadratic form Q, we have $\{\langle \xi, Q\xi\rangle : |\xi| = 1\} = \{\sum_{i=1}^n \lambda_i \eta_i^2 : |\eta| = 1\}$, where the $\lambda_i \geq 0$, $i = 1,\ldots,n$ are the characteristic values of Q. Hence if $\lambda_1(t)$ denotes the largest characteristic value of R(t) and $\lambda_n(t)$ the smallest, we have

$$|u(t)|^2 \lambda_n(t) \leq \langle u(t), R(t)u(t)\rangle \leq |u(t)|^2 \lambda_1(t)$$

for all t in [a,b]. Since λ_n is continuous and $\lambda_n(t) > 0$, it is bounded away from zero on [a,b]. The result now follows.

Note that we have shown that if $\mathscr{X} = E^n$, then the set of admissible controls is $L_2[a,b]$.

(b) Show that under the hypotheses of this exercise, that if there exists an admissible pair, then there exists an admissible pair that minimizes the functional $\hat{J}$ defined in II (3.3).

Hint: One can apply Exercise 6.3. An alternate procedure independent of Exercise 6.3 is the following. In a minimizing sequence we have $\int_a^b |u_k|^2 dt \leq M$ for an appropriate constant M. The variation of parameters formula and the Cauchy-Schwartz inequality then give the equi-absolute continuity of the trajectories $\{\phi_k\}$ in a minimizing sequence. Since the initial point is the same for all trajectories, they must all lie in a compact set. Exercise 6.1 again gives the weak Cesari property.

(c) Is the result true if we replace the requirement that $R(t)$ be positive definite for all t by the requirement that $R(t)$ be positive semi-definite for all t?

7. Proof of Theorem 4.1

Theorem 7.1, which follows, is important in the proof of Theorem 4.1 and elsewhere in optimal control theory.

If T is a measure space and Z is a Hausdorff space then a mapping Γ from T to Z is said to be measurable if the inverse image under Γ of a compact set in Z is a measurable set in T.

THEOREM 7.1. Let T be a measure space, let Z be a Hausdorff space and let D be a topological space that is the union of a countable number of compact metrizable subsets. Let Γ be a measurable map from T to Z and let Φ be a continuous map from D to Z such that $\Gamma(T) \subseteq \Phi(D)$. Then there exists a measurable map μ from T to D such that $\Phi * \mu = \Gamma$, where the symbol $*$ denotes the composition of two maps.

Theorem 7.1 will be proved in Section 10.

We also shall need the following elementary result, whose proof we leave as an exercise.

EXERCISE 7.1. Let $\{s_n\}$ be a sequence of vectors in E^p converging to a limit s. Let $\{n_j\}$ be a subsequence of the integers and let

$$\sigma_{n_j} = \sum_{i=1}^{k} \alpha_{ij} s_{n_j+i}$$

where k depends on n_j and $\alpha_{ij} \geq 0$, $\sum_i \alpha_{ij} = 1$. Then $\sigma_{n_j} \to s$.

Since the graphs of all admissible trajectories are contained in a compact set $\mathscr{R}_0$, there is a compact interval $\mathscr{I} = [a,b]$ such that all intervals $[t_0,t_1]$ of definition of admissible pairs (ϕ,u) are contained in $\mathscr{I}$. If ϕ is an admissible trajectory defined on $[t_0,t_1]$, let $\tilde{\phi}$ be defined as follows. If $t \in [t_0,t_1]$, then $\tilde{\phi}(t) = \phi(t)$; if $a \leq t \leq t_0$, then $\tilde{\phi}(t) = \phi(t_0)$; if $t_1 \leq t \leq b$, then $\tilde{\phi}(t) = \phi(t_1)$. Thus $\tilde{\phi}$ is the extension of ϕ used to construct the metric space $\mathscr{X}_\rho$, except that we take the domain of $\tilde{\phi}$ to be $[a,b]$ instead of $(-\infty,\infty)$. Since all admissible trajectories are defined on intervals contained in $\mathscr{I}$ we may suppose that the metric space $\mathscr{X}_\rho$ is restricted to functions defined on $\mathscr{I}$.

The compactness of $\mathscr{R}_0$ also implies that all the endpoints $(t_0,\phi(t_0),t_1,\phi(t_1))$ of admissible trajectories lie in a bounded subset of $\mathscr{B}$. Since $\mathscr{B}$ is closed, all the end points therefore lie in a compact subset of $\mathscr{B}$. Hence we may assume that $\mathscr{B}$ itself is compact.

We note the following facts which will be used in the proof of Theorem 4.1. By virtue of Lemma 3.1, Assumption 4.1 (iii) is equivalent to the assumption that the set $\mathscr{D}$ is closed. Also the definition of upper semicontinuity and 4.1 (iii) imply that for each (t,x) in $\mathscr{R}$, the set $\Omega(t,x)$ is closed.

In the proof of Theorem 4.1 we shall select subsequences of various sequences. Unless stated otherwise, we shall relabel the subsequence with the labeling of the original sequence. We break the proof up into several steps.

STEP 1. There is a sequence $\{(\phi_k,u_k)\}$ of elements in $\mathscr{A}_0$, a real number γ, points t_0 and t_1 in $\mathscr{I}$ with $t_1 > t_0$ and points x_0 and x_1 in E^n such that for $i = 0,1$, $(t_i,x_i) \in \mathscr{R}_0$,

$(t_0, x_0, t_1, x_1) \in \mathscr{B}$ and

$$t_{ik} \to t_i \qquad \phi_k(t_{ik}) \to x_i \qquad i = 0,1 \tag{7.1}$$
$$I(\phi_k, u_k) \to \gamma.$$

Since f^0 is bounded below on $\mathscr{D}$ and all intervals of definition of admissible pairs are contained in $\mathscr{I}$ it follows that $\{I(\phi,u)\colon (\phi,u) \in \mathscr{A}_0\}$ is bounded below. By hypothesis, this set of numbers is bounded above. Hence there is a real number γ and a sequence $\{(\phi_k, u_k)\}$ of elements in $\mathscr{A}_0$ such that $I(\phi_k, u_k) \to \gamma$. Since $\mathscr{B}$ is compact there is a subsequence $\{(\phi_k, u_k)\}$ such that $\{(t_{0k}, \phi_k(t_{0k}), t_{1k}, \phi_k(t_{1k}))\}$ converges to a point (t_0, x_0, t_1, x_1) in $\mathscr{B}$. Since $t_{1k} > t_{0k}+\delta$ and since $(t_{ik}, \phi_k(t_{ik})) \in \mathscr{R}_0$, it follows that $t_1 \geq t_0+\delta$ and $(t_i, x_i) \in \mathscr{R}_0$.

STEP 2. There exists an absolutely continuous function ϕ defined on $[t_0, t_1]$ and a subsequence $\{\phi_k\}$ such that $\phi_k \to \phi$ in $\mathscr{X}_\rho$ and the extended functions $\tilde{\phi}_k'$ converge weakly to $\tilde{\phi}'$ in $L_1[\mathscr{I}]$. Moreover, ϕ satisfies (i) of Definition II.3.1 and (iii) of Definition II.3.2.

Since the graphs of all admissible trajectories lie in the compact set $\mathscr{R}_0$, the functions ϕ_k are uniformly bounded and so are their extensions $\tilde{\phi}_k$. Since the functions ϕ_k are equi-absolutely continuous, the same is true of their extensions $\tilde{\phi}_k$. It therefore follows from Ascoli's theorem that there exists a subsequence $\{\tilde{\phi}_k\}$ and a function $\tilde{\phi}$ defined on $\mathscr{I}$ such that $\tilde{\phi}_k$ converges uniformly to $\tilde{\phi}$ on $\mathscr{I}$. Moreover, the function $\tilde{\phi}$ is absolutely continuous, so that $\tilde{\phi}'$ exists and is in L_1. Thus

$$\tilde{\phi}(t) = \tilde{\phi}(a) + \int_a^t \tilde{\phi}'(s)ds \qquad a \leq t \leq b. \tag{7.2}$$

Let

$$\phi(t) = x_0 + \int_{t_0}^{t} \tilde{\phi}'(s)ds \qquad t_0 \leq t \leq t_1. \tag{7.3}$$

We assert that $\tilde{\phi}$ is the extension of ϕ. To see this first suppose that $t_0 > a$ and let $t_2 < t_0$. Since $t_{0k} \to t_0$ it follows that for k sufficiently large $t_{0k} > t_2$. Hence $\tilde{\phi}_k(t_2) = \phi_k(t_{0k})$. Since $\tilde{\phi}_k(t_2) \to \tilde{\phi}(t_2)$ and $\phi_k(t_{0k}) \to x_0$ we get that $\tilde{\phi}(t_2) = x_0$ for all $a \leq t_2 < t_0$. From the continuity of $\tilde{\phi}$ we get $\tilde{\phi}(t_0) = x_0$. If $t_0 = a$, then $\tilde{\phi}_k(t_0) = \tilde{\phi}_k(a) = \phi_k(t_{0k})$ for all k. If we now let $k \to \infty$ we get $\tilde{\phi}(t_0) = x_0$. Thus for all t in $[a,t_0]$, $\tilde{\phi}(t) = \tilde{\phi}(t_0) = x_0$. A similar argument shows that for $t_1 \leq t \leq b$, $\tilde{\phi}(t) = \tilde{\phi}(t_1) = x_1$. Thus $\tilde{\phi}'(t) = 0$ on $\mathscr{I} - [t_0,t_1]$. From (7.3) we have $\phi(t_0) = x_0$, and so $\tilde{\phi}(t) = \phi(t_0)$ for $a \leq t \leq t_0$. Since $\tilde{\phi}(a) = x_0$, we have from (7.2) and the vanishing of $\tilde{\phi}'$ outside of $[t_0,t_1]$ that for t in $[t_0,t_1]$

$$\tilde{\phi}(t) = x_0 + \int_a^t \tilde{\phi}'(s)ds = x_0 + \int_{t_0}^t \tilde{\phi}'(s)ds = \phi(t).$$

For $t \geq t_1$ we have

$$\tilde{\phi}(t) = x_0 + \int_{t_0}^{t_1} \tilde{\phi}'(s)ds = \phi(t_1).$$

Hence $\tilde{\phi}$ is the extension of ϕ. Therefore, for $t \in [t_0,t_1]$ we have $\phi'(t) = \tilde{\phi}'(t)$ and $\phi(t) = \tilde{\phi}(t)$.

Since $t_{ik} \to t_i$, $i = 0,1$, we have shown that $\phi_k \to \phi$ in $\mathscr{X}_\rho$. From the relation

$$\tilde{\phi}_k(t) = \phi_k(t_{0k}) + \int_a^t \tilde{\phi}_k'(s)ds \qquad a \leq t \leq b,$$

from (7.2) with $\tilde{\phi}(a)$ replaced by x_0, from (7.1), and from the convergence of $\tilde{\phi}_k$ to $\tilde{\phi}$ it follows that for all t in $[a,b]$

$$\int_a^t \tilde{\phi}_k'(s)ds \to \int_a^t \tilde{\phi}'(s)ds.$$

Since the functions $\tilde{\phi}_k$ are equi-absolutely continuous, their derivatives $\tilde{\phi}'_k$ have equi-absolutely continuous integrals. Hence by Theorem 3.1, $\tilde{\phi}'_k \to \tilde{\phi}'$ weakly in $L_1[\mathscr{I}]$.

Since $\tilde{\phi}$ is the uniform limit of functions for which (i) of Definition II.3.1 and (iii) of Definition II.3.2 hold, these conditions hold for $\tilde{\phi}$ and hence for ϕ.

STEP 3. There exists a real valued function λ that is integrable on $[t_0,t_1]$ such that $(\lambda(t),\phi'(t)) \in \mathscr{Q}^+(t,\phi(t))$ a.e. on $[t_0,t_1]$ and such that

$$\int_{t_0}^{t_1} \lambda(s)ds \le \gamma. \tag{7.4}$$

Since $\tilde{\phi}'_k \to \tilde{\phi}'$ weakly in L_1 we obtain the following statement from Theorem 3.2. For each integer j there exists an integer n_j, a set of integers $i = 1,\ldots,k$, where $k = k(j)$ depends on j and a set of numbers $\alpha_{1j},\ldots,\alpha_{kj}$ satisfying

$$\alpha_{ij} \ge 0 \qquad i = 1,\ldots,k \qquad \sum_{i=1}^{k} \alpha_{ij} = 1 \tag{7.5}$$

such that $n_{j+1} > n_j + k(j)$ and

$$\int_a^b |\tilde{\phi}' - \sum_{i=1}^{k} \alpha_{ij}\tilde{\phi}'_{n_j+i}|dt \to 0 \tag{7.6}$$

as $j \to \infty$. Let

$$\psi_j = \sum_{i=1}^{k} \alpha_{ij}\tilde{\phi}'_{n_j+i}.$$

Recall that for every integer q, if $t \notin [t_{0q},t_{1q}]$ then $\tilde{\phi}'_q(t) = 0$ and that u_q and ϕ_q are only defined on $[t_{0q},t_{1q}]$. If for $t \notin [t_{0q},t_{1q}]$ we define $f(t,\phi_q(t),u_q(t)) = 0$ we can write ψ_j as follows:

$$\psi_j(t) = \sum_{i=1}^{k} \alpha_{ij} f(t,\phi_{n_j+i}(t),u_{n_j+i}(t)) . \tag{7.7}$$

In terms of ψ_j, (7.6) says that $\psi_j \to \tilde{\phi}'$ in $L_1[\mathscr{I}]$. Hence there is a subsequence $\{\psi_j\}$ such that

$$\psi_j(t) \to \tilde{\phi}'(t) \quad \text{a.e. in } \mathscr{I}. \tag{7.8}$$

We suppose that (7.7) is now this subsequence. Corresponding to the sequence (7.7) we define a sequence $\{\lambda_j\}$ as follows:

$$\lambda_j(t) = \sum_{i=1}^{k} \alpha_{ij} f^0(t,\phi_{n_j+i}(t),u_{n_j+i}(t)), \tag{7.9}$$

where if $t \notin [t_{0q},t_{1q}]$ we set $f^0(t,\phi_q(t),u_q(t)) = 0$ and where for each j the numbers α_{ij}, the indices n_j+i and the functions ϕ_{n_j+i} and u_{n_j+i} are as in (7.7). Note that if $t \notin [t_0,t_1]$ there exists a $j_0 = j_0(t)$ such that if $j > j_0$ then $\psi_j(t) = 0$ and $\lambda_j(t) = 0$.

Define

$$\lambda(t) = \liminf \lambda_j(t). \tag{7.10}$$

Since $f^0 \geq -K$, it follows that $\lambda_j \geq -K$. Hence $\lambda \geq -K$. Moreover, if $t \notin [t_0,t_1]$ then $\lambda(t) = 0$. Therefore, if we set $f_q^0(t) \equiv f^0(t,\phi_q(t),u_q(t))$ and use Fatou's theorem we get

$$\begin{aligned}
\int_{t_0}^{t_1} \lambda\, dt = \int_a^b \lambda\, dt &\leq \liminf_{j\to\infty} \left[\sum_{i=1}^{k} \alpha_{ij} \int_a^b f_{n_j+i}^0\, dt\right] \\
&= \liminf_{j\to\infty} \left[\sum_{i=1}^{k} \alpha_{ij} \int_{t_{0,n_j+i}}^{t_{1,n_j+i}} f_{n_j+i}^0\, dt\right] \\
&= \liminf_{j\to\infty} \left[\sum_{i=1}^{k} \alpha_{ij} I(\phi_{n_j+i},u_{n_j+i})\right].
\end{aligned}$$

From (7.1) we have that $I(\phi_{n_j+i},u_{n_j+i}) \to \gamma$ as $j \to \infty$. It then follows from (7.5) and Exercise 7.1 that

$$\lim_{j\to\infty} \sum_{i=1}^{k} \alpha_{ij} I(\phi_{n_j+i},u_{n_j+i}) = \gamma.$$

Thus (7.4) is proved. Since λ is bounded from below it follows that

λ is in $L_1[\mathscr{I}]$ and is finite almost everywhere on $[t_0,t_1]$.

We now show that $(\lambda(t),\phi'(t)) \in \mathscr{Q}^+(t,\phi(t))$ a.e. on $[t_0,t_1]$. Let T_1 denote the set of points in $[t_0,t_1]$ at which $\lambda(t)$ is finite and $\psi_j(t) \to \phi'(t)$. For each integer k define a set E_k as follows:

$$E_k = \{t\colon t \in [t_{0k},t_{1k}],\ u_k(t) \notin \Omega(t,\phi_k(t))\}.$$

Then by Definition II.3.2 (ii), meas $E_k = 0$. Let E denote the union of the sets E_k and let T_2 denote the set of points in $[t_0,t_1]$ that do not belong to E. Let $T' = T_1 \cap T_2$. Clearly, meas $T' = t_1 - t_0$.

Let t be a fixed element in T', $t \neq t_i$, $i = 0,1$. There exists a subsequence $\{\lambda_j(t)\}$, which in general depends on t, such that $\lambda_j(t) \to \lambda(t)$. For the corresponding subsequence $\{\psi_j(t)\}$ we have from (7.8) and the fact that t is interior to (t_0,t_1) that $\psi_j(t) \to \phi'(t)$. Since t is interior to (t_0,t_1) and $t_{ik} \to t_i$, $i = 0,1$, it follows that there exists a j_0 such that if $j > j_0$ then $t \in (t_{0,n_j+i},t_{1,n_j+i})$. For each $\delta > 0$ there exists an integer k_0, depending on δ, such that if $k > k_0$, then $|\phi_k(t)-\phi(t)| < \delta$. Hence for $k > k_0$

$$(t,\phi_k(t)) \in N_{x\delta}(t,\phi(t)).$$

Therefore, for j sufficiently large

$$\hat{f}(t,\phi_{n_j+i}(t),u_{n_j+i}(t)) \in \mathscr{Q}^+(N_{x\delta}(t,\phi(t))),$$

where $\hat{f} = (f^0,f)$. Therefore, by (7.7), (7.9), and (7.5),

$$(\lambda_j(t),\psi_j(t)) \in \text{co}\ \mathscr{Q}^+(N_{x\delta}(t,\phi(t)).$$

Since $\lambda_j(t) \to \lambda(t)$ and $\psi_j(t) \to \phi'(t)$, we have that

$$(\lambda(t),\phi'(t)) \in \text{cl co } \mathcal{Q}^+(N_{x\delta}(t,\phi(t)))$$

Since δ is arbitrary, $(\lambda(t),\phi'(t))$ is in $\text{cl co } \mathcal{Q}^+(N_{x\delta}(t,\phi(t)))$ for every $\delta > 0$ and hence in the intersection of these sets. Since the weak Cesari property holds, we get that $(\lambda(t),\phi'(t)) \in \mathcal{Q}^+(t,\phi(t))$. Since t was an arbitrary point in T' different from t_0 or t_1, we get the desired result.

STEP 4. There exists a measurable function u defined on $[t_0,t_1]$ such that for almost all t: (i) $\phi'(t) = f(t,\phi(t),u(t))$; (ii) $u(t) \in \Omega(t,\phi(t))$; (iii) $\lambda(t) \geq f^0(t,\phi(t),u(t))$.

The existence of a function v satisfying the conclusion of Step 4 is a restatement of $(\lambda(t),\phi'(t)) \in \mathcal{Q}^+(t,\phi(t))$. The problem is to show that there is a measurable function u with this property. This will be done using Theorem 7.1.

Let $T = \{t: (\lambda(t),\phi'(t)) \in \mathcal{Q}^+(t,\phi(t))\}$, let $Z = E^1 \times E^n \times E^n \times E^1$, and let $D = \{(t,x,z,\eta): (t,x,z) \in \mathcal{D}, \eta \geq f^0(t,x,z)\}$. The set T is Lebesgue measurable and thus is a measure space with Lebesgue measure. The space Z is clearly Hausdorff. In the discussion preceding the statement of Step 1 we noted that the set $\mathcal{D}$ is closed. Since f^0 is lower semicontinuous it follows that the set D is also closed. Hence D can be written as a countable union of compact sets D_i, where D_i is the intersection of D with the closed ball of radius i centered at the origin in E^{2n+2}.

Let Γ denote the mapping from T to Z defined by $t \to (t,\phi(t),\phi'(t),\lambda(t))$. Since each of the functions ϕ, ϕ', λ is measurable, Γ is a measurable map. Let Φ denote the mapping from D to Z defined by $(t,x,z,\eta) \to (t,x,f(t,x,z),\eta)$. Since f is continuous, Φ is continuous. From Step 3 we obtain that $\Gamma(T) \subseteq \Phi(D)$. Thus, all of the hypotheses of Theorem 7.1 are fulfilled. Hence there exists a measurable map μ from T to D, say

$$\mu: t \to (\tau(t), x(t), u(t), \eta(t))$$

such that for all t in T

$$\Phi(\mu(t)) = (\tau(t), x(t), f(\tau(t), x(t), u(t)), \eta(t)) = \Gamma(t)$$
$$= (t, \phi(t), \phi'(t), \lambda(t)).$$

From this the conclusion of Step 4 follows.

STEP 5. Completion of Proof.

Let $(\phi^*, u^*) = (\phi, u)$, where ϕ is the function obtained in Step 2 and u is the function obtained in Step 4. Then we already showed in Step 2 that there is a sequence $\{(\phi_k, u_k)\}$ such that $\phi_k \to \phi^*$ in $\mathscr{X}_\rho$ and such that (i) of Definition II.3.1 and (iii) of Definition II.3.2 holds. Statement (i) of Step 4 asserts that (ϕ^*, u^*) satisfies (ii) of Definition II.3.1. Statement (ii) of Step 4 asserts that (ϕ^*, u^*) satisfies (ii) of Definition II.3.2. Thus to prove that (ϕ^*, u^*) is admissible it remains to show that $t \to f^0(t, \phi(t), u(t))$ is Lebesgue integrable on $[t_0, t_1]$. Since ϕ and u are measurable and f^0 is lower semicontinuous $t \to f^0(t, \phi(t), u(t))$ is measurable. Since λ is integrable and f^0 is bounded from below it follows from (iii) of Step 4 that $t \to f^0(t, \phi(t), u(t))$ is integrable.

From (iii) of Step 4 and from (7.4) we get

$$I(\phi^*, u^*) = \int_{t_0}^{t_1} f^0(t, \phi^*(t), u^*(t))dt \le \gamma.$$

But from (7.1) we have that $I(\phi_k, u_k) \to \gamma$. Combining these last two statements, we obtain (4.3) and thereby complete the proof of Theorem 4.1.

In Remark 6.2 we asserted that if (6.1) and (6.2) hold with $H^i = f^0 + K$ for all (ϕ_k, u_k) in a minimizing sequence then (6.1) and (6.2) hold for (ϕ^*, u^*). We now prove this assertion.

Let $\{(\phi_k,u_k)\}$ be a minimizing sequence. Then if $\mu = \inf\{J(\phi,u): (\phi,u) \in \mathscr{A}\}$, we have $I(\phi_k,u_k) \to \mu$. Hence if we take $\mathscr{A}_0 = \{(\phi_k,u_k)\}$, then from Step 1 we have $\mu = \gamma$. From (6.1) and the assumption that $H^i = f^0 + K$ it follows that $I(\phi_k,u_k) + K(t_{1k}-t_{0k}) \leq A$. Since $I(\phi^*,u^*) = \mu$, and for an appropriate subsequence $t_{ik} \to t_i$, $i = 0,1$, we get that $I(\phi^*,u^*) + K(t_1-t_0) \leq A$. But this says that

$$\int_{t_0}^{t_1}[f^0(t,\phi^*(t),u^*(t))+K]dt \leq A,$$

and so (ϕ^*,u^*) satisfies (6.1) with $H^i = f^0 + K$.

If (6.2) holds, then from (7.7), (7.5) and the relation $H^i = f^0 + K$ we have

$$|\psi_j^i(t)| \leq \sum_{q=1}^{k} \alpha_{qj}|f^i(t,\phi_{n_j+q}(t),u_{n_j+q}(t)|$$

$$\leq F_\varepsilon^i(t) + \varepsilon \sum_{q=1}^{k} \alpha_{qj}[f^0(t,\phi_{n_j+q}(t),u_{n_j+q}(t))+K].$$

It now follows from (7.9) that

$$|\psi_j^i(t)| \leq F_\varepsilon^i(t) + \varepsilon[\lambda_j(t)+K].$$

Let t be a point in T_1, defined in Step 3 to be the set of points in $[t_0,t_1]$ at which $\lambda(t)$ is finite and $\psi_j(t) \to \phi'(t)$. There exists a subsequence $\{\lambda_j(t)\}$, depending on t such that $\lambda_j(t) \to \lambda(t)$. Hence for t in T_1

$$|\phi^{i'}(t)| \leq F_\varepsilon^i(t) + \varepsilon[\lambda(t)+K]. \tag{7.11}$$

Since meas $T_1 = t_1 - t_0$, the inequality (7.11) holds a.e. on $[t_0,t_1]$.

From (7.4), the equality $\mu = \gamma$, and (iii) of Step 4 we have

$$I(\phi,u) = \int_{t_0}^{t_1} f^0(t,\phi(t),u(t))dt \leq \int_{t_0}^{t_1}\lambda(t)dt \leq \mu .$$

Since $(\phi^*,u^*) = (\phi,u)$ and $I(\phi^*,u^*) = \mu$, we have that

$$\int_{t_0}^{t_1} [\lambda(t)-f^0(t,\phi(t),u(t))]dt = 0.$$

It now follows from (iii) of Step 4 that $\lambda(t) = f^0(t,\phi^*(t),u^*(t))$ a.e. If we make this substitution for λ in (7.11) and write $\phi^{*i'}$ in (7.11) we have the result that (6.2) holds for (ϕ^*,u^*).

EXERCISE 7.2. Show that in Theorem 4.1 we may replace Assumption 4.1-(v) by the following weaker assumption. There exists a real valued non-negative function ψ in $L_1[\mathscr{I}]$ and an n-vector b such that for all (t,x,z) in $\mathscr{D}$ $f^0(t,x,z) \geq -\psi(t) + \langle b,f(t,x,z)\rangle$.

EXERCISE 7.3. Show that in Theorem 4.1 we may replace Assumption 4.1-(vi) by the following weaker assumption. Let $\mathscr{I}$ be a compact interval in E^1, let $\mathscr{R} = \mathscr{I} \times E^n$ and let $\mathscr{U} = E^m$. For each t in $\mathscr{I}$ let the function $\hat{f} = (f^0,f)$ be a continuous function of (x,z) on E^{n+m} and for each (x,z) in E^{n+m} let f be measurable with respect to t in $\mathscr{I}$.

Hint: From a theorem that appears to have been discovered independently by Scorza-Dragoni [55] and Vainberg ([57]; Theorem 18.2, p. 148) we obtain the following statement. For every $\varepsilon > 0$ there exists an open set $G \subset \mathscr{I}$ such that meas $(G) < \varepsilon$ and such that $\hat{f} = (f^0,f)$ is continuous on $(\mathscr{I}-G) \times E^n \times E^m$. In Step 3 we first obtain the existence of measurable u in $T-(G \cap T)$. Since ε is arbitrary, we obtain the existence of a measurable u on $\mathscr{I}$ with the desired properties.

EXERCISE 7.4. Use the results of Exercise 7.3 to extend the results of Exercise 6.3 to the case in which A and h are integrable on $\mathscr{I}$ and B is bounded and measurable on $\mathscr{I}$.

8. Existence Without the Cesari Property

In this section we shall state and prove two existence theorems in which it is not assumed that the weak Cesari property holds. The sets $\mathcal{Q}^+(t,x)$, however, are still assumed to be convex. In both existence theorems of this section we shall assume that the constraint mapping Ω depend only on t. In one of the theorems we assume that the function $\hat{f}$ satisfies a generalized Lipschitz condition. In the other we assume that the controls in a minimizing sequence all lie in a ball of some L_p space, $1 \leq p \leq \infty$.

In the first theorem the following hypotheses, listed as Assumption 8.1, will be made in addition to Assumption 4.1. Note that (i) of Assumption 8.1 supersedes (vi) of Assumption 4.1.

ASSUMPTION 8.1. (i) The function $\hat{f} = (f^0,f)$ is continuous. (ii) The sets $\Omega(t,x)$ are independent of x: i.e. for a given t, $\Omega(t,x) = \Omega(t,x') \equiv \Omega(t)$ for all x and x' such that (t,x) and (t,x') are in $\mathcal{R}$. (iii) There exists a non-decreasing function ω defined on $[0,\infty)$ such that $\lim_{\delta \to 0} \omega(\delta) = 0$ and a non-negative function L defined on $E^1 \times \mathcal{U}$ such that

$$|\hat{f}(t,x,z)-\hat{f}(t,x',z)| \leq L(t,z)\omega(|x-x'|) \tag{8.1}$$

for all (t,x,z) and (t,x',z) in $\mathcal{D}$.

Note that if $\hat{f}$ is uniformly continuous on $\mathcal{D}$, which occurs if $\mathcal{D}$ is compact, then (8.1) holds with $L \equiv 1$ and ω the modulus of continuity. If $\hat{f}$ is Lipschitz in x, then (8.1) holds with $\omega(\delta) = \delta$ and L equal to the Lipschitz constant.

The following theorem takes the place of Theorem 4.1. We shall give the proof after the statement of Theorem 8.3.

THEOREM 8.1. Let Assumptions 4.1 and 8.1 hold. Let $\mathcal{A}_0$ be a set of admissible pairs (ϕ,u) such that the trajectories ϕ are equi-absolutely continuous, such that $\sup\{I(\phi,u): (\phi,u) \in \mathscr{A}_0\}$ is finite, and such that for all admissible controls u belonging to pairs (ϕ,u) in $\mathscr{A}_0$

$$\int_{t_0}^{t_1} L(t,u(t))dt \leq A, \tag{8.2}$$

where $[t_0,t_1]$ is the interval of definition of (ϕ,u) and A is a constant independent of u. Then there exists a sequence $\{(\phi_k,u_k)\}$ in $\mathscr{A}_0$ and an admissible pair (ϕ^*,u^*) in $\mathscr{A}$ with the following properties: (i) $\phi_k \to \phi^*$ in $\mathscr{X}_\rho$, (ii)

$$\lim_{k\to\infty} I(\phi_k,u_k) \geq I(\phi^*,u^*). \tag{8.3}$$

We now give an example in which the weak Cesari property fails to hold, but the hypotheses of Theorem 8.1 hold.

EXAMPLE 8.1. Let $x = (x^1,x^2)$, let z be a real number, let $\Omega(t,x) = E^1$, let $f^0 \equiv 0$ and let $f(t,x,z) = (z,x^1z)$. Let the end conditions be as follows: $t_0 = 0$, $t_1 = 1$, $x_0^1 = 0$, $0 \leq x_0^2 \leq 1$, x_1^1 and x_1^2 free. Let $\mathscr{R} = \{(t,x): 0 \leq t \leq 1, |x^i| < M, i = 1,2\}$, where M is a large positive constant.

For each (t,x) in $\mathscr{R}$

$$\mathscr{Q}^+(t,x) = \{(\eta,\xi) = (\eta,\xi^1,\xi^2): \eta \geq 0, \xi^1 = z, \xi^2 = x^1z\}. \tag{8.4}$$

The set $\mathscr{Q}^+(t,x)$ is clearly closed and convex. It is also readily verified that the other conditions in Assumption 4.1 and (i) and (ii) of Assumption 8.1 hold. Also $\hat{f} = (f^0,f)$ satisfies (iii) of Assumption 8.1 with $L(t,z) = |z|$ and $\omega(|x-x'|) = |x-x'|$.

Let $\mathscr{A}$ be the set of admissible pairs (ϕ,u). The set $\mathscr{A}$ is not void. Let $\mathscr{A}_0$ be any subset of $\mathscr{A}$ such that the trajectories ϕ belonging to admissible pairs (ϕ,u) in $\mathscr{A}_0$ are equi-absolutely continuous. From the relation $\phi^{1\prime}(t) = u(t)$ a.e. it follows that the functions u, and hence the functions $|u|$, have equi-absolutely continuous integrals. Hence since the interval of integration is $[0,1]$ there is a constant A such that $\int_0^1 |u|dt \leq A$ for all u such that $(\phi,u) \in \mathscr{A}_0$. Thus, the hypotheses of Theorem 8.1 are fulfilled.

On the other hand, for all $\delta > 0$, cl co $\mathscr{Q}^+(N_{x\delta}(t,x)) = \{(\eta,\xi): \eta \geq 0, \xi \in R^2\}$. From this and from (8.4) we see that the weak Cesari property fails.

Theorem 8.2 below is an existence theorem for Problem 2 corresponding to Theorem 4.2 and is obtained from Theorem 8.1 in the same way that Theorem 4.2 is obtained from Theorem 4.1.

THEOREM 8.2. Let Assumptions 4.1 and 8.1 hold. Let G be lower semicontinuous and bounded below on $\mathscr{A}_T$. Let there exist a minimizing sequence $\{(\phi_k,u_k)\}$ such that the functions ϕ_k are equi-absolutely continuous and such that (8.2) holds. Then there exists a (ϕ^*,u^*) in $\mathscr{A}$ such that $\hat{J}(\phi^*,u^*) \leq \hat{J}(\phi,u)$ for all (ϕ,u) in $\mathscr{A}$.

EXERCISE 8.1. Prove Theorem 8.2.

A slightly weakened version of Theorem 5.1, in the sense that the hypotheses are slightly more stringent, can be obtained from Theorem 8.2. This is taken up in the next exercise.

EXERCISE 8.2. Let the hypotheses of Theorem 5.1 hold, and let the additional hypotheses be made that Ω is independent of x and that f^0 is continuous rather than lower semicontinuous. Show that the conclusion of Theorem 5.1 follows from Theorem 8.2.

Existence in the case of non-compact constraints is discussed in the next theorem, whose proof is similar to that of Theorem 6.1, except that we now use Theorem 8.2 instead of Theorem 4.2.

THEOREM 8.3. Let the hypotheses of Theorem 6.1 hold, except for the statement that $\mathscr{Q}^+$ satisfies the weak Cesari property. Let Assumption 8.1 hold and let the controls u_k in the minimizing sequence satisfy

$$\int_{t_{0k}}^{t_{1k}} L(t,u_k(t))dt \leq A,$$

where A is a constant. Then there exists an optimal pair (ϕ^*,u^*) in $\mathscr{A}$.

We now prove Theorem 8.1. The proof proceeds exactly as the proof of Theorem 4.1 up to and including the definition of λ_j in (7.9). The rest of the argument used to prove Step 3 proceeds differently. The reader is urged to keep in mind the order in which various subsequences are chosen.

Define sequences of functions σ_j and θ_j corresponding to ψ_j and λ_j as follows

$$\begin{aligned} \sigma_j(t) &= \sum_{i=1}^{k} \alpha_{ij} f(t,\phi(t),u_{n_j+i}(t)) \\ \theta_j(t) &= \sum_{i=1}^{k} \alpha_{ij} f^0(t,\phi(t),u_{n_j+i}(t)), \end{aligned} \tag{8.5}$$

where if $t \notin [t_{0q},t_{1q}]$ we set $\hat{f}(t,\phi(t),u_q(t)) = 0$, and where for each j the numbers α_{ij}, the indices n_j+i and the functions u_{n_j+i} are as in (7.7). The functions σ_j and θ_j are measurable.

Let $M_k = \max\{|\tilde{\phi}_k(t)-\tilde{\phi}(t)| : t \in \mathscr{I}\}$. Since $\tilde{\phi}_k$ converges uniformly to $\tilde{\phi}$ on $\mathscr{I}$, $M_k \to 0$ as $k \to \infty$. Let $f_q^*(t) = f(t,\phi(t),u_q(t))$ and let $f_q(t) = f(t,\phi_q(t),u_q(t))$. Note that $f_q^*(t) = f_q(t) = 0$ for $t \notin [t_{0q},t_{1q}]$. Since the σ_j are measurable we get, using (8.5), (8.1) and (8.2) that

$$\begin{aligned}\int_a^b |\sigma_j - \psi_j|\,dt &\leq \sum_{i=1}^{k} \alpha_{ij} \int_a^b |f^*_{n_j+i} - f_{n_j+i}|\,dt \\ &= \sum_{i=1}^{k} \alpha_{ij} \int_{t_{0,n_j+i}}^{t_{1,n_j+i}} |f^*_{n_j+i} - f_{n_j+i}|\,dt \\ &\leq \sum_{i=1}^{k} \alpha_{ij}\,\omega(M_{n_j+i}) \int_{t_{0,n_j+i}}^{t_{1,n_j+i}} L(t,u_{n_j+i}(t))\,dt \\ &\leq A \sum_{i=1}^{k} \alpha_{ij}\,\omega(M_{n_j+i}).\end{aligned}$$

Thus $\sigma_j - \psi_j$ is in $L_1[\mathscr{I}]$. Since $M_k \to 0$ and $\omega(\delta) \to 0$ as $\delta \to 0$ we get that $\sigma_j - \psi_j \to 0$ in $L_1[\mathscr{I}]$. A similar argument shows that $\theta_j - \lambda_j$ is in $L_1[\mathscr{I}]$ and that $\theta_j - \lambda_j \to 0$ in $L_1[\mathscr{I}]$. Hence there exist subsequences such that

$$\sigma_j(t) - \psi_j(t) \to 0 \qquad \theta_j(t) - \lambda_j(t) \to 0 \quad \text{a.e.} \; . \tag{8.6}$$

We henceforth take the functions in (8.5), (7.7) and (7.9) to be the functions in these subsequences.

We now define λ as in (7.10) and show as we did in the paragraph following (7.10) that λ is in $L_1[\mathscr{I}]$ and that (7.4) holds.

As in the proof of Theorem 4.1 let T' denote the set of points in $[t_0, t_1]$ at which $\lambda(t)$ is finite, $\psi_j(t) \to \phi'(t)$ and for which $u_k(t) \in \Omega(t)$ for all k. Recall that Ω depends only on t. This set has measure $t_1 - t_0$. Let T denote the set of points in T' such that (8.6) holds. Then meas $T = t_1 - t_0$.

Let t be a fixed but arbitrary point in T. Since $\psi_j(t) \to \phi'(t)$ it follows from (8.6) that $\sigma_j(t) \to \phi'(t)$. From the definition of λ it follows that there is a subsequence $\{\lambda_j(t)\}$, which in general depends on t such that $\lambda_j(t) \to \lambda(t)$. From (8.6) we get that $\theta_j(t) \to \lambda(t)$. For the corresponding subsequence $\sigma_j(t)$ we still have $\sigma_j(t) \to \phi'(t)$. Since $T \subset T'$ and Ω is independent of x it follows that for all j and i, $u_{n_j+i}(t) \in \Omega(t)$. Hence

$$\hat{f}(t,\phi(t),u_{n_j+i}(t)) \in \mathscr{Q}^+(t,\phi(t))$$

Since $\mathscr{Q}^+(t,\phi(t))$ is convex the points $(\theta_j(t),\sigma_j(t))$ belong to $\mathscr{Q}^+(t,\phi(t))$. Since $\mathscr{Q}^+(t,\phi(t))$ is closed and $(\theta_j(t),\sigma_j(t)) \to (\lambda(t),\phi'(t))$ we get that $(\lambda(t),\phi'(t)) \in \mathscr{Q}^+(t,\phi(t))$. Since t is an arbitrary point in T, we have that $(\lambda(t),\phi'(t)) \in \mathscr{Q}^+(t,\phi(t))$ a.e. on $[t_0,t_1]$.

The remainder of the proof is now exactly the same as the proof of Theorem 4.1.

EXERCISE 8.3. Use Theorem 8.2 to obtain the existence theorem of Exercise 6.5 for the "linear plant quadratic integral cost criterion" problem.

We now take up our second existence theorem. The following theorem takes the place of Theorem 4.1.

THEOREM 8.4. Let Assumption 4.1 hold, except for statement (vi). Let $\hat{f} = (f^0,f)$ be continuous and let the sets $\Omega(t,x)$ be independent of x: i.e. for a given t, $\Omega(t,x) = \Omega(t,x') = \Omega(t)$ for all x and x' such that (t,x) and (t,x') are in $\mathscr{R}$. Let $\mathscr{A}_0$ be a set of admissible pairs (ϕ,u) such that the trajectories ϕ are equi-absolutely continuous, such that $\sup\{I(\phi,u): (\phi,u) \in \mathscr{A}_0\}$ is finite and such that all admissible controls u belonging to pairs (ϕ,u) in $\mathscr{A}_0$ satisfy

$$||u||_p \leq M, \tag{8.7}$$

where $1 \leq p \leq \infty$ and $M > 0$ are fixed and $||u||_p$ denotes the $L_p[t_0,t_1]$ norm of u. Then there exists a sequence $\{(\phi_k,u_k)\}$ in $\mathscr{A}_0$ and an admissible pair (ϕ^*,u^*) in $\mathscr{A}$ with the following properties: (i) $\phi_k \to \phi^*$ in $\mathscr{X}_\rho$, (ii)

$$\lim_{k\to\infty} I(\phi_k,u_k) \geq I(\phi^*,u^*).$$

The proof of this theorem will be given after the statement of Theorem 8.6.

REMARK 8.1. In Example 8.1 the hypotheses of Theorem 8.4 are satisfied.

Theorem 8.5 below is an existence theorem for Problem 2 corresponding to Theorems 4.2 and 8.2 and is obtained from Theorem 8.4 in the same way that Theorems 4.2 and 8.2 were obtained from Theorems 4.1 and 8.1 respectively.

THEOREM 8.5. Let Assumption 4.1 hold, except for statement (vi). Let $\hat{f}$ and Ω be as in Theorem 8.4. Let G be lower semicontinuous and bounded below on $\mathscr{A}_T$. Let there exist a minimizing sequence $\{(\phi_k,u_k)\}$ such that the functions ϕ_k are equi-absolutely continuous and such that (8.7) holds. Then there exists a (ϕ^*,u^*) in $\mathscr{A}$ such that $\hat{J}(\phi^*,u^*) \leq \hat{J}(\phi,u)$ for all (ϕ,u) in $\mathscr{A}$.

The theorem corresponding to Theorems 6.1 and 8.3 is Theorem 8.6, which follows. Its proof is similar to that of Theorem 6.1 except that we now use Theorem 8.5 instead of Theorem 4.2.

THEOREM 8.6. Let the hypotheses of Theorem 6.1 hold, except for the statement that $\mathscr{Q}^+$ satisfies the weak Cesari property and the statement that (vi) of Assumption 4.1 holds. Let $\hat{f}$ and Ω be as in Theorem 8.4 and let the controls u_k in the minimizing sequence satisfy (8.7). Then there exists an optimal pair (ϕ^*,u^*) in $\mathscr{A}$.

We now prove Theorem 8.4. The following lemma is crucial to the proof.

LEMMA 8.1. Let $h\colon (t,\xi) \to h(t,\xi)$ be a continuous mapping from $[\alpha,\beta] \times R^r \to R^1$. Let $\{v_k\}$ and $\{w_k\}$ be sequences in $L_p[\alpha,\beta]$,

$1 \leq p \leq \infty$, such that $||v_k||_p \leq M$ and $||w_k|| \leq M$ for some $M > 0$ and such that $(v_k - w_k) \to 0$ in measure on $[\alpha,\beta]$. Then

$$h(t,v_k(t)) - h(t,w_k(t)) \to 0$$

in measure on $[\alpha,\beta]$.

We postpone the proof of Lemma 8.1 and proceed to sketch the proof of Theorem 8.5. The proof proceeds as does the proof of Theorem 4.1 through Steps 1 and 2. Step 3, however, is modified as follows.

Let p be as in (8.7). Since $\tilde{\phi}_k \to \tilde{\phi}$ uniformly on $[a,b]$ and all trajectories lie in a fixed compact set it follows that there exists an M' such that $||\tilde{\phi}_k||_p \leq M'$ and $||\tilde{\phi}||_p \leq M'$, where $||\ ||_p$ denotes the $L_p[a,b]$ norm. Let $\tilde{u}_k$ be the extension of u_k from $[t_{0k},t_{1k}]$ to $[a,b]$ obtained by setting $\tilde{u}_k(t) = 0$ if $t \notin [t_{0k},t_{1k}]$. Since by (8.7), the elements u_k lie in a ball of radius M in $L_p[t_{0k},t_{1k}]$ we get that there exists a constant $A > 0$ such that for all k the functions $v_k = (\tilde{\phi}_k,\tilde{u}_k)$ and $w_k = (\tilde{\phi},u_k)$ lie in a ball of radius A in $L_p[a,b]$. Also note that $v_k(t)-w_k(t) \to 0$ at all points of $[a,b]$.

Let

$$\hat{\Delta}_k(t) = \hat{f}(t,\phi(t),u_k(t)) - \hat{f}(t,\phi_k(t),u_k(t)), \qquad (8.8)$$

where we set $\hat{\Delta}_k(t) = 0$ if $t \notin [t_{0k},t_{1k}]$. It is then a consequence of Lemma 8.1 with $\xi = (x,z)$ and of the convergence of t_{ik} to t_i, $i = 0,1$, that $\hat{\Delta}_k \to 0$ in measure on $[a,b]$. Since $\hat{\Delta}_k \to 0$ in measure there exists a subsequence $\{(\phi_k,u_k)\}$ such that

$$\hat{\Delta}_k(t) \to 0 \qquad \text{a.e.} \qquad (8.9)$$

in $[a,b]$.

The functions ψ_j, λ_j, and λ are next defined as in Step 4 of the proof of Theorem 4.1 and it is shown that (7.4) holds.

Sequences σ_j and θ_j are then defined as in (8.5). If we denote the first component of the vector $\hat{\Delta}_k(t)$ by $\Delta_k^0(t)$ and the vector consisting of the remaining n components by $\Delta_k(t)$ we get, using (8.5), (8.8), (7.7) and (7.9) that

$$\sigma_j(t) - \psi_j(t) = \sum_{i=1}^{k} \alpha_{ij} \Delta_{n_j+i}(t)$$

$$\theta_j(t) - \lambda_j(t) = \sum_{i=1}^{k} \alpha_{ij} \Delta_{n_j+i}^{0}(t).$$

It then follows from (8.9) and Exercise 7.1 that (8.6) holds.

The rest of the proof is a verbatim repetition of the last four paragraphs of the proof of Theorem 8.1.

We now prove Lemma 8.1. We must show that for arbitrary $\eta > 0$ and $\varepsilon > 0$ there exists an integer N such that if $n > N$ then

$$\text{meas}\{t: |f(t,v_n(t))-f(t,w_n(t))| \geq \eta\} < \varepsilon. \tag{8.10}$$

Let

$$A = M(2/\varepsilon)^{1/p}, \tag{8.11}$$

where we interpret $1/\infty$ as zero. Let G_A denote the set of points ξ in R^r such that $|\xi| \leq A$. Since f is uniformly continuous on $[\alpha,\beta] \times G_A$ it follows that there exists a $\delta > 0$ such that if $|\xi-\xi'| < \delta$ and ξ and ξ' belong to G_A then

$$|f(t,\xi) - f(t,\xi')| < \eta \tag{8.12}$$

for all t in $[\alpha,\beta]$.

Let I_n denote the set of points in $[\alpha,\beta]$ at which either $|v_n(t)| > A$ or $|w_n(t)| > A$. Let

$$G_n = \{t: |v_n(t) - w_n(t)| \geq \delta\}.$$

From (8.12) we have that for $t \notin I_n \cup G_n$,

$$|f(t,v_n(t)) - f(t,w_n(t))| < \eta.$$

Therefore to establish the lemma we must show that for n sufficiently large $\text{meas}\,(I_n \cup G_n) < \varepsilon$.

For $p < \infty$ we have

$$M \geq \left(\int_\alpha^\beta |v_n|^p dt\right)^{1/p} > \left(\int_{I_n} A^p dt\right)^{1/p} = A(\text{meas } I_n)^{1/p}.$$

From this and from (8.11) it follows that $\text{meas } I_n < \varepsilon/2$. For $p = \infty$ we have from (8.11) that $\text{meas } I_n = 0$. Since $v_n - w_n \to 0$ in measure, there exists an integer N such that for $n > N$ $\text{meas } G_n < \varepsilon/2$. Hence $\text{meas}\,(I_n \cup G_n) < \varepsilon$, and the lemma is proved.

EXERCISE 8.4. Show that in Theorem 8.4 we may replace the assumption that $\hat{f}$ is continuous by the weaker assumption made in Exercise 7.4.

EXERCISE 8.5. Use Theorem 8.6 to obtain the Nagumo-Tonelli theorem, Theorem 6.2.

EXERCISE 8.6. Use Theorems 8.4-8.6 to obtain the existence theorems for the "linear plant convex integral cost criterion and unbounded controls" given in Exercise 6.3. Do the same for the "linear plant quadratic integral cost criterion problem" discussed in Exercises 6.5 and 6.6.

EXERCISE 8.7. In some applications an isoperimetric constraint of the form

$$\int_a^b |u|^2 dt < M, \tag{8.13}$$

where M is a positive constant, is present in the linear plant quadratic integral criterion problem. This constraint arises where there are limitations on the energy available. Obtain an existence theorem for the problem posed in Exercise 6.5 in which the additional constraint (8.13) is present.

9. Behavior of Controls in a Minimizing Sequence

In the statements and proofs of the various existence theorems there were no assertions made nor conclusions drawn concerning the behavior of the controls belonging to a minimizing sequence. In this section we give an example showing that it is possible for the trajectories of a minimizing sequence to converge in $\mathscr{X}_\rho$ to an optimal trajectory while no subsequence of the corresponding controls converges in any of the usual senses to the optimal control.

EXAMPLE 9.1. Let the state equations be

$$\begin{aligned} \frac{dx^1}{dt} &= p^1(t)v^1(t) + p^2(t)v^3(t) \\ \frac{dx^2}{dt} &= p^1(t)v^2(t) + p^2(t)v^4(t) \\ \frac{dx^3}{dt} &= 1. \end{aligned} \tag{9.1}$$

Let $f^0(t,x,z) = (x^1)^2 + (x^2)^2$, let $\mathscr{T}_0 = \{(t_0,x_0): x_0 = 0,\ t_0 = 0\}$, and let $\mathscr{T}_1 = \{(t_1,x_1): t_1 = x_1^3 = 1,\ x_1^1 = x_1^2 = 0\}$. The control function is $u = (v^1,v^2,v^3,v^4,p^1,p^2)$. The constraints on the controls are:

$$\begin{aligned} &(v^1(t))^2 + (v^2(t))^2 = 1 \qquad && (v^3(t))^2 + (v^4(t))^2 = 1 \\ &p^1(t) \geq 0 \qquad p^2(t) \geq 0 \qquad && p^1(t) + p^2(t) = 1. \end{aligned} \tag{9.2}$$

The problem is to minimize

$$J(\phi,u) = \int_0^1 [(\phi^1(t))^2 + (\phi^2(t))^2]dt.$$

It is readily verified that all of the hypotheses of theorem 5.1 are satisfied. In particular, to see that the sets $\mathscr{Q}^+(t,x)$

are convex note that for fixed t the set of vectors of the form $(y^1(t), y^2(t))$, where $y^1(t)$ is given by the right hand side of the first equation in (9.1), $y^2(t)$ is given by the right hand side of the second equation in (9.1), and the conditions (9.2) are fulfilled, can be written in the form

$$(y^1(t), y^2(t)) = p^1(t)(v^1(t), v^2(t)) + p^2(t)(v^3(t), v^4(t)).$$

Thus this set is the convex hull of the unit circle, which is the unit disc. From this the convexity of $\mathscr{Q}^+(t,x)$ is evident.

Let $\phi^*(t) = (0,0,t)$, $0 \le t \le 1$. Let θ be a fixed real number in $[0,2\pi)$ and let u_θ^* be a control defined as follows:

$$\begin{aligned}(v^1(t), v^2(t)) &= (\cos\theta, \sin\theta)\\ (v^3(t), v^4(t)) &= (-\cos\theta, -\sin\theta)\\ p^1(t) = p^2(t) &= 1/2.\end{aligned}$$

Then (ϕ^*, u_θ^*) is admissible and $J(\phi^*, u_\theta^*) = 0$. Since $J(\phi,u) \ge 0$ for all admissible (ϕ,u) it follows that (ϕ^*, u_θ^*) is optimal. Note that there are infinitely many optimal pairs (ϕ^*, u_θ^*), each one corresponding to a value of θ in $[0,2\pi)$. Moreover, all of the optimal pairs have the same trajectory. In fact it is clear from the form of J and the end conditions that ϕ^* is the only possible optimal trajectory. Thus the optimal trajectory is unique, while the optimal control is not.

We now construct a minimizing sequence. Let

$$\begin{aligned}v_k^1(t) &= v_k^3(t) = \cos 2\pi kt\\ v_k^2(t) &= v_k^4(t) = \sin 2\pi kt\\ p_k^1(t) &= p_k^2(t) = 1/2 \qquad k = 1,2,3,\ldots\end{aligned}$$

Then (9.1) becomes

$$\frac{dx^1}{dt} = \cos 2\pi kt \qquad \frac{dx^2}{dt} = \sin 2\pi kt$$

$$\frac{dx^3}{dt} = 1.$$

As in Example 2.3 we see that the corresponding sequence $\{(\phi_k, u_k)\}$ of admissible pairs is a minimizing sequence and $\phi_k \to \phi^*$ uniformly on $[0,1]$.

Let f be any function in $L_1[0,1]$. Then by the Riemann-Lebesgue Lemma

$$\int_0^1 f(t)\sin 2\pi kt\, dt \to 0 \qquad \int_0^1 f(t)\cos 2\pi kt\, dt \to 0. \tag{9.3}$$

Suppose that for some θ in $[0,2\pi)$ a subsequence of $v_k = (v_k^1, \ldots, v_k^4)$ converges in L_p, $1 \leq p \leq \infty$, or in measure, or almost everywhere, to the function $\tilde{u}_\theta^* = (\cos\theta, \sin\theta, -\cos\theta, -\sin\theta)$. Note that $\tilde{u}_\theta^*$ is obtained by taking the first four components of u_θ^*. Since the functions v_k are uniformly bounded we would in all cases have

$$\int_0^1 v_k dt \to \int_0^1 \tilde{u}_\theta^* dt = (\cos\theta, \sin\theta, -\cos\theta, -\sin\theta)$$

for the subsequence in question. Since $\tilde{u}_\theta^* \neq 0$ for all θ, we obtain a contradiction by taking $f \equiv 1$ in (9.3).

If in (9.3) we let f range over the bounded measurable functions on $[0,1]$ we see that the sequence $\{v_k\}$ converges to zero in the weak topology of $L_1[0,1]$. Thus, no subsequence of $\{v_k\}$ can converge to $\tilde{u}_\theta^*$ in this topology. Similarly, if we let f range over all functions in $L_q[0,1]$, $1 < q < \infty$, we get that no subsequence of $\{v_k\}$ can converge to $\tilde{u}_\theta^*$ in the weak topology of $L_p[0,1]$. Finally, we note that if we let f range over all functions in $L_1[0,1]$, then we get that no subsequence of $\{v_k\}$ can converge to $\tilde{u}_\theta^*$ in the weak*-topology induced on $L_\infty[0,1]$ by $L_1[0,1]$.

10. Proof of Theorem 7.1

We first prove the theorem under the assumption that D is a closed subset L of the open interval $(0,\infty)$.

For each positive integer k we can partition $\Phi(L)$, the range of Φ, into an at most countable union of sets B_j^k, $j = 1,2,\ldots$ as follows. Let

$$\Lambda_j^k = \Phi(L \cap [0,j2^{-k}]) \quad j = 1,2,\ldots$$

and let

$$B_j^k = \Lambda_j^k - \Lambda_{j-1}^k.$$

Some of the sets B_j^k may be empty. Each non empty set B_j^k consists of those points in $\Phi(L)$ that are images of points in $L \cap ((j-1)2^{-k},j2^{-k}]$ but are not images of points in $L \cap [0,(j-1)2^{-k}]$. Thus the non empty sets B_j^k are disjoint and their union is all of $\Phi(L)$. Note that the sets Λ_j^k are compact. Also note that

$$B_j^k = B_{2j}^{k+1} \cup B_{2j-1}^{k+1}. \tag{10.1}$$

For each positive integer k we define a function μ_k from T to L as follows. Since $\Gamma(T) \subseteq \Phi(L)$ it follows that for each t in T there is a unique integer j such that $\Gamma(t) \in B_j^k$. Let

$$\mu_k(t) = \inf \Phi^{-1}(B_j^k),$$

where B_j^k is the set in $\Phi(L)$ that contains $\Gamma(t)$. The function μ_k is measurable since it assumes an at most countable set of values, $\inf \Phi^{-1}(B_j^k)$, $j = 1,2,\ldots,$ and each set $\mu_k^{-1}(\inf \Phi^{-1}(B_j^k))$ is measurable, as the following argument shows. For each j we have $\mu_k^{-1}(\inf \Phi^{-1}(B_j^k)) = \Gamma^{-1}(B_j^k)$. But

$$\Gamma^{-1}(B_j^k) = \Gamma^{-1}(\Lambda_j^k - \Lambda_{j-1}^k) = \Gamma^{-1}(\Lambda_j^k)-\Gamma^{-1}(\Lambda_{j-1}^k).$$

The sets Λ_j^k and Λ_{j-1}^k are compact. Therefore, since Γ is measurable, the sets $\Gamma^{-1}(\Lambda_j^k)$ and $\Gamma^{-1}(\Lambda_{j-1}^k)$ are measurable, and so is their difference $\Gamma^{-1}(B_j^k)$.

We show that the sequence $\{\mu_k\}$ of measurable mappings from T to L is an increasing sequence. If t is such that $\Gamma(t) \in B_j^k$, then it follows from (10.1) that $\Gamma(t) \in B_i^{k+1}$ with $i = 2j$ or $i = 2j-1$. Therefore $B_i^{k+1} \subseteq B_j^k$ and therefore $\inf \Phi^{-1}(B_j^{k+1}) \geq \inf \Phi^{-1}(B_j^k)$. Hence $\mu_{k+1}(t) \geq \mu_k(t)$.

For t such that $\Gamma(t) \in B_j^k$ it also follows that for all $q \geq k$,

$$\mu_q(t) \leq j2^{-k}. \tag{10.2}$$

In particular, if $\Gamma(t) \in B_j^0$, we have that $\mu_k(t) \leq j$ for all k. Hence the sequence $\{\mu_k(t)\}$ is bounded above, and the functions μ_k converge to a measurable function μ. Since L is closed and the numbers $\mu_k(t)$ are in L, the range of μ is contained in L.

We now show that for all t in T, $\Phi(\mu(t)) = \Gamma(t)$. Suppose this statement were false. Then since Z is Hausdorff there exists a t in T and an open set $\mathcal{O}$ in Z such that $\Phi(\mu(t)) \in \mathcal{O}$ and $\Gamma(t) \notin \mathcal{O}$. Since Φ is continuous there exists a neighborhood of $\mu(t)$ in L that is contained in $\Phi^{-1}(\mathcal{O})$. Therefore there exist integers k and j_0 such that

$$\begin{aligned} &\mu(t) \in L \cap ((j_0-1)2^{-k}, j_0 2^{-k}) \\ &L \cap [(j_0-1)2^{-k}, j_0 2^{-k}] \subseteq \Phi^{-1}(\mathcal{O}). \end{aligned} \tag{10.3}$$

There exists an integer j such that $\Gamma(t) \in B_j^k$. If $j < j_0$, then from (10.2) we get that $\mu_q(t) \leq j2^{-k} \leq (j_0-1)2^{-k}$ for all $q \geq k$. Hence $\mu(t) \leq (j_0-1)2^{-k}$. But by the first line of (10.3) we have $\mu(t) > (j_0-1)2^{-k}$ and so $j \geq j_0$. If $j > j_0$ then $\Gamma(t)$ is the image under Φ of some point λ in $((j-1)2^{-k}, j2^{-k}]$ and is not the image of any point in $[0,(j-1)2^{-k}]$. Hence there exists a $q \geq k$

such that $\mu_q(t) > j_0 2^{-k} \geq \mu(t)$. But since $\mu_q(t)$ is an increasing sequence converging to $\mu(t)$, this cannot be. Hence $j = j_0$. Therefore $\Gamma(t)$ is the image under Φ of some point in $L \cap ((j_0-1)2^{-k}, j_0 2^{-k}]$. Therefore, using the second line of (10.3) we have $\Gamma(t) \in \Phi(L \cap [(j_0-1)2^{-k}, j_0 2^{-k}]) \subseteq \mathcal{O}$ which contradicts the assumption that $\Gamma(t) \notin \mathcal{O}$. Hence the Theorem is proved for the case in which $D = L$, a closed subset of $(0,\infty)$.

We next prove the theorem for the case in which D is the continuous image of a closed set L in $(0,\infty)$. Thus $D = \psi(L)$, where ψ is a continuous mapping from L onto D. The state of affairs is pictured in Figure 4.

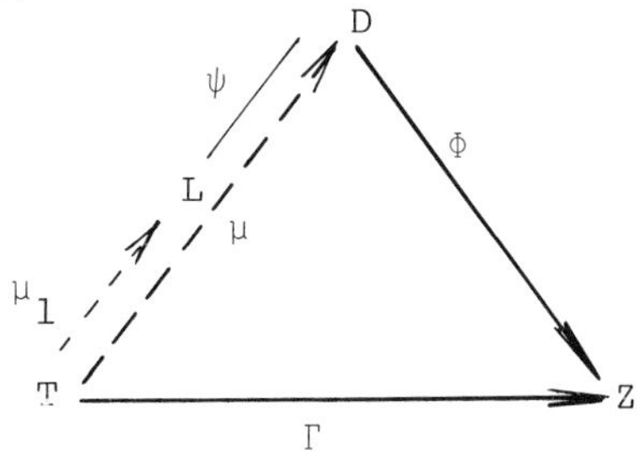

Figure 4

The mapping $\Phi*\psi$, where $*$ denotes composition, is a continuous mapping from L to Z. Since $\Gamma(T) \subseteq \Phi(D)$ and $\Phi(D) = \Phi(\psi(L))$, it follows that $\Gamma(T) \subseteq (\Phi*\psi)(L)$. Therefore, since we have established the theorem for the case $D = L$, there exists a measurable function μ_1 from T to L such that $(\Phi*\psi)*\mu_1 = \Gamma$. Let $\mu = (\psi*\mu_1)$. Then μ is a mapping from T to D such that $\Phi*\mu = \Gamma$. We assert that μ is measurable. Let C be a compact set in D. Then $\psi^{-1}(C)$ is closed in L and is therefore the union of the countable collection of sets $L \cap [0,k] \cap \psi^{-1}(C)$, $k = 1,2,\ldots$. Hence $\mu_1^{-1}(\psi^{-1}(C))$ is measurable, and therefore so is $\mu^{-1}(C)$.

Now let $D = \bigcup_{i=1}^{\infty} C_i$, where the C_i are compact metric spaces. By a theorem of Urysohn, every compact metric space is the continuous image of the Cantor set. For a proof of this the reader is referred to ([29] Theorem 3.28, p. 127). Hence for each integer k there is a translate of the Cantor set L_k contained in $[2k-1,2k]$ and a continuous mapping ψ_k from L_k onto C_k. Note that each L_k is closed. Let $L = \cup L_k$ and let ψ be a mapping from L to D defined by the formula $\psi(t) = \psi_k(t)$ if $t \in L_k$, $k = 1,2,\ldots$. Then L is a closed subset of $(0,\infty)$, the function ψ is continuous on L and $\psi(L) = D$. Hence the hypotheses of the previous case are fulfilled and there is a measurable μ such that $\Phi^*\mu = \Gamma$.

11. Existence of Controls in Systems Linear in the State

We consider systems with state equations

$$\frac{dx}{dt} = A(t)x + h(t,u(t)), \tag{11.1}$$

and shall use Theorem 7.1 to show that under reasonable hypotheses on the system a control exists.

LEMMA 11.1. Let $\mathscr{I}$ be a fixed closed interval in E^1 and let $\Omega: t \to \Omega(t)$ be a mapping from $\mathscr{I}$ to the closed subsets of E^m that is upper semicontinuous on $\mathscr{I}$. Then there exists a measurable function $u: \mathscr{I} \to E^m$ such that $u(t) \in \Omega(t)$ a.e.

<u>Proof</u>: Let

$$\Delta = \{(t,z): t \in \mathscr{I},\ z \in \Omega(t)\}. \tag{11.2}$$

Since Ω is upper semicontinuous it follows from Lemma 3.1 that Δ is closed. The set Δ can therefore be written as the countable union of compact sets $\Delta_i = \Delta \cap B_i$, where B_i is the closed ball of radius i in E^{m+1}.

Let $Z = \mathscr{I} \times E^1$ and let $T = \mathscr{I}$. Then Z is a Hausdorff space and T is a measure space with Lebesgue measure. Let Φ be the mapping from Δ to Z defined by the formula

$$\Phi(t,z) = (t,|z|).$$

Then Φ is clearly continuous. For each t in $\mathscr{I}$ let

$$d(t) = \inf\{|z|: \quad z \in \Omega(t)\}.$$

Let Γ be the map from T to Z defined by the formula $\Gamma(t) = (t,d(t))$. Since for each t the set $\Omega(t)$ is closed, there exists at least one $z \in \Omega(t)$ such that $|z| = d(t)$. Thus $\Gamma(T) \subseteq \Phi(D)$.

We next show that the mapping d is lower semicontinuous on $\mathscr{I}$ by showing that for any real number α, the set $E_\alpha = \{t: d(t) \le \alpha\}$ is closed. Suppose that for some α, the set E_α were not closed. Then there would exist a sequence of points $\{t_n\}$ in E_α converging to a point t_0 in $\mathscr{I}$ such that $d(t_0) = \beta$, where $\beta > \alpha$. We show that this leads to a contradiction.

Since Ω is upper semicontinuous,

$$\Omega(t_0) \supseteq \bigcap_{\delta>0} \operatorname{cl}\Big(\bigcup_{|t-t_0|<\delta} \Omega(t)\Big). \tag{11.3}$$

Let $\varepsilon = (\beta-\alpha)/2$. Since $d(t_n) \le \alpha$, it follows that in each set $\Omega(t_n)$ there is a point z_n such that $|z_n| < \alpha+\varepsilon$. Select a subsequence $\{z_n\}$ that converges to some point z_0 in E^m. Clearly,

$$|z_0| \le \alpha+\varepsilon < \beta = d(t_0). \tag{11.4}$$

The corresponding sequence $\{t_n\}$ still converges to t_0. Hence for each $\delta > 0$ there exists an $n_0(\delta)$ such that for $n > n_0(\delta)$ we have $|t_n-t_0| < \delta$. Consequently, for all $n > n_0(\delta)$,

$$z_n \in \bigcup_{|t-t_0|<\delta} \Omega(t). \tag{11.5}$$

Therefore, for each $\delta > 0$, z_0 belongs to the closure of the right hand side of (11.5). Therefore, z_0 belongs to the right hand side of (11.3). But by (11.4), $|z_0| < d(t_0)$, and so z_0 cannot belong to $\Omega(t_0)$. This, however, contradicts (11.3).

Since d is lower semicontinuous on $\mathscr{I}$, the mapping Γ is certainly measurable. The hypotheses of Theorem 7.1 are thus satisfied with $D = \Delta$. Hence there is a measurable map $\mu: t \to (\tau(t),u(t))$ such that $u(t) \in \Omega(\tau(t))$ and

$$(\Phi*\mu)(t) = (\tau(t),|u(t)|) = \Gamma(t) = (t,d(t)).$$

Hence $\tau(t) = t$ and $u(t) \in \Omega(t)$. The mapping $u: t \to u(t)$ is the desired mapping. Note that

$$|u(t)| = \inf\{|z|: z \in \Omega(t)\}.$$

LEMMA 11.2. Let $\mathscr{I}$ be a closed interval in E^1 and let A be an $n \times n$ continuous matrix on $\mathscr{I}$. Let $\mathscr{U}$ be a region in E^m and let h be a continuous map from $\mathscr{I} \times \mathscr{U}$ to E^n. Let $\Omega: t \to \Omega(t)$ be a mapping from $\mathscr{I}$ to the subsets of E^m that is upper semicontinuous on $\mathscr{I}$. Let there exist a nonnegative integrable function μ defined on $\mathscr{I}$ such that

$$|h(t,z)| \leq \mu(t) \tag{11.6}$$

for all (t,z) in Δ, where Δ is as in (11.2). Then there exists a control u on the interval $\mathscr{I}$ satisfying $u(t) \in \Omega(t)$.

<u>Proof</u>: By Lemma 11.1 there exists a measurable function u defined on $\mathscr{I}$ such that $u(t) \in \Omega(t)$. Since h is continuous the mapping $t \to h(t,u(t))$ is measurable. Therefore, by (11.6) it is integrable. Since the resulting system (11.1) is linear in x, a solution exists on all of $\mathscr{I}$.

REMARK 11.1. If the mapping Ω is assumed to be u.s.c.i. and the constraint sets $\Omega(t)$ are assumed to be compact, then by Lemma 5.2 the set Δ is compact and (11.6) is satisfied.

CHAPTER IV

EXISTENCE WITHOUT CONVEXITY

1. Introduction

The existence theorems in the preceding chapter required certain regularity in the behavior of the data of the problem and required that certain sets $\mathscr{Q}^+(t,x)$ be convex. In the case of noncompact constraints it was also required that the trajectories be equi-absolutely continuous, and reasonable growth conditions to ensure this were formulated. All of the conditions placed on the problem can be justified except, perhaps, the requirement that the sets $\mathscr{Q}^+(t,x)$ be convex. This requirement essentially restricts us to systems whose state equations are linear in the control, whose payoff function f^0 is convex in z, and for which the constraint sets $\Omega(t,x)$ are convex. In this chapter we investigate systems in which the sets $\mathscr{Q}^+(t,x)$ need not be convex.

In Section 2 we do not require the sets $\mathscr{Q}^+(t,x)$ to be convex. Instead, we essentially place restrictions on the rate of change of the controls. This is a mathematical idealization of systems in which the controls are assumed to possess inertia. It turns out that for such controls the assumption that the sets $\mathscr{Q}^+(t,x)$ are convex is not needed to obtain existence.

In Section 3 we replace the original problem with the "relaxed problem" in which the convexity assumption is satisfied. All trajectories of the original problem are also relaxed trajectories. The state of affairs sought for by introducing the relaxed problem is the following. Although the absence of convexity in the sets $\mathscr{Q}^+(t,x)$ may prevent us from asserting that the original problem has a solution, we can conclude that the relaxed problem does. It is then hoped that a further characterization of the solution of the relaxed problem will yield the information that the putative relaxed solution is actually a solution of the original problem. Of course, this need not be the case and the relaxed problem may have a solution which the original problem does not. In Section 4, however, we show that under reasonable hypotheses on the system, the trajectories of the original problem are dense in the set of trajectories of the relaxed problem. Thus relaxed trajectories can be uniformly approximated by ordinary trajectories. On the other hand, the infimum of the relaxed functional can be strictly less than the infimum of the original functional. The material in Section 4 will also be used in the proof of the maximum principle in Sections 2-7, Chapter VI.

Some writers base their investigation of the optimal control problem on the properties of the attainable set. We reverse the procedure and in Section 5 we use Theorem III 4.1 to obtain the basic properties of the attainable set.

In Section 6 we study problems in which f^0 and f are linear in the state variable and the constraint set is independent of the state variable x. We show that for such systems the requirement that the sets $\mathscr{Q}^+(t,x)$ are convex can be dropped. The basic fact here is that the relaxed attainable set and the ordinary attainable set are equal and both are compact and convex. This fact is also used to obtain the bang-bang principle for linear systems. Except for the need to refer to Definition 5.2 and the related discussion, Section 6 can be read independently.

The reader who is primarily interested in applications can at first reading confine himself to Section 2, Theorem 6.3, the statement of Theorem 6.4 and the discussion following Theorem 6.4. In reading Theorem 6.4 he will, however, have to refer back to Definitions 5.2 and 3.2 and the accompanying discussions.

2. Inertial Controllers

In Chapter 3 the controls u are measurable functions. Thus, there are no restrictions on the rate at which the controls can be varied. Physically this means that the controls are assumed to have no inertia. While this may be a reasonable model for electrical controls it is not so reasonable for mechanical systems or economic systems. We therefore consider controls that do have inertia. Continuous functions u with piecewise continuous derivatives u' such that $|u'(t)| \leq K$, where K is a constant, appear to be reasonable models of inertial controllers. At points of discontinuity of u' we interpret the inequality $|u'(t)| \leq K$ to hold for both the right and left hand limits $u'(t+0)$ and $u'(t-0)$. We now have a bound on the rate at which a control can be changed. It turns out that the class of functions u just described is too restrictive to enable us to prove an existence theorem for problems with inertial controllers. For this purpose it is necessary to take our model of an inertial controller to be an absolutely continuous function u such that $|u'(t)| \leq K$ a.e., where K is a constant independent of u.

We now state the minimization problem for inertial controllers.

PROBLEM 3. Minimize

$$J(\phi,u) = G(\phi) + \int_{t_0}^{t_1} f^0(t,\phi(t),u(t))\,dt$$

subject to:

$$dx/dt = f(t,x,u(t))$$

$$u(t) \in \Omega(t,x)$$

$$(t_0,\phi(t_0),t_1,\phi(t_1)) \in \mathscr{B}$$

u is absolutely continuous on $[t_0,t_1]$

$$|u'(t)| \leq K \quad \text{a.e. on} \quad [t_0,t_1],$$

where K is a pre-assigned constant and G is a functional defined on $\mathscr{X}_\rho$.

An admissible pair (ϕ,u) for Problem 3 is an absolutely continuous function ϕ and an absolutely continuous function u such that the requirements of Definition II 3.1 and II 3.2 are fulfilled and such that $|u'(t)| \leq K$ a.e.

THEOREM 2.1. Let the class $\mathscr{A}$ of admissible pairs for Problem 3 be non-empty and let the following hypotheses hold. (i) There exists a compact set $\mathscr{R}_0 \subset \mathscr{R}$ such that for all admissible trajectories ϕ we have $(t,\phi(t)) \in \mathscr{R}_0$ for all t in $[t_0,t_1]$. (ii) The set $\mathscr{B}$ is closed. (iii) The mapping Ω is u.s.c.i. on $\mathscr{R}_0$. (iv) For each (t,x) in $\mathscr{R}_0$ the set $\Omega(t,x)$ is compact. (v) The function f^0 is lower semicontinuous on $\mathscr{G} = \mathscr{R} \times \mathscr{U}$ and the function f is continuous on $\mathscr{G}$. Let G be lower semicontinuous and bounded below on $\mathscr{A}_T$. Then there exists an optimal pair (ϕ^*,u^*) in $\mathscr{A}$ for Problem 3.

REMARK 2.1. Note that the hypotheses of this theorem are the same as those of Theorem III 5.1 except that we now <u>do not</u> require the set $\mathscr{Q}^+(t,x)$ to be convex.

We shall prove Theorem 2.1 by rewriting Problem 3 as an equivalent problem having the form of Problem 2 and then applying Theorem III 5.1.

We take z to be a state variable and take the derivative u' to be the control. The system equations then become

$$\frac{dx}{dt} = f(t,x,z)$$
$$\frac{dz}{dt} = v(t). \tag{2.1}$$

We shall denote solutions of (2.1) by (ϕ,ψ); thus, $\phi'(t) = f(t,\phi(t),\psi(t))$ and $\psi'(t) = v(t)$ a.e. Note that since f, ϕ, and ψ are continuous, $\phi'(t) = f(t,\phi(t),\psi(t))$ everywhere. Let

$\mathscr{C}^m = \{w \in E^m: |w| \leq K\}$. Let

$$\tilde{\mathscr{R}} = \{(t,x,z): (t,x) \in \mathscr{R}, \quad z \in \Omega(t,x)\}$$
$$\tilde{\mathscr{B}} = \{(t_0,x_0,z_0,t_1,x_1,z_1): (t_0,x_0,t_1,x_1) \in \mathscr{B},$$
$$z_i \in \Omega(t_i,x_i), \ i = 0,1\}$$
$$\tilde{\Omega}(t,x,z) = \{w \in E^m: |w| \leq K\} = \mathscr{C}^m.$$

Consider Problem 3', defined as follows. Minimize

$$\tilde{J}(\phi,\psi,v) = G(\phi) + \int_{t_0}^{t_1} f^0(t,\phi(t),\psi(t))dt$$

subject to (2.1), $(t,\phi(t),\psi(t)) \in \tilde{\mathscr{R}}$, $v(t) \in \mathscr{C}^m$ a.e., and $(t_0,\phi(t_0),\psi(t_0), t_1,\phi(t_1),\psi(t_1)) \in \tilde{\mathscr{B}}$. To each admissible pair (ϕ,ψ,v) of Problem 3' there corresponds a unique admissible pair $(\phi,u) = (\phi,\psi)$ of Problem 3 such that $J(\phi,u) = \tilde{J}(\phi,\psi,v)$. Conversely, to each admissible pair (ϕ,u) of Problem 3 there corresponds the admissible pair (ϕ,ψ,v) of Problem 3' with $\psi(t) = u(t)$ and such that $J(\phi,u) = \tilde{J}(\phi,\psi,v)$. Hence Problem 3' is equivalent to Problem 3. In the next paragraph we shall show that Problem 3' satisfies the hypotheses of Theorem III 5.1. Hence Problem 3' has a solution (ϕ^*,ψ^*,v^*); hence Problem 3 has a solution (ϕ^*,u^*) with $u^*(t) = \psi^*(t)$.

We now verify that the hypotheses of Theorem III 5.1 are satisfied in Problem 3'. First note that since u is continuous and Ω is u.s.c.i. that $u(t) \in \Omega(t,\phi(t))$ a.e. implies $u(t) \in \Omega(t,\phi(t))$ everywhere. Hence since $\psi(t) = u(t)$ we have that $\psi(t) \in \Omega(t,\phi(t))$ for all values of t. Since the graphs of all trajectories ϕ of Problem 3 lie in a compact set $\mathscr{R}_0 \subset \mathscr{R}$, it follows that if we set $\tilde{\mathscr{R}}_0 = \mathscr{D}$, where $\mathscr{D}$ is defined in III (4.1), then the graphs of all trajectories (ϕ,ψ) of Problem 3' lie in $\tilde{\mathscr{R}}_0$. Clearly, $\tilde{\mathscr{R}}_0 \subset \tilde{\mathscr{R}}$. Moreover, since Ω is u.s.c.i. and each $\Omega(t,x)$ is compact it follows from Lemma III 5.2 that the set $\tilde{\mathscr{R}}_0$ is compact. Thus all trajectories of Problem 3' lie in a compact set $\tilde{\mathscr{R}}_0 \subset \tilde{\mathscr{R}}$.

To see that $\tilde{\mathscr{B}}$ is closed, let $\{(t_{0n},x_{0n},z_{0n},t_{1n},x_{1n},z_{1n})\}$ be a sequence of points in $\tilde{\mathscr{B}}$ converging to a point $(t_0,x_0,z_0,t_1,x_1,z_1)$. Since $\mathscr{B}$ is closed, it follows that $(t_0,x_0,t_1,x_1) \in \mathscr{B}$. Let $\varepsilon > 0$ be given. Since Ω is u.s.c.i. it follows that for n sufficiently large

$$z_{in} \in \Omega(t_{in},x_{in}) \subseteq [\Omega(t_i,x_i)]_\varepsilon \qquad i = 0,1.$$

Hence $z_i \in [\Omega(t_i,x_i)]_\varepsilon$ for arbitrary $\varepsilon > 0$ and $i = 0,1$. Since each set $\Omega(t_i,x_i)$, $i = 0,1$, is closed we get that $z_i \in \Omega(t_i,x_i)$, $i = 0,1$, and so $\tilde{\mathscr{B}}$ is closed. Since for each (t,x,z) in $\tilde{\mathscr{R}}_0$ we have $\tilde{\Omega}(t,x,z) = \mathscr{C}^m$, it follows that $\tilde{\Omega}$ is u.s.c.i. and each set $\tilde{\Omega}(t,x,z)$ is compact. The sets $\tilde{\mathscr{Q}}^+(t,x,z)$ for Problem 3' are defined as follows:

$$\tilde{\mathscr{Q}}^+(t,x,z) = \{(y^0,y,\eta): y^0 \geq f^0(t,x,z),\ y = f(t,x,z),$$
$$\eta = w,\ w \in \mathscr{C}^m\},$$

and are clearly convex. Finally, the functions f^0, f and G satisfy the required continuity hypotheses.

EXERCISE 2.1. Prove Theorem 2.1 directly, without appealing to Theorem III 5.1 and without using Theorem III 4.1. Hint: If $\{(\phi_k,u_k)\}$ is a minimizing sequence then the functions u_k have equi-absolutely continuous integrals.

3. The Relaxed Problem

In this section we formulate the relaxed problem corresponding to Problem 2 and prove an existence theorem for the relaxed problem. We shall refer to Problem 2 as the original problem. As noted in the introduction, the relaxed problem is a problem related to the original problem and is one in which the convexity assumption of Chapter 3 is satisfied even though the original problem does not satisfy the convexity assumption. Thus, under appropriate hypotheses, we can guarantee that the relaxed problem will have a solution in situations where we cannot guarantee that the original problem will have a solution. In some situations we can use the knowledge of the existence of a solution to the relaxed problem to show that the original problem has a solution. For example, when a solution is characterized by the necessary conditions of Chapter 5 it may turn out that the relaxed solution is actually an ordinary solution. Another example of the use of the relaxed problem to obtain the existence of a solution to the original problem is found in Theorem 6.2 below. Of course, the relaxed problem may have a solution while the original problem does not.

Let q be a positive integer and let S be any set. The symbol $[S]^q$ will denote the q-fold Cartesian product of S with itself. Let $\tilde{z}$ denote a vector in $[E^m]^{n+2}$. Thus, $\tilde{z} = (z_1,\ldots,z_{n+2})$, where $z_i = (z_i^1,\ldots,z_i^m) \in E^m$ for $i = 1,\ldots,n+2$. Let

$$\Gamma = \{\pi : \pi = (\pi^1,\ldots,\pi^{n+2}),\quad \pi^i \geq 0,\ \Sigma\pi^i = 1\}$$

$$\tilde{\Omega}(t,x) = [\Omega(t,x)]^{n+2} \times \Gamma$$

$$\tilde{f}^0(t,x,\tilde{z},\pi) = \sum_{i=1}^{n+2} \pi^i f^0(t,x,z_i) \tag{3.1}$$

$$\tilde{f}(t,x,\tilde{z},\pi) = \sum_{i=1}^{n+2} \pi^i f(t,x,z_i),$$

where the vectors z_i are in $\mathscr{U}$. Let $\mathscr{B}$ and $\mathscr{R}$ be as in Problem 2 of Chapter 2. If $u_1,\ldots,u_{n+2}$ is a set of measurable functions defined on a common interval $[t_0,t_1]$ and if each u_i has range in E^m, then we define a measurable mapping $\tilde{u}$ from $[t_0,t_1]$ to $[E^m]^{(n+2)}$ as follows: $\tilde{u} = (u_1,\ldots,u_{n+2})$.

DEFINITION 3.1. A measurable function v of the form

$$v = (\tilde{u},p) = (u_1,\ldots,u_{n+2}, p^1,\ldots,p^{n+2})$$

defined on an interval $[t_0,t_1]$ is said to be a relaxed control if the following hold. For each $i = 1,\ldots,n+2$ the function u_i has range contained in $\mathscr{U}$. For each $i = 1,\ldots,n+2$ the function p^i is real valued, $p^i(t) \geq 0$ a.e. on $[t_0,t_1]$ and $\sum_{i=1}^{n+2} p^i(t) = 1$ a.e. on $[t_0,t_1]$. There exists an absolutely continuous function $\psi = (\psi^1,\ldots,\psi^n)$ defined on $[t_0,t_1]$ such that

(i) $(t,\psi(t)) \in \mathscr{R}$ for all $t \in [t_0,t_1]$,

(ii) ψ is a solution of the system of differential equations

$$\frac{dx}{dt} = \tilde{f}(t,x,\tilde{u}(t),p(t)) = \sum_{i=1}^{n+2} p^i(t) f(t,x,u_i(t)); \qquad (3.2)$$

that is $\psi'(t) = \tilde{f}(t,\psi(t),\tilde{u}(t),p(t))$ a.e. on $[t_0,t_1]$. The function ψ is called a relaxed trajectory corresponding to v. The system of differential equations (3.2) is called the system of relaxed state equations.

DEFINITION 3.2. A relaxed control v is said to be an admissible relaxed control if there exists a relaxed trajectory ψ corresponding to v such that

(i) The mapping $t \to \tilde{f}^0(t,\psi(t),\tilde{u}(t),p(t))$

$$= \sum_{i=1}^{n+2} p^i(t) f^0(t,\psi(t),u_i(t)) \text{ is in } L_1[t_0,t_1],$$

(ii) $v(t) = (\tilde{u}(t),p(t)) \in \tilde{\Omega}(t,\psi(t))$ a.e. on $[t_0,t_1]$,

(iii) $(t_0,\psi(t_0),t_1,\psi(t_1)) \in \mathscr{B}$.

The trajectory ψ is called an admissible relaxed trajectory. The pair (ψ,v) is called an admissible relaxed pair.

We now state the <u>relaxed problem</u> corresponding to Problem 2.

PROBLEM 2R. Let $\mathscr{A}_R$ denote the class of all admissible relaxed pairs and let $\mathscr{A}_R$ be non-empty. Let

$$\tilde{J}(\psi,v) = G(\psi) + \int_{t_0}^{t_1} \tilde{f}^0(t,\psi(t),\tilde{u}(t),p(t))dt,$$

where $(\psi,v) \in \mathscr{A}_R$ and G is a functional defined on $\mathscr{X}_\rho$. Let $\mathscr{A}_{1R}$ be a non-empty subset of $\mathscr{A}_R$. Find an optimal admissible relaxed pair (ψ^*,v^*) in $\mathscr{A}_{1R}$; i.e. find an element (ψ^*,v^*) in $\mathscr{A}_{1R}$ such that $\tilde{J}(\psi^*,v^*) \leq \tilde{J}(\psi,v)$ for all (ψ,v) in $\mathscr{A}_{1R}$.

The control problem in Example III 9.1 is the relaxed problem corresponding to Example III 2.3.

Let (ϕ,u) be an admissible pair for Problem 2. Let $v = (u_1,\ldots,u_{n+2}, 1,0,\ldots,0)$, where $u_i = u$ for $i = 1,\ldots,n+2$. Then if we take $\psi = \phi$, the pair (ψ,v) is an admissible relaxed pair. Thus, if $\mathscr{A}$ is non-empty then $\mathscr{A}_R$ is non empty. Speaking loosely, we say that every admissible pair is a relaxed admissible pair. The following remark is now obvious.

REMARK 3.1. If $\tilde{\mu} = \inf\{\tilde{J}(\psi,v): (\psi,v) \in \mathscr{A}_R\}$ and $\mu = \inf\{J(\phi,u): (\phi,u) \in \mathscr{A}\}$, then $\tilde{\mu} \leq \mu$.

In Exercise 3.1 we consider an example in which both the original and relaxed problems have solutions and $\tilde{\mu}$ is strictly less than μ. In Exercise 4.1 of this Chapter we shall give criteria guaranteeing that $\tilde{\mu} = \mu$.

We now state the principal result of this section. We shall only consider compact constraint sets $\Omega(t,x)$. We leave the formulation and proof of the corresponding results for non compact constraint sets to the reader.

THEOREM 3.1. Let there exist a compact set $\mathscr{R}_0 \subset \mathscr{R}$ such that all admissible relaxed trajectories lie in $\mathscr{R}_0$. Let the original problem satisfy hypotheses (ii), (iii), (iv) of Theorem III 5.1 with $\mathscr{R}_0$ as in the present theorem. Let f^0 and f satisfy the hypotheses (vi) of Theorem III 5.1 and let G be as in Remark III 5.1. Then the relaxed problem has a solution in $\mathscr{A}_R$.

It is a straight forward matter to verify that the relaxed problem satisfies all of the hypotheses of Theorem III 5.1, with the possible exception of (v). For (v) to hold we must show that for every (t,x) in $\mathscr{R}_0$ the set

$$\tilde{\mathscr{Q}}^+(t,x) = \{(y^0,y): y^0 \geq \tilde{f}^0(t,x,\tilde{z},\pi),\ y = \tilde{f}(t,x,\tilde{z},\pi),$$
$$(\tilde{z},\pi) \in \tilde{\Omega}(t,x)\}$$

is convex. Let

$$\tilde{\mathscr{Q}}(t,x) = \{(y^0,y): y^0 = \tilde{f}^0(t,x,\tilde{z},\pi),\ y = \tilde{f}(t,x,\tilde{z},\pi),$$
$$\tilde{z} \in [\Omega(t,x)]^{n+2},\quad \pi \in \Gamma\}$$

and let

$$\mathscr{Q}(t,x) = \{(y^0,y): y^0 = f^0(t,x,z),\ y = f(t,x,z),\ z \in \Omega(t,x)\}.$$

Then $\tilde{\mathscr{Q}}(t,x) \subset \text{co } \mathscr{Q}(t,x)$. By Caratheodory's theorem (Theorem III 5.2), every point in co $\mathscr{Q}(t,x)$ can be written as a convex combination of at most $(n+2)$-points in $\mathscr{Q}(t,x)$. From the definition of $\tilde{\mathscr{Q}}(t,x)$ and from (3.1) we see that we therefore have co $\mathscr{Q}(t,x) \subseteq \tilde{\mathscr{Q}}(t,x)$. Hence $\tilde{\mathscr{Q}}(t,x) = \text{co } \mathscr{Q}(t,x)$ and so $\tilde{\mathscr{Q}}(t,x)$ is convex. If $\tilde{\mathscr{Q}}(t,x)$ is convex, then so is $\tilde{\mathscr{Q}}^+(t,x)$. Therefore (v) of Theorem III 5.1 holds for the relaxed problem. Thus all of the hypotheses of Theorem III 5.1 hold for the relaxed problem, and so the relaxed problem has a solution.

EXERCISE 3.1. Consider the problem with state equations

$$dx^1/dt = (x^2)^2 - (u(t))^2$$
$$dx^2/dt = u(t)$$
$$dx^3/dt = (x^2)^4,$$

where u is a real valued function subject to the constraint $|u(t)| \leq 1$. Let $\mathscr{B} = \{(t_0,x_0,t_1,x_1): t_0 = 0,\ x_0 = 0,\ t_1 = 1,\ x_1^3 = 0\}$. Let $G(\phi) = \phi^1(1)$ and let $f^0 \equiv 0$. Show that the convexity hypothesis of Theorem III 5.1 is not satisfied, yet an optimal pair exists. Find an optimal pair for the relaxed problem and show that $\tilde{\mu} < \mu$. Find a sequence of ordinary trajectories ϕ_k and corresponding controls u_k such that all constraints except the end conditions are satisfied and such that $G(\phi_k) \to \tilde{\mu}$ and $\phi_k \to \psi^*$ uniformly on $[0,1]$, where ψ^* is the optimal relaxed trajectory. Note that it is not possible to do this with pairs (ϕ_k, u_k) that are admissible for the original problem.

EXERCISE 3.2. Let the hypotheses of Theorem 3.1 hold except that f^0 is taken to be continuous. Then there is no loss of generality in assuming $f^0 \equiv 0$. Under these assumptions the following definition of relaxed trajectory is equivalent to the one given in the text. A relaxed trajectory is an absolutely continuous function ψ such that

$$\psi'(t) \in \text{co } f(t,\psi(t),\Omega(t,\psi(t))) \text{ a.e.,}$$

where

$$f(t,x,\Omega(t,x)) = \{y: y = f(t,x,z),\ z \in \Omega(t,x)\}.$$

4. The Chattering Lemma; Approximations to Relaxed Controls

Theorem 4.1 will be used in Chapter 6 in the derivation of the maximum principle and will be used in Theorem 4.3 of this section to show that under reasonable hypotheses the ordinary trajectories of a control system are dense in the relaxed trajectories of the system.

In both of these applications of Theorem 4.1 we shall see that the hypothesis below that the f_i are only measurable in t is imposed on us by the applications. Theorem 4.1 is sometimes called the "Chattering Lemma" for reasons which will be discussed in Remark 4.4 below.

THEOREM 4.1. Let $\mathscr{I}$ be a finite closed interval and let $\mathscr{X}$ be a compact set in E^n. Let $f_1,\ldots,f_q$ be functions defined on $\mathscr{I} \times \mathscr{X}$ with range in E^n and possessing the following properties:

(i) Each f_i is a measurable function on $\mathscr{I}$ for each x in $\mathscr{X}$.

(ii) Each f_i is continuous on $\mathscr{X}$ for each t in $\mathscr{I}$.

(iii) There exists an integrable function μ defined on $\mathscr{I}$ such that for all (t,x) and (t,x') in $\mathscr{I} \times \mathscr{X}$ and $i = 1,\ldots,q$:

$$
\begin{aligned}
&|f_i(t,x)| \leq \mu(t) \\
&|f_i(t,x)-f_i(t,x')| \leq \mu(t)|x-x'|.
\end{aligned}
\tag{4.1}
$$

Let p^i, $i = 1,\ldots,q$ be real valued non negative measurable functions defined on $\mathscr{I}$ and satisfying

$$\sum_{i=1}^{q} p^i(t) = 1 \quad \text{a.e.} \tag{4.2}$$

Then for every $\varepsilon > 0$ there exists a subdivision of $\mathscr{I}$ into a finite collection of non-overlapping intervals E_j, $j = 1,\ldots,k$ and an assignment of one of the functions $f_1,\ldots,f_q$ to each E_j such that the following holds. If f_{E_j} denotes the function assigned to E_j and if f is a function that agrees with f_{E_j} on the interior E_j^0 of each E_j i.e.,

$$f(t,x) = f_{E_j}(t,x) \quad \text{if} \quad t \in E_j^0 \qquad j = 1,\ldots,k,$$

then for every t', t'' in $\mathcal{I}$ and all x in $\mathcal{X}$

$$\left| \int_{t'}^{t''} \left(\sum_{i=1}^{q} p^i(t) f_i(t,x) - f(t,x) \right) dt \right| < \varepsilon. \tag{4.3}$$

REMARK 4.1. Let $E_j = [\tau_j, \tau_{j+1}]$, $j = 1,\ldots,k$. If we set $f(\tau_j,x) = f_{E_j}(\tau_j,x)$, $j = 1,\ldots,k$ and set $f(\tau_{k+1},x) = f_{E_k}(\tau_{k+1},x)$, then (4.3) will still hold and the following statements will be true. The function f satisfies (4.1). If the functions f_i are of class $C^{(r)}$ on $\mathcal{X}$ for some values of t, then f is of class $C^{(r)}$ for the same values of t.

For an $n \times m$ matrix M we take $|M|$ to be the norm of the linear transformation that M determines relative to the standard bases in E^n and E^m. If for each t in $\mathcal{I}$ the f_i are of class $C^{(1)}$ in $\mathcal{X}$ and if for each i and x in $\mathcal{X}$

$$|\partial f_i(t,x)/\partial x| \le \mu(t),$$

where $\partial f_i/\partial x$ denotes the Jacobian matrix of first partial derivatives of f_i, then $|\partial f(t,x)/\partial x| \le \mu(t)$.

The first step in our proof is to establish the following lemma.

LEMMA 4.1. Let $\mathcal{I}$ and $\mathcal{X}$ be as in the theorem and let f be a function from $\mathcal{I} \times \mathcal{X}$ to E^n having the same properties as the functions $f_1,\ldots,f_q$ of the theorem. Then for every $\varepsilon > 0$ there exists a continuous function g, depending on ε, from $\mathcal{I} \times \mathcal{X}$ to E^n such that for every x in $\mathcal{X}$

$$\int_{\mathcal{I}} |f(t,x) - g(t,x)| dt < \varepsilon . \tag{4.4}$$

<u>Proof</u>. It follows from (4.1) that for x and x' in $\mathcal{X}$

$$\int_{\mathcal{I}} |f(t,x) - f(t,x')| dt \le |x - x'| \int_{\mathcal{I}} \mu(t) dt.$$

Hence for arbitrary $\varepsilon > 0$, we have

$$\int_{\mathscr{I}} |f(t,x)-f(t,x')|dt < \varepsilon/2 \tag{4.5}$$

whenever $|x-x'| < \varepsilon/2 \int_{\mathscr{I}} \mu(t)dt$. Since $\mathscr{X}$ is compact, there exists a finite open cover $\mathscr{O}_1,\ldots,\mathscr{O}_k$ of $\mathscr{X}$ such that if x and x' are in the same $\mathscr{O}_i$, then (4.5) holds.

Let $x_1,\ldots,x_k$ be a finite set of points such that $x_i \in \mathscr{O}_i$. For each $i = 1,\ldots,k$ there exists a continuous function h_i defined on $\mathscr{I}$ such that

$$\int_{\mathscr{I}} |f(t,x_i)-h_i(t)|dt < \varepsilon/2. \tag{4.6}$$

Let $\gamma_1,\ldots,\gamma_k$ be a partition of unity corresponding to the finite open cover $\mathscr{O}_1,\ldots,\mathscr{O}_k$ of $\mathscr{X}$. That is, let $\gamma_1,\ldots,\gamma_k$ be continuous real valued functions on $\mathscr{X}$ such that:

$$\begin{aligned} &\text{(i)} \quad \gamma_i(x) \geq 0 \quad \text{for all } x \in \mathscr{X} \\ &\text{(ii)} \quad \gamma_i(x) = 0 \quad \text{if } x \notin \mathscr{O}_i \\ &\text{(iii)} \quad \sum_{i=1}^{k} \gamma_i(x) = 1. \end{aligned} \tag{4.7}$$

For a proof of the existence of partitions of unity corresponding to finite open covers of compact subsets of locally compact Hausdorff spaces see Rudin ([54], p. 40).

Define

$$g(t,x) = \sum_{i=1}^{k} \gamma_i(x)h_i(t).$$

Then g is continuous on $\mathscr{I} \times \mathscr{X}$. We now show that g satisfies (4.4) and therefore is the desired function.

$$\int_{\mathscr{I}} |g(t,x)-f(t,x)|dt \leq \int_{\mathscr{I}} \left|\sum_{i=1}^{k} \gamma_i(x)h_i(t) - \sum_{i=1}^{k} \gamma_i(x)f(t,x_i)\right|dt$$

$$+ \int_{\mathscr{I}} \left|\sum_{i=1}^{k} \gamma_i(x)f(t,x_i) - \sum_{i=1}^{k} \gamma_i(x)f(t,x)\right|dt$$

$$\leq \sum_{i=1}^{k} \gamma_i(x) \int_{\mathscr{I}} |h_i(t) - f(t,x_i)|dt$$

$$+ \sum_{i=1}^{k} \gamma_i(x) \int_{\mathscr{I}} |f(t,x_i) - f(t,x)|dt.$$

By virtue of (4.6) each of the integrals in the first sum on the right is less than $\varepsilon/2$. From this and from (4.7)-(iii) it follows that the first sum on the right is less than $\varepsilon/2$. We now examine the i-th summand in the second sum on the right. If $x \notin \mathscr{O}_i$ then by (4.7)-(ii), $\gamma_i(x) = 0$ and so the summand is zero. If $x \in \mathscr{O}_i$, then by (4.5) the integral is less than $\varepsilon/2$ and therefore by (4.7)-(i) the summand is less than $\varepsilon\gamma_i(x)/2$. Therefore, each summand in the second sum is less than $\varepsilon\gamma_i(x)/2$. It now follows from (4.7)-(iii) that the second sum is less than $\varepsilon/2$. Hence g satisfies (4.4) and the lemma is proved.

Let $\bar{\varepsilon} > 0$ be given and let

$$\varepsilon = \bar{\varepsilon}/2(2 + q + |\mathscr{I}|), \tag{4.8}$$

where $|\mathscr{I}|$ denotes the measure of $\mathscr{I}$. Henceforth if A is a measurable set we shall use $|A|$ to denote the measure of A. From Lemma 4.1 we get that for each $i = 1,\ldots,q$ there is a continuous function g_i defined on $\mathscr{I} \times \mathscr{X}$ with range in E^n such that

$$\int_{\mathscr{I}} |f_i(t,x) - g_i(t,x)|dt < \varepsilon. \tag{4.9}$$

Since each g_i is continuous on $\mathscr{I} \times \mathscr{X}$ and $\mathscr{I}$ and $\mathscr{X}$ are compact, each g_i is uniformly continuous on $\mathscr{I} \times \mathscr{X}$. Therefore there exists a $\delta > 0$ such that if $|t-t'| < \delta$ then

$$|g_i(t,x) - g_i(t',x)| < \varepsilon. \tag{4.10}$$

Moreover, we may suppose that δ is such that if E is a measurable subset of $\mathscr{I}$ with $|E| < \delta$, then

$$\int_E \mu(t)dt < \varepsilon. \tag{4.11}$$

Let $\{I_k\}$ be a subdivision of $\mathscr{I}$ into a finite number of non-overlapping intervals with $|I_k| < \delta$ for each interval I_k. Moreover, suppose that $I_k = [t_k, t_{k+1}]$ and that $\ldots < t_{k-1} < t_k < t_{k+1} < t_{k+2} < \ldots$. For each I_k we can construct a subdivision of I_k into non-overlapping subintervals $E_{ki}, \ldots, E_{kq}$ such that

$$|E_{ki}| = \int_{I_k} p^i(t)dt. \tag{4.12}$$

This is possible since

$$\sum_{i=1}^{q} |E_{ki}| = \sum_{i=1}^{q} \int_{I_k} p^i(t)dt = \int_{I_k} \Big(\sum_{i=1}^{q} p^i(t)\Big)dt = |I_k|,$$

the last equality following from (4.2).

Define

$$f(t,x) = f_i(t,x) \qquad t \in E_{ki}^0, \tag{4.13}$$

where E_{ki}^0 denotes the interior of E_{ki}. Thus f is defined at all points of $\mathscr{I}$ except the end points of the intervals E_{ki}. At these points f can be defined as in Remark 4.1 or in any arbitrary manner. Let

$$\lambda(t,x) = \sum_{i=1}^{q} p^i(t)f_i(t,x) - f(t,x). \tag{4.14}$$

The collection of intervals $\{E_{ki}\}$ where k ranges over the same index set as do the intervals I_k and i ranges over the set $1,\ldots,q$, constitutes a subdivision of $\mathscr{I}$ into a finite number of non-overlapping subintervals. This subdivision, relabelled as $\{E_j\}$, is the subdivision whose existence is asserted in the theorem. If an interval E_j was originally the interval E_k then the function f_{E_j} assigned to E_j is f_i. If we now compare the definition of λ in (4.14) with (4.3) and note (4.13) we see that to prove the theorem we must show that for arbitrary t' and t'' in $\mathscr{I}$ and all x in $\mathscr{X}$

$$\left|\int_{t'}^{t''} \lambda(t,x)dt\right| < \bar{\varepsilon}. \tag{4.15}$$

There is no loss of generality in assuming that $t' < t''$. The point t' will belong to some interval I_α of the subdivision $\{I_k\}$ and the point t'' will belong to some interval I_β. If $I_\alpha \neq I_\beta$, let s_1 denote the right hand end point $t_{\alpha+1}$ of I_α and let s_2 denote the left hand end point t_β of I_β. Then if J denotes the set of indices $\{\alpha+1, \alpha+2,\ldots,\beta\}$ we have

$$[s_1,s_2] \equiv [t_{\alpha+1},t_\beta] = \bigcup_{j\varepsilon J} I_j.$$

See figure 1.

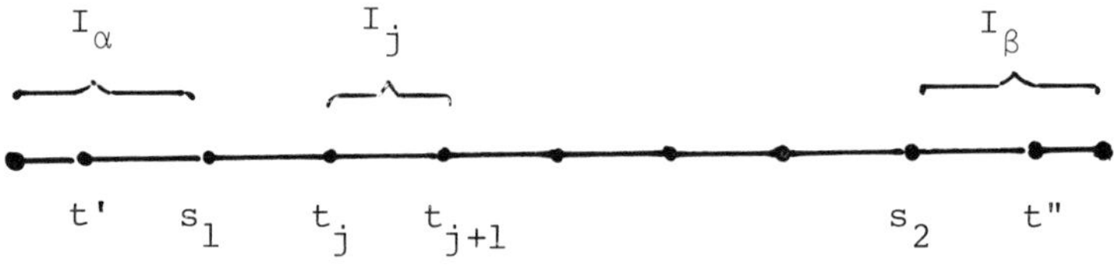

Figure 1

Hence we have

$$\left|\int_{t'}^{t''} \lambda\, dt\right| \le \left|\int_{t'}^{s_1} \lambda\, dt\right| + \left|\int_{s_1}^{s_2} \lambda\, dt\right| + \left|\int_{s_2}^{t''} \lambda\, dt\right|$$

$$\equiv A + B + C.$$

It follows from (4.14) Remark 4.1, (4.2), (4.11), and the fact that t' and s_1 are in an interval I_α with $|I_\alpha| < \delta$ that:

$$A \le \int_{t'}^{s_1} \left(\sum_{i=1}^{q} |p^i f_i| + |f|\right) dt = \int_{t'}^{s_1} \sum_{i=1}^{q} p^i |f_i| dt + \int_{t'}^{s_1} |f| dt$$

$$\le \int_{t'}^{s_1} \left(\sum_{i=1}^{q} p^i\right) \mu\, dt + \int_{t'}^{s_1} \mu\, dt = 2\int_{t'}^{s_1} \mu\, dt < 2\varepsilon.$$

Note that if t' and t'' are in the same interval I_α then the preceding estimate and (4.8) combine to give (4.15).

An argument similar to the preceding one gives $C < 2\varepsilon$.

We now estimate B. Recall that $I_k = [t_k, t_{k+1}]$. Then

$$B = \left|\int_{s_1}^{s_2} \lambda \, dt\right| \leq \sum_{j\varepsilon J} \left|\int_{t_j}^{t_{j+1}} \lambda \, dt\right|.$$

Let $g(t,x) = g_i(t,x)$ for $t \,\varepsilon\, E_{ji}$, where $i = 1,\ldots,q$ and $j \,\varepsilon\, J$; then we can estimate each summand on the right as follows.

$$\left|\int_{t_j}^{t_{j+1}} \lambda \, dt\right| \leq \left|\int_{t_j}^{t_{j+1}} \left(\sum_{i=1}^{q} p^i (f_i - g_i)\right) dt\right|$$
$$+ \left|\int_{t_j}^{t_{j+1}} \left(\sum_{i=1}^{q} p^i g_i - g\right) dt\right| + \left|\int_{t_j}^{t_{j+1}} (g-f) dt\right|$$
$$\equiv A_j + B_j + C_j.$$

Hence

$$B \leq \sum_{j\varepsilon J} (A_j + B_j + C_j). \tag{4.16}$$

From the non negativeness of the p^i and (4.2) we get

$$A_j \leq \int_{t_j}^{t_{j+1}} \left(\sum_{i=1}^{q} p^i |f_i - g_i|\right) dt \leq \sum_{i=1}^{q} \int_{t_j}^{t_{j+1}} |f_i - g_i| dt.$$

From the definitions of f and g we get

$$C_j \leq \int_{t_j}^{t_{j+1}} |g-f| dt \leq \sum_{i=1}^{q} \int_{t_j}^{t_{j+1}} |f_i - g_i| dt.$$

Therefore

$$\sum_{j\varepsilon J} (A_j + C_j) \leq 2 \sum_{i=1}^{q} \int_{s_1}^{s_2} |f_i - g_i| dt \leq 2 \sum_{i=1}^{q} \int_{\mathcal{J}} |f_i - g_i| dt \tag{4.17}$$

$$< 2q\varepsilon,$$

where the last inequality follows from (4.9).

We now consider B_j.

$$B_j = \left|\int_{t_j}^{t_{j+1}} \left(\sum_{i=1}^{q} p^i g_i - g\right) dt\right| = \left|\sum_{i=1}^{q} \int_{t_j}^{t_{j+1}} p^i g_i \, dt - \sum_{i=1}^{q} \int_{E_{ji}} g_i \, dt\right|.$$

In each set E_{ji} select a point t_{ji}. Since $E_{ji} \subset I_j$ and $|I_j| < \delta$ it follows from (4.10) that for all t in I_j and all x in $\mathscr{X}$ and all $i = 1,\ldots,q$

$$g_i(t,x) = g_i(t_{ji},x) + \eta_i(t,x),$$

where $|\eta_i(t,x)| < \varepsilon$. Therefore, using (4.12), we get

$$\begin{aligned} B_j &= \left| \sum_{i=1}^{q} \left(\left(\int_{t_j}^{t_{j+1}} (p^i(t) g_i(t_{ji},x) + p^i(t)\eta_i(t,x))dt \right.\right.\right. \\ &\qquad \left.\left.\left. - \int_{E_{ji}} (g_i(t_{ji},x) + \eta_i(t,x))dt \right) \right| \\ &= \left| \sum_{i=1}^{q} \left(g_i(t_{ji},x)|E_{ji}| - g_i(t_{ji},x)|E_{ji}| \right.\right. \\ &\qquad \left.\left. + \int_{t_j}^{t_{j+1}} p^i(t)\eta_i(t,x)dt - \int_{E_{ji}} \eta_i(t,x)dt \right) \right| \\ &< \sum_{i=1}^{q} \left(\varepsilon \int_{t_j}^{t_{j+1}} p^i dt + \varepsilon|E_{ji}| \right) = 2\varepsilon|I_j|. \end{aligned} \tag{4.18}$$

Hence

$$\sum_{j\varepsilon J} B_j < 2\varepsilon|s_2 - s_1| \leq 2\varepsilon|\mathscr{I}|.$$

Combining this with (4.17) and (4.16) gives $B < 2\varepsilon(q+|\mathscr{I}|)$. If we now combine this estimate with the estimates on A and C and use (4.8) we get that

$$\left| \int_{t'}^{t''} \lambda dt \right| < 2\varepsilon(2+q+|\mathscr{I}|) = \bar{\varepsilon},$$

which is (4.15), as required. This completes the proof of Theorem 4.1.

In the next theorem we show that (4.3) remains true if we replace the vectors x in $\mathscr{X}$ by functions ψ from an equi-continuous family.

THEOREM 4.2. Let $f_1,\ldots,f_q$ and $p^1,\ldots,p^q$ be as in Theorem

4.1. Let Ψ be a family of equicontinuous functions defined on $\mathscr{I}$ with range in $\mathscr{X}$. Then for every $\varepsilon > 0$ there exists a subdivision of $\mathscr{I}$ into a finite number of disjoint intervals E_j and an assignment of one of the functions $f_1, \ldots, f_q$ to each interval E_j such that the following holds. If f_{E_j} denotes the function assinged to E_j and if f is a function that agrees with f_{E_j} on E_j^0, the interior of E_j, i.e.

$$f(t,x) = f_{E_j}(t,x) \qquad t \in E_j^0,$$

then for every t' and t'' in $\mathscr{I}$ and every function ψ in Ψ

$$\left| \int_{t'}^{t''} \left(\sum_{i=1}^{q} p^i(t) f_i(t,\psi(t)) - f(t,\psi(t)) \right) dt \right| < \varepsilon. \tag{4.19}$$

<u>Proof</u>. Let $\varepsilon > 0$ be given. Since the functions in Ψ are equicontinuous and $\mathscr{I}$ is compact, there exists a partition of $\mathscr{I}$ into a finite number of non-overlapping subintervals $\{I_j\} = \{[t_j, t_{j+1}]\}$, $j = 1, \ldots, k$ such that $\ldots < t_{j-1} < t_j < t_{j+1} < t_{j+2} < \ldots$ and such that for all ψ in Ψ and all t in I_j, $j = 1, \ldots, k$.

$$|\psi(t) - \psi(t_j)| < \varepsilon \left(2(q+1) \int_I \mu dt \right)^{-1} \equiv \varepsilon'. \tag{4.20}$$

We now apply Theorem 4.1 to $f_1, \ldots, f_q$ and $p^1, \ldots, p^q$ with ε replaced by $\varepsilon/2k$. Then there exists a function f as described in Theorem 4.1 such that for all x in $\mathscr{X}$ and all t', t'' in $\mathscr{I}$,

$$\left| \int_{t'}^{t''} \lambda(t,x) dt \right| < \varepsilon/2k, \tag{4.21}$$

where λ is defined in (4.14). We shall prove the present theorem by showing that for all ψ in Ψ and t', t'' in $\mathscr{I}$,

$$\left| \int_{t'}^{t''} \lambda(t,\psi(t)) dt \right| < \varepsilon.$$

Define

$$\hat{\lambda}(t) = \lambda(t, \psi(t_j)) \qquad t_j \leq t < t_{j+1}, \quad j = 1, \ldots, k$$

and let $\hat{\lambda}(t_{k+1}) = \lambda(t_{k+1},\psi(t_k))$. Then

$$\left|\int_{t'}^{t''} \lambda(t,\psi(t))dt\right| \leq \left|\int_{t'}^{t''} (\lambda(t,\psi(t))-\hat{\lambda}(t))dt\right| + \left|\int_{t'}^{t''} \hat{\lambda}(t)dt\right| \equiv A + B. \tag{4.22}$$

Let $t' < t''$, let $t' \in I_\alpha = [t_\alpha,t_{\alpha+1}]$, and let $t'' \in I_\beta = [t_\beta,t_{\beta+1}]$. Let J now denote the index set $\{\alpha,\alpha+1,\alpha+2,\ldots,\beta\}$. Then

$$A \leq \int_{t'}^{t''}\left|\lambda(t,\psi(t))-\hat{\lambda}(t)\right|dt \leq \int_{t_\alpha}^{t_{\beta+1}}|\lambda(t,\psi(t))-\hat{\lambda}(t)|dt$$

$$= \sum_{j\in J}\int_{t_j}^{t_{j+1}}|\lambda(t,\psi(t))-\lambda(t,\psi(t_j))|dt$$

$$= \sum_{j\in J}\int_{t_j}^{t_{j+1}}\Big|\sum_{i=1}^{q} p^i(t)\{f_i(t,\psi(t))-f_i(t,\psi(t_j))\}$$

$$+ f(t,\psi(t_j))-f(t,\psi(t))\Big|dt$$

$$\leq \sum_{j\in J}\int_{t_j}^{t_{j+1}}\left(\sum_{i=1}^{q} p^i(t)\mu(t)|\psi(t)-\psi(t_j)|+\mu(t)|\psi(t)-\psi(t_j)|\right)dt,$$

where the last inequality follows from (4.1) and Remark 4.1. From the relation $0 \leq p^i(t) \leq 1$ and from (4.20) we see that the last sum in turn is less than

$$\sum_{j\in J}(q+1)\varepsilon'\int_{t_j}^{t_{j+1}}\mu dt \leq \varepsilon'(q+1)\int_{\mathcal{I}}\mu dt = \varepsilon/2.$$

We have thus shown that $A < \varepsilon/2$.

To estimate B we write

$$B \leq \left|\int_{t'}^{t_{\alpha+1}}\lambda(t,\psi(t_\alpha))dt\right| + \sum_{j=\alpha+1}^{\beta-1}\left|\int_{t_j}^{t_{j+1}}\lambda(t,\psi(t_j))dt\right| + \left|\int_{t}^{t''}\lambda(t,\psi(t_\beta))dt\right|.$$

By (4.21) each summand on the right is $< \varepsilon/2k$. Since there are at most k summands (the number of intervals I_j) it follows that $B < \varepsilon/2$. If we combine this estimate with the estimate for A and substitute into (4.22), then we get the desired result.

The proof of our next theorem requires an inequality that is known in the literature as Gronwall's Inequality. This inequality is very useful in the study of differential equations.

LEMMA 4.2. Let ρ and μ be non-negative real valued functions continuous on $[0,\infty)$ such that

$$\rho(t) \leq \alpha + \int_{t_0}^{t} \mu(s)\rho(s)ds \qquad \alpha \geq 0 \tag{4.23}$$

for all t_0, t in $[0,\infty)$. Then

$$\rho(t) \leq \alpha \exp\left(\int_{t_0}^{t} \mu(s)ds\right). \tag{4.24}$$

<u>Proof</u>. Suppose that $\alpha > 0$. Then the right hand side of (4.23) is strictly positive and we get that

$$\rho(t)\mu(t)\left[\alpha + \int_{t_0}^{t} \mu(s)\rho(s)ds\right]^{-1} \leq \mu(t).$$

Integrating both sides of this inequality from t_0 to t and using (4.23) gives

$$\log \rho(t) \leq \log\left[\alpha + \int_{t_0}^{t} \mu\rho ds\right] \leq \log \alpha + \int_{t_0}^{t} \mu ds.$$

From this we get (4.24).

If $\alpha = 0$, then (4.23) holds for all $\alpha_1 > 0$. Hence (4.24) holds for all $\alpha_1 > 0$. Letting $\alpha_1 \to 0$ now yields $\rho(t) \equiv 0$. Hence (4.24) is trivially true.

REMARK 4.2. The proof shows that if $\alpha > 0$ and strict inequality holds in (4.23), then strict inequality holds in (4.24).

THEOREM 4.3. Let $\mathcal{I}$ be a compact interval in E^1, let $\mathcal{X}$ be a compact interval in E^n, and let $\mathcal{R} = \mathcal{I} \times \mathcal{X}$. Let $\mathcal{G} = \mathcal{R} \times \mathcal{U}$, where $\mathscr{U}$ is a region of E^m, and let f be a continuous mapping from $\mathscr{G}$ to E^n. Let Ω be a mapping from $\mathscr{R}$ to subsets of E^m that is independent of x; i.e. $\Omega(t,x') = \Omega(t,x) \equiv \Omega(t)$ for all x and x' in $\mathscr{X}$. Let $\mathscr{D}$ be as in III (4.1). Let there exist an integrable function μ defined on $\mathscr{I}$ such that for all (t,x,z) in $\mathscr{D}$

$$|f(t,x,z)| \leq \mu(t)$$

and for all (t,x,z) and (t,x',z) in $\mathscr{D}$

$$|f(t,x,z)-f(t,x',z)| \leq \mu(t)|x-x'|. \tag{4.25}$$

Let $\mathscr{I}_1 = [t_0,t_1]$ be a compact interval contained in the interior of $\mathscr{I}$ and let $\mathscr{X}_1$ be a compact interval in the interior of $\mathscr{X}$. Let $\mathscr{R}_1 = \mathscr{I}_1 \times \mathscr{X}_1$. Let $v = (\tilde{u},p) = (u_1,\ldots,u_{n+2}, p^1,\ldots,p^{n+2})$ be a relaxed control on $\mathscr{I}_1$ for the relaxed system

$$\frac{dx}{dt} = \sum_{i=1}^{n+2} p^i(t)f(t,x,u_i(t))$$

corresponding to the control system

$$\frac{dx}{dt} = f(t,x,u(t)).$$

Let both systems have initial point (t_0,x_0). Let ψ be a relaxed trajectory corresponding to v on $\mathscr{I}_1$ and let $\psi(t) \in \mathscr{X}_1$ for all t in $[t_0,t_1]$. Then there exists an $\varepsilon_0 > 0$ such that for each ε satisfying $0 < \varepsilon < \varepsilon_0$ there is a control u_ε defined on $\mathscr{I}_1$ with the following properties. For a.e. t in $\mathscr{I}_1$ $u_\varepsilon(t) \in \Omega(t)$, the trajectory ϕ_ε corresponding to u_ε lies in $\mathscr{I}_1 \times \mathscr{X}$, and for all t in $\mathscr{I}_1$, $|\phi_\varepsilon(t)-\psi(t)| < \varepsilon$.

REMARK 4.3. Theorem 4.3 states that under appropriate hypotheses the ordinary trajectories of a system are dense in the set of relaxed trajectories in the uniform topology on $[t_0,t_1]$. Thus, for any relaxed trajectory ψ on $[t_0,t_1]$ there is a sequence of controls

$\{u_k\}$ and a sequence of corresponding trajectories $\{\phi_k\}$ such that $u_k(t) \in \Omega(t)$ a.e. and $\phi_k \to \psi$ uniformly on $[t_0, t_1]$. We caution the reader that with reference to a specific control problem, if ψ is an admissible relaxed trajectory the pairs (ϕ_k, u_k) need not be admissible for the original problem in that either $t \to f^0(t, \phi_k(t), u_k(t))$ may not be integrable or the end points of the ϕ_k may not satisfy the end condition. Recall the distinction between a control (Definition II 3.1) and an admissible control (Definition II 3.2).

Note that no assumption is made concerning the nature of the constraint sets $\Omega(t)$.

Proof. Let ε_0 denote the distance between $\partial\mathscr{X}$ and $\partial\mathscr{X}_1$, where for any set A the symbol ∂A denotes the boundary of A. Then $\varepsilon_0 > 0$. Let

$$K = \int_{t_0}^{t_1} \mu dt \tag{4.26}$$

and let ε be any number satisfying $0 < \varepsilon < \varepsilon_0$. For (t,x) in $\mathscr{I}_1 \times \mathscr{X}$ and $i = 1,\ldots,n+2$ let

$$f_i(t,x) = f(t,x,u_i(t)). \tag{4.27}$$

It is readily verified that as a consequence of the hypotheses of the present theorem, the functions f_i satisfy the hypotheses of Theorems 4.1 and 4.2. In particular note that since f is continuous on $\mathscr{D}$ and each u_i is measurable, the functions f_i are measurable on $\mathscr{I}_1$ for each fixed x in $\mathscr{X}_1$.

Let $\varepsilon' = \varepsilon e^{-K}$. We next apply Theorem 4.2 to the functions $f_1,\ldots,f_{n+2}$ just defined, the functions $p^1,\ldots,p^{n+2}$ in the relaxed control, the family Ψ consisting of one element - the relaxed control ψ, and the value of epsilon equal to ε'. We obtain the existence of a function $\hat{f}$ such that for $x \in \mathscr{X}_1$ and $t \in \mathscr{I}_1$

$$\hat{f}(t,x) = f_{E_j}(t,x) \qquad t \in E_j^0 \tag{4.28}$$

and

$$\left| \int_{t'}^{t''} \left(\sum_{i=1}^{n+2} p^i(t) f_i(t,\psi(t)) - \hat{f}(t,\psi(t)) \right) dt \right| < \varepsilon' \tag{4.29}$$

for arbitrary t' and t'' in $\mathscr{I}_1$.

It follows from the definition of the f_i and from (4.28) that

$$\hat{f}(t,x) = f_{E_j}(t,x) = f(t,x,u_{E_j}(t)) \qquad t \in E_j^0. \tag{4.30}$$

Define

$$u_\varepsilon(t) = u_{E_j}(t) \quad \text{if} \quad t \in E_j^0.$$

Then since u_{E_j} is one of the $u_1,\ldots,u_{n+2}$ and each u_i satisfies $u_i(t) \in \Omega(t)$ a.e. on $\mathscr{I}_1$ it follows that $u_\varepsilon(t) \in \Omega(t)$ on $\mathscr{I}_1$ a.e. From the definition of u_ε and (4.30) we get

$$\hat{f}(t,x) = f(t,x,u_\varepsilon(t)).$$

Consider the system

$$\frac{dx}{dt} = f(t,x,u_\varepsilon(t)) = \hat{f}(t,x) \tag{4.31}$$

with initial point (t_0,x_0). Since f satisfies (4.25) it follows that through each point (t_2,x_2) in the interior of $\mathscr{I} \times \mathscr{X}$, there passes a unique solution of (4.31), provided we extend u_ε to be defined and measurable on $\mathscr{I}$. In particular there exists a unique solution ϕ_ε of (4.31) with initial point (t_0,x_0). This solution will be defined on some open interval containing t_0 in its interior. Let $\mathscr{I}_{max} = (a,b)$ denote the maximal interval on which ϕ_ε is defined. If $[a,b] \subset \mathscr{I}_1$, then $\limsup_{t\to b} \phi_\varepsilon(t)$ must be a boundary point of $\mathscr{X}$, for otherwise we could extend the solution ϕ_ε to an interval containing $\mathscr{I}_{max}$ in its interior. This would contradict the maximality of $\mathscr{I}_{max}$. We shall show that for all t in $\mathscr{I}_{max}$, the inequality $|\phi_\varepsilon(t)-\psi(t)| < \varepsilon$ holds. Since $\psi(t) \in \mathscr{X}_1$ for all t in $[t_0,t_1]$ and since $\varepsilon < \varepsilon_0 = \text{dist}(\partial\mathscr{X}_1, \partial\mathscr{X})$ it will follow that $[a,b] = \mathscr{I}_1$ and ϕ_ε is defined in all of $\mathscr{I}_1$. Moreover, we shall have $|\phi_\varepsilon(t)-\psi(t)| < \varepsilon$ on all of $\mathscr{I}_1$.

Since ψ is defined on all of $\mathscr{I}_1$ and $\psi(t_0) = \phi_\varepsilon(t_0) = x_0$, we have for all t in $[t_0,b]$

$$\begin{aligned}
|\psi(t)-\phi_\varepsilon(t)| &= \left|\int_{t_0}^{t} (\psi'(s)-\phi_\varepsilon'(s))ds\right| \\
&= \left|\int_{t_0}^{t} \left(\sum_{i=1}^{n+2} p^i(s)f_i(s,\psi(s))-\hat{f}(s,\phi_\varepsilon(s))\right) ds\right| \\
&\le \left|\int_{t_0}^{t} \left(\sum_{i=1}^{n+2} p^i(s)f_i(s,\psi(s))-\hat{f}(s,\psi(s))\right) ds\right| \\
&\quad + \left|\int_{t_0}^{t} (\hat{f}(s,\psi(s))-\hat{f}(s,\phi_\varepsilon(s)))ds\right| \\
&< \varepsilon' + \int_{t_0}^{t} |\hat{f}(s,\psi(s))-\hat{f}(s,\phi_\varepsilon(s))|ds,
\end{aligned}$$

where the last inequality follows from (4.29). It now follows from (4.30) and (4.25) that

$$\int_{t_0}^{t} |\hat{f}(s,\psi(s))-\hat{f}(s,\phi_\varepsilon(s))|ds \le \int_{t_0}^{t} \mu(s)|\psi(s)-\phi_\varepsilon(s)|ds.$$

Combining this with the preceding inequality gives

$$|\psi(t)-\phi_\varepsilon(t)| < \varepsilon' + \int_{t_0}^{t} \mu(s)|\psi(s)-\phi_\varepsilon(s)|ds.$$

From Lemma 4.2, Remark 4.2, (4.26), and the definition of ε' we now conclude that

$$|\psi(t)-\phi_\varepsilon(t)| < \varepsilon' \exp\left(\int_{t_0}^{t} \mu ds\right) \le \varepsilon' e^K = \varepsilon,$$

and the theorem is proved.

REMARK 4.4. From the proof of Theorem 4.3 we see why we must assume that the functions f_i of Theorem 4.1 are measurable in t and continuous in x, rather than continuous in (t,x). Since controls u are only assumed to be measurable, we can only guarantee that the functions f_i defined in (4.27) will be measurable in t, no matter how regular we assume the behavior of f to be.

The reason for calling Theorem 4.1 the "Chattering Lemma" can now be given. In most applications the functions $f_1,\ldots,f_q$ are obtained as in Theorem 4.3. That is, we have a system with state equations $dx/dt = f(t,x,u(t))$, we choose q controls $u_1,\ldots,u_q$, and define functions $f_1,\ldots,f_q$ by means of equations (4.27). The function f of Theorem 4.1 is obtained in the same fashion as the function $\hat{f}$ of the present theorem. That is, the basic interval $\mathscr{I}$ is divided up into a large number of small intervals and on each subinterval we choose one of the controls $u_1,\ldots,u_q$ to build the control u_ε. In a physical system the control u_ε corresponds to a rapid switching back and forth among the various controls $u_1,\ldots,u_q$. In the engineering vernacular the system is said to "chatter". The control u_ε is therefore sometimes called a <u>chattering control</u>.

From the proof of Theorem 4.3 we learn more than just the fact that a relaxed trajectory can be approximated as close as we please by an ordinary trajectory. We learn that the approximation can be effected through the use of a chattering control built from the controls used to define the relaxed control in question.

REMARK 4.5. The theorem remains valid if we take $\mathscr{X} = \mathscr{X}_1 = E^n$.

EXERCISE 4.1. Consider Problem 2 with f^0, f, and Ω satisfying the hypotheses of Theorem 4.3 and with $\mathscr{X}_1 = \mathscr{X} = E^n$. Let the terminal set $\mathscr{B}$ be given as follows: t_0, x_0, t_1, fixed, and x_1 an arbitrary element in $\mathscr{X} = E^n$. Let G be a continuous functional on $C[t_0,t_1]$. Show that if Problem 2 has a solution (ϕ^*,u^*), then

(ϕ^*,u^*) is also a solution of the corresponding relaxed problem, Problem 2R.

5. The Attainable Set

Stated somewhat imprecisely, the attainable set at time t_1 of the control system

$$\frac{dx}{dt} = f(t,x,u(t)) \tag{5.1}$$

with initial point (t_0,x_0) is the set of points that can be reached at time t_1 by the trajectories of the system. We shall obtain the properties of the attainable sets from Theorem III 4.1 and from Theorem III 8.1. Therefore, in studying the trajectories of (5.1) it will be useful for us to suppose that the system (5.1) is the system of state equations for a control problem with $f^0 \equiv 0$. Henceforth in this section we make this supposition. It will also be useful to introduce a certain metric on the space of compact subsets of a metric space.

DEFINITION 5.1. Let $\mathscr{S}$ be a metric space and let A and B be compact subsets of $\mathscr{S}$. Let

$$h(A,B) = \frac{1}{2}\{\max_{a\varepsilon A} d(a,B) + \max_{b\varepsilon B} d(A,b)\},$$

where d denotes the metric on $\mathscr{S}$.

It can be shown that h defines a metric on the space $\mathscr{T}$ consisting of compact subsets of $\mathscr{S}$. The metric h is called the Hausdorff metric.

EXERCISE 5.1. Prove the assertion that h defines a metric on $\mathscr{T}$.

We now give a precise definition of the attainable set. Let $\mathscr{A}^+$ denote the set of all pairs (ϕ,u), where u is a control and

ϕ is a trajectory corresponding to u, that satisfy all of the requirements in the definition of admissible pair except the requirement that $(t_0, \phi(t_0), t_1, \phi(t_1)) \in \mathscr{B}$. Note that since $f^0 \equiv 0$, the requirement that the mapping $t \to f^0(t,\phi(t),u(t))$ is in $L_1[t_0,t_1]$ is always satisfied.

DEFINITION 5.2. The attainable set at time $t \geq t_0$ for the system (5.1) with initial point (t_0,x_0) is defined to be the set of all points x such that for some trajectory ϕ belonging to a pair (ϕ,u) in $\mathscr{A}^+$ and satisfying $\phi(t_0) = x_0$, the relation $\phi(t) = x$ holds. We shall denote the attainable set by $\mathscr{K}(t,t_0,x_0)$.

Let (t_0,x_0) be fixed and let $t_1 > t_0$ be fixed. Let $\mathscr{A}^+(t_0,x_0,t_1)$ denote the set of pairs (ϕ,u) in $\mathscr{A}^+$ that are defined on $[t_0,t_1]$ and satisfy $\phi(t_0) = x_0$. We henceforth suppose that $\mathscr{A}^+(t_0,x_0,t_1)$ is not empty, and we shall not repeat this assumption in various statements. For systems that are linear in the state, conditions ensuring that $\mathscr{A}^+(t_0,x_0,t_1)$ is not empty were given in Section 2 of Chapter III. For each t in the interval $t_0 < t < t_1$ we can define a set $\mathscr{A}^+(t_0,x_0,t)$ of pairs (ϕ,u) in the same way as we defined $\mathscr{A}^+(t_0,x_0,t_1)$. If $t < t'$, then $\mathscr{A}^+(t_0,x_0,t) \supset \mathscr{A}^+(t_0,x_0,t')$. Since the sets $\mathscr{A}^+(t_0,x_0,t)$, $t_0 < t \leq t_1$ are not empty, the attainable sets $\mathscr{K}(t,t_0,x_0)$ for $t_0 \leq t \leq t_1$ are not empty.

If we henceforth restrict our attention to the interval $[t_0,t_1]$ and do not consider trajectories or portions of trajectories defined outside of this interval then the following definition of $\mathscr{A}^+$ is consistent with our previous one.

$$\mathscr{A}^+ = \bigcup_{t_0 \leq t \leq t_1} \mathscr{A}^+(t_0,x_0,t).$$

We shall denote the set of trajectories ϕ belonging to pairs (ϕ,u)

in $\mathscr{A}^+$ by the symbol $\mathscr{A}_T^+$. We shall denote the set of trajectories ϕ belonging to pairs (ϕ,u) in $\mathscr{A}^+(t_0,x_0,t)$ by $\mathscr{A}_T^+(t_0,x_0,t)$.

The next two theorems are general theorems dealing with the structure of the attainable sets. In Theorem 5.1 the hypotheses correspond to those of Theorem III 4.1. As with Theorem III 4.1 in the study of existence problems, we shall use Theorems 5.1 and 5.2 below to obtain information about the attainable sets in problems where the hypotheses involve conditions that are easier to verify than those of Theorems 5.1 and 5.2 below.

THEOREM 5.1. Let the function f in (5.1) be continuous and let the constraint mapping Ω be upper semicontinuous on $\mathscr{R}$. For each (t,x) in $\mathscr{R}$ let the set

$$\mathscr{Q}(t,x) = \{y: y = f(t,x,z), \quad z \in \Omega(t,x)\} \tag{5.2}$$

be closed and convex and let the mapping $\mathscr{Q}$ satisfy the weak Cesari property at every point of $\mathscr{R}$. Let the set of trajectories $\mathscr{A}_T^+$ be equi-absolutely continuous. Then for each t in $[t_0,t_1]$ the set $\mathscr{K}(t,t_0,x_0)$ is compact. Moreover, if for each t in $[t_0,t_1]$ each trajectory in $\mathscr{A}_T^+(t_0,x_0,t)$ is a restriction of a trajectory in $\mathscr{A}_T^+(t_0,x_0,t_1)$ then the mapping $t \to \mathscr{K}(t,t_0,x_0)$ is a continuous mapping from $[t_0,t_1]$ into the space of compact subsets of E^n endowed with the Hausdorff metric.

We first show that the sets $\mathscr{K}(t,t_0,x_0)$ are compact. If $t = t_0$, then $\mathscr{K}(t_0,t_0,x_0) = x_0$ and so is compact. We now suppose $t > t_0$. Since $\phi(t_0) = x_0$ for all ϕ in $\mathscr{A}_T^+$ and since the functions in $\mathscr{A}_T^+$ are equi-absolutely continuous it follows that all trajectories ϕ in $\mathscr{A}_T^+$ lie in a compact set $\mathscr{R}_0 \subseteq \mathscr{R}$. Hence the sets $\mathscr{K}(t,t_0,x_0)$ are bounded and to prove that they are compact we need only show that they are closed.

Let x be a limit point of $\mathscr{K}(t,t_0,x_0)$. Then there exists a sequence $\{x_k\}$ such that $x_k \in \mathscr{K}(t,t_0,x_0)$ and $x_k \to x$. Hence there exists a sequence $\{(\phi_k,u_k)\}$ in $\mathscr{A}^+$ such that $\phi_k(t) = x_k$. Let us suppose that in addition to (5.1) being the state equations for a problem $f^0 \equiv 0$, that the terminal conditions for this problem are determined by the set $\hat{\mathscr{B}} = \{(t_0',x_0',t_1',x_1'): t_0' = t_0,\ t_1' = t,\ x_0' = x_0$ and x_1' an arbitrary vector in $E^n\}$. Then the pairs (ϕ,u) in $\mathscr{A}^+(t_0,x_0,t)$ are admissible pairs for this problem. In particular, the pairs (ϕ_k,u_k) in the sequence above are admissible. It is readily verified that under the hypotheses of the present theorem, Theorem III 4.1 can be applied with the sequence $\{(\phi_k,u_k)\}$ taken to be set $\mathscr{A}_0$ of Theorem III 4.1. We obtain the existence of a pair (ϕ^*,u^*) in $\mathscr{A}^+(t_0,x_0,t)$ such that $\phi_k \to \phi^*$ uniformly on $[t_0,t]$. Hence $\phi^*(t) = x$, and so $\mathscr{K}(t,t_0,x_0)$ is closed.

To prove the continuity of the mapping $t \to \mathscr{K}(t,t_0,x_0)$ we proceed as follows. Let $\varepsilon > 0$ be given. Since the functions ϕ in $\mathscr{A}_T^+$ are equi-absolutely continuous, there exists a $\delta > 0$ such that for all t and t' in $[t_0,t_1]$ satisfying $|t-t'| < \delta$ the inequality $\left|\int_t^{t'} \phi'(s)ds\right| < \varepsilon$ holds. Let x be a point of $\mathscr{K}(t,t_0,x_0)$. Then there is a pair (ϕ,u) in $\mathscr{A}^+(t_0,x_0,t)$, and also in $\mathscr{A}^+(t_0,x_0,t_1)$, such that $\phi(t) = x$. Let t' be a point of $[t_0,t_1]$ such that $|t'-t| < \delta$. Then

$$|\phi(t')-\phi(t)| = \left|\int_t^{t'} \phi'(s)ds\right| < \varepsilon.$$

Hence $d(x,\mathscr{K}(t',t_0',x_0)) < \varepsilon$, where d denotes the euclidean distance. Since x is an arbitrary point of $\mathscr{K}(t,t_0,x_0)$, we have $\max\{d(x,\mathscr{K}(t',t_0,x_0)); x \in \mathscr{K}(t,t_0,x_0)\} < \varepsilon$. Similarly, $\max\{d(x,\mathscr{K}(t,t_0,x_0)): x \in \mathscr{K}(t',t_0,x_0)\} < \varepsilon$. Hence $h(\mathscr{K}(t,t_0,x_0), \mathscr{K}(t',t_0,x_0)) < \varepsilon$ for all t and t' in $[t_0,t_1]$ and $|t'-t| < \delta$.

COROLLARY 5.1. For each t in $[t_0,t_1]$ the sets $\mathscr{A}_T^+(t_0,x_0,t)$ are compact subsets of $C[t_0,t]$, the space of continuous functions on $[t_0,t]$ in the uniform topology.

Proof. Since the trajectories ϕ in $\mathscr{A}_T^+(t_0,x_0,t)$ are equi-absolutely continuous they are uniformly bounded and equi-continuous. Hence by Ascoli's theorem it follows that if $\{\phi_k\}$ is a sequence of trajectories in $\mathscr{A}_T^+(t_0,x_0,t)$, then there is a subsequence, again written as $\{\phi_k\}$ and a continuous function ϕ in $C[t_0,t]$ such that $\phi_k \to \phi$ in $C[t_0,t]$. As in the proof that $\mathscr{K}(t,t_0,x_0)$ is closed, we use Theorem III 4.1 to show that ϕ is in $\mathscr{A}_T^+(t_0,x_0,t)$.

THEOREM 5.2. Let the hypotheses of Theorem 5.1 hold, except the assumption that the mapping $\mathscr{Q}^+$ satisfies the weak Cesari property. Let (ii) and (iii) of Assumption III 8.1 hold and let III (8.2) hold for all (ϕ,u) in $\mathscr{A}^+$. Then the conclusion of Theorem 5.1 holds.

The proof of Theorem 5.2 is similar to that of Theorem 5.1 except that Theorem III 8.1 now plays the role played by Theorem III 4.1 in the proof of Theorem 5.1. A corollary similar to Corollary 5.1 can also be stated. We leave the details to the reader.

Theorems similar to Theorem 5.1 and based on Theorem III 8.1 and on Theorem 8.4 can also be formulated. We leave this to the reader.

We now consider the properties of the attainable sets when the sets $\Omega(t,x)$ are compact. The hypotheses of the next theorem, Theorem 5.3, and its corollary imply that the hypotheses of Theorem 5.1 hold. In specific examples it is usually easier to check the validity of the hypotheses of Theorem 5.3 than it is to check the validity of the hypotheses of Theorem 5.1. Theorem 5.3 stands in relation to Theorem 5.1 as Theorem III 5.1 stands to Theorem III 4.1.

THEOREM 5.3. Let the function f in (5.1) be continuous. Let there exist a compact set $\mathscr{R}_0 \subset \mathscr{R}$ such that all trajectories ϕ in $\mathscr{A}_T^+$ lie in $\mathscr{R}_0$. Let the mapping Ω be u.s.c.i. on $\mathscr{R}_0$ and for each (t,x) in $\mathscr{R}_0$ let the set $\Omega(t,x)$ be compact. For each (t,x) in $\mathscr{R}_0$, let the set $\mathscr{Q}(t,x)$ defined by (5.2) be convex. Then for each t in $[t_0,t_1]$ the set $\mathscr{K}(t,t_0,x_0)$ is compact in E^n and the set $\mathscr{A}_T^+(t_0,x_0,t)$ is compact in $C[t_0,t]$. Moreover, if for each t in $[t_0,t_1]$ the trajectories in $\mathscr{A}_T^+(t_0,x_0,t)$ are restrictions of trajectories in $\mathscr{A}_T^+(t_0,x_0,t_1)$, then the mapping $t \to \mathscr{K}(t,t_0,x_0)$ is a continuous mapping from $[t_0,t_1]$ into the space of compact subsets of E^n endowed with the Hausdorff metric.

The proof is similar to the proof of Theorem III 5.1 and consists of showing that the hypotheses of the present theorem imply the hypotheses of Theorem 5.1. We leave the details to the reader.

In Section 5 of Chapter III we showed that many special classes of problems that are of importance in applications satisfy the hypotheses of Theorem III 5.1. Since the hypotheses of Theorem 5.3 concerning f, $\mathscr{R}_0$, Ω and $\mathscr{Q}$ are the same as those hypotheses of Theorem III 5.1 that deal with the state equations, it follows that Theorem 5.3 is applicable to the special problems of Section 5, Chapter III. In particular, we have the following corollary to Theorem 5.3, which is analogous to Corollary III 5.1 and which embraces many problems of interest in applications.

COROLLARY 5.3.1. Let the system (5.1) have the form

$$\frac{dx}{dt} = h(t,x) + B(t,x)u(t) \tag{5.3}$$

where h and B are continuous on $\mathscr{R}$. Let there exist a compact set $\mathscr{R}_0 \subset \mathscr{R}$ such that all trajectories ϕ in $\mathscr{A}_T^+$ lie in $\mathscr{R}_0$. Let the mapping Ω be u.s.c.i. on $\mathscr{R}_0$ and for each (t,x) in $\mathscr{R}_0$

let the set $\Omega(t,x)$ be compact and convex. Then the conclusions of Theorem 5.3 hold.

For linear systems we have the following statement.

COROLLARY 5.3.2. Let the system (5.1) have the form

$$\frac{dx}{dt} = A(t)x + B(t)u(t), \tag{5.4}$$

where A and B are continuous on $[t_0,t_1]$. Let the mapping Ω depend on t alone and let Ω be u.s.c.i. on $[t_0,t_1]$. Let each set $\Omega(t)$ be compact and convex. Then the attainable sets are compact and convex and the mapping $t \to \mathscr{K}(t,t_0,x_0)$ is continuous. For each t in $[t_0,t_1]$ the sets $\mathscr{A}_T^+(t_0,x_0,t)$ are convex and compact in $C[t_0,t]$.

Proof. Since Ω is u.s.c.i. and each set $\Omega(t)$ is compact, it follows from Lemma III 5.2 that any measurable function u defined on $[t_0,t_1]$ and satisfying $u(t) \in \Omega(t)$ a.e. is essentially bounded. Since the system is linear in x it follows that any measurable function u defined on $[t_0,t_1]$ and satisfying $u(t) \in \Omega(t)$ will give rise to a unique solution ϕ of (5.4) satisfying $\phi(t_0) = x_0$. Moreover, this solution will be defined on all of $[t_0,t_1]$. By Lemma III 11.1 there exists at least one measurable function u defined on $[t_0,t_1]$ such that $u(t) \in \Omega(t)$ a.e. Therefore any measurable function u defined on a subinterval $[t_0,t_2]$ of $[t_0,t_1]$ and satisfying the constraint $u(t) \in \Omega(t)$ can be extended to a measurable function that is defined on $[t_0,t_1]$ and satisfies the constraint on $[t_0,t_1]$. Thus, for each t in $[t_0,t_1]$ the trajectories in $\mathscr{A}_T^+(t_0,x_0,t)$ are restrictions of trajectories in $\mathscr{A}_T^+(t_0,x_0,t_1)$. From Exercise III 5.1 it follows that under the hypotheses of the present corollary, all trajectories ϕ in $\mathscr{A}_T^+$ lie in a compact set. Corollary 5.3.2, with the exception of the statements that the sets $\mathscr{K}(t,t_0,x_0)$ and $\mathscr{A}_T^+(t_0,x_0,t)$ are convex, follows from

Corollary 5.3.1. The convexity of the sets $\mathscr{K}(t,t_0,x_0)$ and $\mathscr{A}^+_T(t_0,x_0,t)$ are consequences of the convexity of the sets $\Omega(t)$ and the variations of parameter formula for solutions ϕ of (5.4); namely

$$\phi(t) = \Phi(t)[x_0 + \int_{t_0}^{t} \Phi^{-1}(s)B(s)u(s)ds],$$

where Φ is the fundamental matrix of the homogeneous system $dx/dt = A(t)x$ such that $\Phi(t_0)$ is the identity matrix.

The convexity of the sets $\mathscr{Q}^+(t,x)$ is required in Theorem 5.3. If these sets are not convex, we consider the relaxed trajectories corresponding to (5.1). The next two theorems are concerned with the structure of the attainable sets for the relaxed trajectories and their relationship with the attainable sets for the original problem.

The relaxed system corresponding to (5.1) is the system

$$\frac{dx}{dt} = \tilde{f}(t,x,v(t)) = \sum_{i=1}^{n+1} p^i(t)f(t,x,u_i(t)). \tag{5.5}$$

Since we are considering (5.1) to be the state equations for a control problem with $f^0 \equiv 0$, the system (5.5) gives the state equations for the corresponding relaxed system. We define $\mathscr{A}^+_R$ for the relaxed system to be the set of all relaxed pairs (ψ,v), where v is a relaxed control and ψ is a relaxed trajectory corresponding to v, that satisfy all of the requirements in the definition of a relaxed admissible pair except the requirement that $(t_0,\psi(t_0),t_1,\psi(t_1)) \varepsilon \mathscr{B}$. The set $\mathscr{A}^+_R$ for the relaxed problem corresponds to the set $\mathscr{A}^+$ for the original problem. In similar fashion we define sets $\mathscr{K}_R(t,t_0,x_0)$, $\mathscr{A}^+_R(t_0,x_0,t)$, $\mathscr{A}^+_{RT}$ and $\mathscr{A}^+_{RT}(t_0,x_0,t)$ for the relaxed system. These sets correspond respectively to the sets $\mathscr{K}(t,t_0,x_0)$, $\mathscr{A}^+(t_0,x_0,t)$, $\mathscr{A}^+_T$ and $\mathscr{A}^+_T(t_0,x_0,t)$ for the original system. The sets $\mathscr{K}_R(t,t_0,x_0)$ will be called <u>relaxed attainable sets</u>.

THEOREM 5.4. Let the function f in (5.1) be continuous. Let there exist a compact set $\mathscr{R}_0 \subset \mathscr{R}$ such that all relaxed trajectories ψ in $\mathscr{A}^+_{RT}$ lie in $\mathscr{R}_0$. Let the mapping Ω be u.s.c.i. on $\mathscr{R}_0$

and for each (t,x) in $\mathscr{R}_0$ let the set $\Omega(t,x)$ be compact. Then for each t in $[t_0,t_1]$ the set $\mathscr{K}_R(t,t_0,x_0)$ is compact in E^n and the set $\mathscr{A}^+_{RT}(t_0,x_0,t)$ is compact in $C[t_0,t]$. Moreover, if for each t in $[t_0,t_1]$ the trajectories in $\mathscr{A}^+_{RT}(t_0,x_0,t)$ are restrictions of trajectories in $\mathscr{A}^+_{RT}(t_0,x_0,t_1)$, then the mapping $t \to \mathscr{K}_R(t,t_0,x_0)$ is a continuous mapping from $[t_0,t_1]$ into the space of compact subsets of E^n endowed with the Hausdorff metric.

Theorem 5.4 is an immediate consequence of Theorem 5.3 and the observation that the sets

$$\tilde{\mathscr{Q}}(t,x) = \{y\colon y = \tilde{f}(t,x,\tilde{z},\pi),\ \tilde{z} \in [\Omega(t,x)]^{n+1},\ \pi \in \Gamma\}$$

are convex. For the definitions of $\tilde{f}$, $\tilde{z}$, π and Γ, see (3.1).

REMARK 5.1. For relaxed systems corresponding to (5.3) we can state a corollary that corresponds to Corollary 5.3.1. The assumptions are that h and B are continuous, all relaxed trajectories lie in a compact set, Ω is u.s.c.i., and each set $\Omega(t,x)$ is compact. Note that the sets $\Omega(t,x)$ are not assumed to be convex. We assert that the conclusion of Theorem 5.4 holds.

REMARK 5.2. For relaxed systems corresponding to the linear system (5.4) we can state a corollary corresponding to Corollary 5.3.2. The assumptions on A, B and Ω are as in Corollary 5.3.2, except that the sets $\Omega(t)$ are not assumed to be convex. We conclude that the relaxed attainable sets are compact and convex and that the mapping $t \to \mathscr{K}_R(t,t_0,x_0)$ is continuous. The sets $\mathscr{A}_{RT}(t_0,x_0,t)$ are compact and convex in $C[t_0,t]$.

EXERCISE 5.2. Prove the assertion that the sets $\mathscr{K}_R(t,t_0,x_0)$ and $\mathscr{A}_{RT}(t_0,x_0,t_1)$ are convex. Hint: Use Theorems III 5.2 and III 7.1 and the variation of parameters formula.

THEOREM 5.5. Let $\mathscr{I}$ be an interval in E^1, let $\mathscr{X}$ be an interval in E^n and let $\mathscr{R} = \mathscr{I} \times \mathscr{X}$. Let the function f in (5.1) be continuous on $\mathscr{G} = \mathscr{R} \times \mathscr{U}$, where $\mathscr{U}$ is a region in E^m. Let Ω be independent of x and let Ω be u.s.c.i. on $\mathscr{I}$. For each t in $\mathscr{I}$ let the set $\Omega(t)$ be compact. Let $\mathscr{D}$ be as in III (4.1) and let there exist an integrable function μ defined on $\mathscr{I}$ such that for all (t,x,z) and (t,x',z) in $\mathscr{D}$

$$|f(t,x,z)-f(t,x',z)| \leq \mu(t)|x-x'|.$$

Let $\mathscr{I}_1 = [t_0,t_1]$ be a compact interval in $\mathscr{I}$ and let x_0 be a fixed point in $\mathscr{X}$. Let there exist a compact interval $\mathscr{X}_1 \subset \mathscr{X}$ such that all trajectories of the relaxed system corresponding to (5.1) with initial point (t_0,x_0) and defined on subintervals of $\mathscr{I}_1$ lie in $\mathscr{I}_1 \times \mathscr{X}_1$. Then for all t in $\mathscr{I}_1$ the sets $\mathscr{K}_R(t,t_0,x_0)$ and $\mathscr{A}^+_{RT}(t_0,x_0,t)$ are compact and

$$\mathscr{K}_R(t,t_0,x_0) = \text{cl } \mathscr{K}(t,t_0,x_0)$$

$$\mathscr{A}^+_{RT}(t_0,x_0,t) = \text{cl } \mathscr{A}^+_T(t_0,x_0,t),$$

where the closure of $\mathscr{A}^+_T(t_0,x_0,t)$ is taken in $C[t_0,t]$.

Since every trajectory of (5.1) is also a relaxed trajectory it follows that

$$\mathscr{K}(t,t_0,x_0) \subseteq \mathscr{K}_R(t,t_0,x_0), \quad \mathscr{A}^+_T(t_0,x_0,t) \subseteq \mathscr{A}^+_{RT}(t_0,x_0,t). \tag{5.6}$$

It is readily seen that we may apply Theorem 5.4 to the present system and obtain that for each t in $[t_0,t_1]$, the set $\mathscr{K}_R(t,t_0,x_0)$ is compact in E^n and the set $\mathscr{A}^+_{RT}(t_0,x_0,t)$ is compact in $C[t_0,t]$. Let $\mathscr{X}_2$ be any compact set such that $\mathscr{X}_1 \subset \mathscr{X}_2 \subset \mathscr{X}$. It follows from Lemma III 5.2, the u.s.c.i. property of Ω and the compactness of $[t_0,t_1]$ that the sets $\mathscr{D}_1 = \{(t,x,z)\colon (t,x) \in \mathscr{I}_1 \times \mathscr{X}_2,$

$z \in \Omega(t)\}$ is compact. Hence f is bounded on $\mathscr{D}_1$ and the hypotheses of Theorem 4.3 are satisfied, after a possible redefinition of the function μ. Hence from Theorem 4.3 we get that $\mathscr{A}^+_{RT}(t_0,x_0,t) \subseteq$ cl $\mathscr{A}^+_T(t_0,x_0,t)$ and $\mathscr{K}_R(t,t_0,x_0) \subseteq$ cl $\mathscr{K}(t,t_0,x_0)$. Since the sets $\mathscr{A}^+_{RT}(t_0,x_0,t)$ and $\mathscr{K}_R(t,t_0,x_0)$ are closed in the appropriate spaces and since (5.6) holds, the conclusion of the theorem follows.

EXERCISE 5.3. Show that Theorem 5.5 remains true if we replace the hypotheses that Ω is u.s.c.i. and that the sets $\Omega(t)$ are compact by the following hypotheses: The mapping $t \to \Omega(t)$ is upper semicontinuous. For each (t,x) in $\mathscr{I} \times \mathscr{X}_1$ the sets

$$\mathscr{Q}(t,x) = \{y: y = f(t,x,z),\ z \in \Omega(t)\}$$

are closed. For all (t,x,z) in $\mathscr{D}$,

$$|f(t,x,z)| \leq \mu(t).$$

Hint: Use Theorem 5.2.

6. Systems Linear in the State Variable

The hypotheses of the existence theorem III 5.1 and of Theorem 5.3 of the present chapter can be justified in problems arising in applications. The assumption concerning the convexity of the sets $\mathscr{Q}^+(t,x)$, however, essentially restricts us to systems that are linear in the control and for which the constraint sets $\Omega(t,x)$ are convex. In this section we show that for systems linear in x we can drop the requirement that the sets $\mathscr{Q}^+(t,x)$ be convex, provided we assume that the mapping Ω depends on t alone. For linear systems this means that we need not require that the sets $\Omega(t)$ be convex. Our analysis will also yield the "bang-bang principle" for linear systems.

We first prove a lemma that can be considered as a special

case of the "bang-bang principle" to be discussed in Theorem 6.4.

LEMMA 6.1. Let E be a measurable subset of the line and let E have finite measure. Let y be a function defined on E with values in E^k and such that $y \in L_1(E)$. Let w be a real valued measurable function defined on E such that $0 \le w \le 1$. Then there exists a measurable subset $F \subset E$ such that

$$\int_E y(t)w(t)dt = \int_F y(t)dt.$$

In the proof of Lemma 6.1 and elsewhere in this section we shall need the Krein-Milman theorem, which we state as Lemma 6.2 below. We refer the reader to Dunford-Schwartz ([20], Theorem V 8.4, p. 440) for a proof of this theorem. Certain basic facts about the weak* topology of a Banach space will also be used. For these topics we refer the reader to Dunford-Schwartz ([20], Chapter V, pp. 420-439). A short, readable treatment of some of the topics used can also be found in Hermes and LaSalle ([27], pp. 1-22).

LEMMA 6.2. (Krein-Milman). Let $\mathscr{C}$ be a compact convex set in a locally convex topological vector space. Then $\mathscr{C}$ is the closed convex hull of its extreme points.

Proof of Lemma 6.1. Define a mapping T as follows. For each real valued function ρ in $L_\infty(E)$ let $T\rho = \int_E y(t)\rho(t)dt$.

The mapping T so defined is a continuous mapping from $L_\infty(E)$ with the weak* topology to E^k with the euclidean topology. Let $a = \int_E y(t)w(t)dt$. Then $T^{-1}(a)$ is a convex, weak* closed set of $L_\infty(E)$. Let Σ denote the intersection of $T^{-1}(a)$ and the unit ball in $L_\infty(E)$. Since $w \in \Sigma$, the set Σ is not empty. The weak* topology is a Hausdorff topology. Therefore, since the unit ball in

$L_\infty(E)$ is weak* compact and $T^{-1}(a)$ is weak* closed, the set Σ is weak* compact. It is also convex. Therefore, by the Krein-Milman theorem, Σ has extreme points. We show that the extreme points of Σ are characteristic functions χ_F of measurable subsets F of E. This will prove the Lemma, for then

$$a = \int_E y\,\chi_F dt = \int_F y\,dt.$$

The proof proceeds by induction on k, the dimension of the range of y. We shall give the general induction step. The proof for the initial step, $k = 1$, is essentially the same as the proof for the general step and will be left to the reader.

Assume that the Lemma is true for $k - 1$. We suppose that θ is an extreme point of Σ and that θ is not a characteristic function of some set F. Then there exists an $\varepsilon > 0$ and a measurable set $E_1 \subset E$ with meas $(E_1) > 0$ such that $\varepsilon \leq \theta(t) \leq 1-\varepsilon$ for a.e. t in E_1. Let E_2 and E_3 be two subsets of E_1 such that E_2 and E_3 have positive measure and $E_3 = E_1 - E_2$. From the inductive hypothesis applied to E_2 and E_3 we obtain the existence of measurable sets F_2 and F_3 such that $F_2 \subset E_2$, $F_3 \subset E_3$ and

$$\frac{1}{2}\int_{E_j} y^i(t)dt = \int_{F_j} y^i(t)dt \qquad \begin{array}{l} i = 1,\ldots,k-1 \\ j = 2,3. \end{array}$$

Let $h_2 = 2\chi_{F_2} - \chi_{E_2}$ and let $h_3 = 2\chi_{F_3} - \chi_{E_3}$. Then h_2 and h_3 are not identically zero on E_1, do not exceed one in absolute value and for $i = 1,\ldots,k-1$ satisfy

$$\int_{E_2} y^i h_2 dt = \int_{E_3} y^i h_3 dt = 0.$$

Let $h(t) = \alpha h_2(t) + \beta h_3(t)$ where α and β are chosen so that $|\alpha| < \varepsilon$, $|\beta| < \varepsilon$, $\alpha^2 + \beta^2 > 0$ and

$$\int_{E_1} y^k h\, dt = \alpha \int_{E_2} y^k h_2 dt + \beta \int_{E_3} y^k h_3 dt = 0.$$

This can always be done. The function h so defined also satisfies $\int_{E_1} y^i h\, dt = 0$ for $i = 1,\ldots,k-1$ and $|h(t)| < \varepsilon$. Hence $0 < \theta \pm h < 1$ and

$$\int_{E_1} y(\theta \pm h)dt = \int_{E_1} y\theta\, dt = a.$$

Hence $\theta + h$ and $\theta - h$ are in $\sum$. But then θ cannot be an extreme point of $\sum$ since it is the midpoint of the segment with end points $\theta + h$ and $\theta - h$. Therefore, θ must be a characteristic function of some set F.

THEOREM 6.1. Let $\mathscr{I}$ be a closed interval in E^1, let A be an $n \times n$ matrix function continuous on $\mathscr{I}$, let $\mathscr{U}$ be a region in E^m and let h be a continuous map from $\mathscr{I} \times \mathscr{U}$ to E^n. Let Ω be independent of x and upper semicontinuous on $\mathscr{I}$. For each t in $\mathscr{I}$ let the set $h(t,\Omega(t))$ be closed. Let

$$\Delta = \{(t,z): t \in \mathscr{I},\ z \in \Omega(t)\}$$

and let there exist a nonnegative integrable function μ defined on $\mathscr{I}$ such that for all (t,z) in Δ

$$|h(t,z)| \leq \mu(t). \tag{6.1}$$

Let the state equations be given by

$$\frac{dx}{dt} = A(t)x + h(t,u(t)) \tag{6.2}$$

and let (t_0,x_0) be the initial point. Then the mapping $t \to \mathscr{K}(t,t_0,x_0)$ is continuous, the sets $\mathscr{K}_R(t,t_0,x_0)$ and $\mathscr{K}(t,t_0,x_0)$ are non-empty and are compact and convex, and $\mathscr{K}_R(t,t_0,x_0) = \mathscr{K}(t,t_0,x_0)$.

Proof. By Lemma III 11.2 there is at least one control and corresponding trajectory defined on $\mathcal{J}$. Therefore, for all $t_1 > t_0$ and all t in $[t_0,t_1]$ the sets $\mathcal{K}(t,t_0,x_0)$ and $\mathcal{K}_R(t,t_0,x_0)$ are not empty. Also, every control u defined on an interval $[t_0,t]$ can be extended to a control that is defined on the entire interval $[t_0,t_1]$.

From the variation of parameters formula for solutions of (6.2) and from (6.1) it follows that the trajectories in $\mathcal{A}_T^+$ are equi-absolutely continuous. The arguments used in Theorem 5.1 to show that the mapping $t \to \mathcal{K}(t,t_0,x_0)$ was continuous there, can therefore be used to show that the mapping $t \to \mathcal{K}(t,t_0,x_0)$ is continuous in the present case.

We next show that the sets $\mathcal{K}_R(t,t_0,x_0)$ are convex and compact. The relaxed system corresponding to (6.2) is

$$\frac{dx}{dt} = A(t)x + \sum_{i=1}^{n+1} p^i(t)h(t,u_i(t)), \tag{6.3}$$

where for almost all t in $\mathcal{J}$

$$\sum_{i=1}^{n+1} p^i(t) = 1 \qquad 0 \leq p^i(t) \leq 1 \quad i = 1,\ldots,n+1 \tag{6.4}$$

and $u_i(t) \in \Omega(t)$. Thus, solutions of (6.3) that satisfy $\psi(t_0) = x_0$ are defined for all t in $\mathcal{J}$ as are the solutions of (6.2) that satisfy $\phi(t_0) = x_0$.

Let Θ denote the set of functions θ in $L_1[t_0,t]$ that satisfy

$$\theta(s) \in \text{co } h(s,\Omega(s)) \quad \text{a.e.} \tag{6.5}$$

We show that the set Θ can also be characterized as the set of all functions θ in $L_1[t_0,t]$ such that

$$\theta(s) = \sum_{i=1}^{n+1} p^i(s)h(s,u_i(s)) \quad \text{a.e.,} \tag{6.6}$$

where $p^1,\ldots,p^{n+1}$ are real valued measurable functions on $[t_0,t]$ satisfying the relations (6.4) and where $u_1,\ldots,u_{n+1}$ are measurable functions satisfying $u_i(s) \in \Omega(s)$. Any function of the form (6.6) satisfies (6.5) and is integrable by virtue of (6.4) and (6.1). Conversely, by Caratheodory's theorem (Theorem III 5.2) any function θ that satisfies (6.5) can be written in the form (6.6) without any assertion about the measurability of the functions p^i and u_i, $i = 1,\ldots,n+1$. From Theorem III 7.1, however, we conclude that the p^i and u_i may be chosen to be measurable. Thus, (6.5) and (6.6) give equivalent characterizations of Θ.

From the preceding discussion it follows that a relaxed trajectory can also be defined as an absolutely continuous function such that

$$\psi'(s) = A(s)\psi(s) + \theta(s)$$

for some function θ in Θ. From the variation of parameters formula we have that

$$\psi(t) = \Psi(t_0)[x_0 + \int_{t_0}^{t} \Psi^{-1}(s)\theta(s)ds],$$

where Ψ is the fundamental matrix solution of the homogeneous system $dx/dt = A(t)x$ and $\Psi(t_0) = I$, the $n \times n$ identity matrix. Hence the set $\mathscr{K}_R(t,t_0,x_0)$ can be characterized as the set of points x in E^n such that

$$x = \Psi(t_0)[x_0 + \int_{t_0}^{t} \Psi^{-1}(s)\theta(s)ds] \tag{6.7}$$

for some θ in Θ. Hence to prove that $\mathscr{K}_R(t,t_0,x_0)$ is convex and compact it suffices to prove that the set of points in E^n of the form

$$\{x:\ x = \int_{t_0}^{t} \Psi^{-1}(s)\theta(s)ds,\quad \theta \in \Theta\} \tag{6.8}$$

is convex and compact.

It is immediate from (6.5) that the set Θ is a convex subset of $L_1[t_0,t]$. The convexity of $\mathscr{K}_R(t,t_0,x_0)$ now follows from (6.7).

Define a mapping T from $L_1[t_0,t]$ to E^n as follows:

$$T\rho = \int_{t_0}^{t} \Psi^{-1}(s)\rho(s)ds \qquad \rho \in L_1[t_0,t].$$

The mapping T is continuous from the strong topology of $L_1[t_0,t]$ to E^n. It is therefore continuous from the weak topology of $L_1[t_0,t]$ to E^n. Hence the image of a weakly compact set in $L_1[t_0,t]$ will be compact in E^n. Therefore, to show that the set (6.8), and hence $\mathscr{K}_R(t,t_0,x_0)$ is compact it suffices to show that Θ is weakly compact. We now do this.

We have already noted that the set Θ is a convex subset of $L_1[t_0,t]$. The set Θ is also closed in $L_1[t_0,t]$, as the following argument shows. Let $\{\theta_k\}$ be a sequence in Θ converging in $L_1[t_0,t]$ to an element θ_0. Then there exists a subsequence, again written as $\{\theta_k\}$, such that $\theta_k \to \theta_0$ at all points of $[t_0,t]$, except those in a subset E_0 of measure zero. Let E_{-1} denote the set of points s at which $\mu(s) = +\infty$, where μ is as in (6.1). For each integer k let E_k denote the set of points at which (6.6) fails. Then $E = \cup_{k=-1}^{\infty} E_k$ has measure zero. Let s be a point of $[t_0,t_1]$ that is not in E. Then there exists a subsequence of the integers, which we again label as $\{k\}$, such that for all $i = 1,\ldots,n+1$ the sequences

$$\{p_k^i(s)\}_{k=1}^{\infty} \quad \text{and} \quad \{h(s,u_{ik}(s))\}_{k=1}^{\infty}$$

converge. Let $p^i(s) = \lim p_k^i(s)$, $i = 1,\ldots,n+1$. Then, $0 \leq p^i(s) \leq 1$ and $\sum p^i(s) = 1$. Since $h(s,\Omega(s))$ is closed, it follows that $\lim_{k\to\infty} h(s,u_{ik}(s))$ is in $h(s,\Omega(s))$ for each $i = 1,\ldots,n+1$. Hence θ_0 satisfies (6.5) at the point s. Since s is an arbitrary point in $[t_0,t]-E$, it follows that (6.5) holds a.e., and Θ is closed in $L_1[t_0,t]$.

Since Θ is a convex, closed set in $L_1[t_0,t]$ it is also weakly closed. From (6.6), (6.4), and (6.1) it follows that $|\theta(s)| \leq \mu(s)$ for all θ in Θ. Hence the functions in Θ have equi-absolutely continuous integrals. Therefore ([20] Corollary IV. 8.11, p. 294) Θ is weakly sequentially conditionally compact. Hence by the Eberlein-Smulian theorem ([20] Theorem V.6.1) the closure of Θ is weakly compact. Since Θ is weakly closed, it is also weakly compact. Hence $\mathscr{K}_R(t,t_0,x_0)$ is compact.

Since a trajectory of the system (6.2) is also a relaxed trajectory, we have that $\mathscr{K}(t,t_0,x_0) \subseteq \mathscr{K}_R(t,t_0,x_0)$. Therefore to complete the proof it suffices to show that $\mathscr{K}_R(t,t_0,x_0) \subseteq \mathscr{K}(t,t_0,x_0)$.

Let x be an element of $\mathscr{K}_R(t,t_0,x_0)$. Then x is given by (6.7). Let

$$a = \Psi^{-1}(t_0)x - x_0.$$

The point a is in the set defined by (6.8). The set $T^{-1}(a)$ is weakly closed and is convex in $L_1[t_0,t]$. Let $\sum$ denote the intersection of $T^{-1}(a)$ and Θ. Then $\sum$ is weakly closed and convex. By the Krein-Milman Theorem, (Lemma 6.2) $\sum$ has an extreme point θ_0. Since $\theta_0 \in \Theta$ it follows that θ_0 has a representation given by the right side of (6.6). We now assert that on no measurable subset E of $[t_0,t]$ with positive measure can we have $0 < \varepsilon \leq p^i(s) \leq 1-\varepsilon$ for some i in the set $i,\dots,n+1$ and some $\varepsilon > 0$. This assertion implies that θ_0 can be written

$$\theta_0(s) = h(s,u(s)) \tag{6.9}$$

where u is a measurable function on $[t_0,t]$ satisfying $u(s) \in \Omega(s)$. Once we establish (6.9), the theorem will be proved, for then we will have that

$$a = \Psi^{-1}(t)x - x_0 = \int_{t_0}^{t} \Psi^{-1}(s)h(s,u(s))ds,$$

which says that $x \in \mathscr{K}(t,t_0,x_0)$.

Let us suppose that θ_0 has the form (6.6) and that there exists an $\varepsilon > 0$ and a measurable set $E \subseteq [t_0,t]$ such that $meas(E) > 0$ and $\varepsilon \leq p^i(t) \leq 1-\varepsilon$ on E. Then for at least one other index i we must also have $\varepsilon \leq p^i(t) \leq 1-\varepsilon$ on E. For definiteness let us suppose that $i = 2$ is one such index. Since a belongs to the set defined by (6.8), we get from (6.6) that

$$a = \int_{t_0}^{t} \{\Psi^{-1}(s) \sum_{i=1}^{n+1} p^i(s)h(s,u_i(s))\}ds$$

$$= \int_{t_0}^{t} \{\sum_{i=1}^{n+1} p^i(s)(\Psi^{-1}(s)h(s,u_i(s))\}ds.$$

We next apply Lemma 6.1 with $w \equiv 1/2$ and y the vector function in $L_1(E)$ with range in E^{2n} defined by

$$y(s) = \begin{pmatrix} \Psi^{-1}(s)h(s,u_1(s)) \\ \Psi^{-1}(s)h(s,u_2(s)) \end{pmatrix} .$$

We obtain the existence of a set $F \subseteq E$ such that

$$\frac{1}{2}\int_E \Psi^{-1}(s)h(s,u_i(s))ds = \int_F \Psi^{-1}(s)h(s,u_i(s))ds \qquad i = 1,2.$$

Let χ_F denote the characteristic function of the set F and let χ_E denote the characteristic function of the set E. Let the function γ be defined as follows:

$$\gamma(s) = 2\chi_F(s) - \chi_E(s).$$

The function γ is equal to one in absolute value on the set E. Let

$$\pi_1^1 = p^1 + \varepsilon\gamma \quad \pi_1^2 = p^2 - \varepsilon\gamma \quad \pi_1^i = p^i, \quad i = 3,\ldots,n+1.$$

$$\pi_2^1 = p^1 - \varepsilon\gamma \quad \pi_2^2 = p^2 + \varepsilon\gamma \quad \pi_2^i = p^i, \quad i = 3,\ldots,n+1.$$

Then $0 \leq \pi_1^i \leq 1$ and $0 \leq \pi_2^i \leq 1$ for $i = 1,\ldots,n+1$. Also, $\sum \pi_1^i = 1$ and $\sum \pi_2^i = 1$.

Let

$$\theta_1(s) = \sum_{i=1}^{n+1} \pi_1^i(s) h(s,u_i(s))$$

$$\theta_2(s) = \sum_{i=1}^{n+1} \pi_2^i(s) h(s,u_i(s)).$$

Then θ_1 and θ_2 are in Θ and

$$a = \int_{t_0}^{t} \Psi^{-1}(s)\theta_1(s)ds = \int_{t_0}^{t} \Psi^{-1}(s)\theta_2(s)ds.$$

Hence θ_1 and θ_2 are in $\sum$. But, $\theta_0 = (\theta_1 + \theta_2)/2$, which contradicts the assumption that θ_0 is an extreme point, and the Theorem is proved.

When the constraint sets $\Omega(t)$ are compact we have the following result, which is a corollary of Theorem 6.1.

THEOREM 6.2. Let $\mathscr{I}$ be a compact interval in E^1 and let A be an $n \times n$ matrix function continuous on $\mathscr{I}$. Let $\mathscr{U}$ be a region in E^m and let h be a continuous map from $\mathscr{I} \times \mathscr{U}$ to E^n. Let Ω be independent of x and u.s.c.i. on $\mathscr{I}$. For each t in $\mathscr{I}$ let the set $\Omega(t)$ be compact. Then for the system (6.2) with initial point (t_0,x_0) and subject to the constraint $u(t) \in \Omega(t)$ the following holds. For all t in $\mathscr{I}$, $t > t_0$, the attainable sets $\mathscr{K}(t,t_0,x_0)$ are non empty, compact and convex and $\mathscr{K}(t,t_0,x_0) = \mathscr{K}_R(t,t_0,x_0)$. The mapping $t \to \mathscr{K}(t,t_0,x_0)$ is continuous on any interval $[t_0,t_1] \subset \mathscr{I}$.

Since $\mathscr{I}$ is compact and Ω is u.s.c.i. on $\mathscr{I}$ it follows from

Lemma III 5.2 that the set Δ defined in the statement of Theorem 6.1 is compact. Since h is continuous, h is bounded on Δ. Hence (6.1) holds and the hypotheses of Theorem 6.1 are satisfied.

For systems that are linear in the state variable we have the following existence theorem that does not require the sets $\mathscr{Q}^+(t,x)$ to be convex.

THEOREM 6.3. Let $\mathscr{I} = [t_0,t_1]$ be a fixed compact interval in E^1, let $\mathscr{U}$ be a region in E^m and let h be a continuous mapping from $\mathscr{I} \times \mathscr{U}$ to E^n. Let A be a continuous $n \times n$ matrix function on $\mathscr{I}$. Let the mappings $a_0\colon \mathscr{I} \to E^n$ and $h_0\colon \mathscr{I} \times \mathscr{U} \to E^1$ be continuous. Let Ω be independent of x, u.s.c.i. on $\mathscr{I}$, and for each t in $\mathscr{I}$ let the set $\Omega(t)$ be compact. Let g be real valued and continuous on E^n and let x_0 be fixed. Then the problem of minimizing the functional

$$J(\phi,u) = g(\phi(t_1)) + \int_{t_0}^{t_1} (\langle a_0(t),\phi(t)\rangle + h_0(t,u(t)))dt$$

subject to the end conditions $\mathscr{B} = \{(t_0',x_0',t_1',x_1')\colon t_0' = t_0,\ x_0' = x_0,\ t_1' = t_1\}$, control constraint Ω, and system equations

$$\frac{dx}{dt} = A(t)x + h(t,u(t))$$

has a solution.

REMARK 6.1. It follows from Lemma III 11.2 and Remark 11.1 that the class of admissible pairs is non-empty.

REMARK 6.2. For linear systems we have $h(t,z) = B(t)z$. Theorem 6.2 enables us to dispense with the requirement imposed up to now that the constraint sets $\Omega(t)$ be convex.

<u>Proof</u>. By introducing an additional coordinate x^0 and state equation $dx^0/dt = \langle a_0(t),x\rangle + h_0(t,u(t))$, we may assume, without loss

of generality, that a_0 and h_0 are identically zero. (See Chapter II, Section 4). By Theorem 6.2 the set $\mathscr{K}(t_1,t_0,x_0)$ is compact. Since g is continuous on the attainable set $\mathscr{K}(t_1,t_0,x_0)$, it attains its minimum at some point of $\mathscr{K}(t_1,t_0,x_0)$, say x_1^*. By definition, $\mathscr{K}(t_1,t_0,x_0)$ consists precisely of those points x that are equal to $\phi(t_1)$ for some admissible trajectory ϕ. Let ϕ^* correspond to x_1^*. Then $J(\phi^*,u^*) = g(\phi^*(t_1)) \leq g(\phi(t_1)) = J(\phi,u)$ for all admissible pairs (ϕ,u), and the Theorem is proved.

Another corollary of Theorem 6.1 is the so-called "bang-bang principle", which is contained in Theorem 6.4 which follows. The reason for the terminology and the significance of the principle in applications will be discussed after the proof of Theorem 6.4 is given.

If $\mathscr{C}$ is a compact convex set in E^m, then we shall denote the set of extreme points of $\mathscr{C}$ by $\mathscr{C}_e$. By the Krein-Milman Theorem, $\mathscr{C}_e$ is non-void and $\mathscr{C} = \text{cl co}(\mathscr{C}_e)$.

THEOREM 6.4. Let $\mathscr{I} = [t_0,t_1]$ be a compact interval in E^1, let A be an $n \times n$ continuous matrix on $\mathscr{I}$ and let B be an $n \times m$ continuous matrix on $\mathscr{I}$. Let $\mathscr{C}$ be a compact convex set in E^m whose set of extreme points $\mathscr{C}_e$ is closed. Let $\mathscr{K}(t_1,t_0,x_0)$ denote the attainable set at t_1 for the system

$$\frac{dx}{dt} = A(t)x + B(t)u(t) \tag{6.10}$$

with initial point (t_0,x_0) and with the control constraint $u(t) \in \mathscr{C}$. Let $\mathscr{K}_e(t_1,t_0,x_0)$ denote the attainable set for the system (6.10) with initial point (t_0,x_0) and with control constraint $u(t) \in \mathscr{C}_e$. Then $\mathscr{K}_e(t_1,t_0,x_0)$ is non-empty and $\mathscr{K}(t_1,t_0,x_0) = \mathscr{K}_e(t_1,t_0,x_0)$.

Proof. Since the function defined by $u(t) = z_0$, where z_0 is any point of $\mathscr{C}_e$ is admissible for the system (6.10) with initial point (t_0,x_0) and control constraint $u(t) \in \mathscr{C}_e$, it follows that $\mathscr{K}_e(t_1,t_0,x_0)$ is non-empty. Since $\mathscr{C}$ is compact and convex $\mathscr{C} =$ cl co $(\mathscr{C}_e)$. Since $\mathscr{C}$ is compact, $\mathscr{C}_e$ is bounded. Since $\mathscr{C}_e$ is closed by hypothesis, the set $\mathscr{C}_e$ is compact. By Theorem III 5.2 every point in co $(\mathscr{C}_e)$ can be written as a convex combination of at most $(n+1)$ points in $\mathscr{C}_e$. Thus co $(\mathscr{C}_e)$ is the continuous image of $\Gamma \times [\mathscr{C}_e]^{n+1}$, where Γ is as in (3.1). Hence co $(\mathscr{C}_e)$ is compact and therefore $\mathscr{C} =$ cl co $(\mathscr{C}_e) =$ co $(\mathscr{C}_e)$. Therefore any control u such that $u(t) \in \mathscr{C}$ can be written as

$$u(t) = \sum_{i=1}^{n+1} p^i(t)u_i(t)$$

where $0 \leq p^i(t) \leq 1$, $\sum p^i(t) = 1$, and $u_i(t) \in \mathscr{C}_e$. By Theorem III 7.1 the functions p^i and u_i can be chosen to be measurable. Hence the set $\mathscr{K}(t_1,t_0,x_0)$ is contained in the relaxed attainable set $\mathscr{K}_{eR}(t_1,t_0,x_0)$ corresponding to $\mathscr{K}_e(t_1,t_0,x_0)$.

Conversely, every relaxed control for the system (6.10) with control constraint $u(t) \in \mathscr{C}_e$ is a control for the system (6.10) with control constraint $u(t) \in \mathscr{C}$. Hence $\mathscr{K}(t_1,t_0,x_0) = \mathscr{K}_{eR}(t_1,t_0,x_0)$. It is readily checked that the system (6.10) with initial point (t_0,x_0) and control constraint $u(t) \in \mathscr{C}_e$ satisfies the hypotheses of Theorem 6.1. Hence $\mathscr{K}_{eR}(t_1,t_0,x_0) = \mathscr{K}_e(t_1,t_0,x_0)$ and the present theorem is established.

In many applications the constraint set $\mathscr{C}$ is a compact convex polyhedron, or even a cube, in E^m. The set of extreme points $\mathscr{C}_e$ is the set of vertices of the polyhedron, and is therefore closed. Theorem 6.4 in this situation states that if a control u with values $u(t) \in \mathscr{C}$ will transfer the system from a point x_0 at time t_0 to a point x_1 at time t_1, then there exists a control u_e with values $u_e(t)$ in $\mathscr{C}_e$ that will do the same thing. Thus in designing a control system the designer need only allow for a finite number of control positions corresponding to the vertices of $\mathscr{C}$. The term "bang-bang" to describe controls with values on the vertices of derives from the case where $\mathscr{C}$ is a one-dimensional interval. In this case controls u_e with $u_e(t) \in \mathscr{C}_e$ are controls that take on the values $+1$ and -1. Such controls represent the extreme positions of the control device and are therefore often referred to in the engineering vernacular as "bang-bang" controls. In the control literature the terminology has been carried over to theorems such as Theorem 6.4.

CHAPTER V

THE MAXIMUM PRINCIPLE AND SOME OF ITS APPLICATIONS

1. Introduction

In this chapter we shall state the maximum principle and shall use it to characterize the optimal controls in several important classes of problems. The proof of the maximum principle will be given in the next chapter.

In Section 2 we use a dynamic programming argument to derive the maximum principle. Although the arguments are mathematically correct, the assumptions are such that most interesting problems are ruled out. The purpose of this section is to make plausible the statements of the theorem and to give some insight and interpretation to the theorem. From the point of view of logical development, Section 2 can be omitted, except for one concept. The concept is that of optimal synthesis, or optimal feedback control, which is introduced in Section 2 and used again in Section 9.

In Section 3 we give a precise statement of the maximum principle for the control problem in Lagrange form. The statements of the maximum principle for other formulations of the problem, such as those discussed in Chapter II, are taken up in the exercises. In special cases of importance more precise characterizations of the optimal pair can often be given. Some of these are also taken up in the exercises. The exercises in this section are an important supplement to the general theory.

In Section 4 we use the maximum principle and one of our existence theorems to determine the optimal pair in a specific example. The purpose here is to illustrate how the maximum principle is used and some of the difficulties that one can expect to encounter in large scale problems.

The remaining sections of the chapter are devoted to applications of the maximum principle to special classes of problems. In Section 5 we show how to obtain the first order necessary conditions of the classical calculus of variations from the maximum principle. In the exercises we take up the relationship between the classical Bolza problem in the calculus of variations and the control problem. In Section 6 we take up control problems that are linear in the state variable. We specialize this in Section 7 to linear problems, and further specialize in Section 8 to the linear time optimal problem. The standard results for these problems are obtained, whenever possible, as relatively simple consequences of the maximum principle. The power of the maximum principle will be apparent to the reader.

In Section 9 we take up the so-called linear plant quadratic cost criterion problem. Here again we obtain the standard characterization of the optimal pair from the maximum principle. We also show that the necessary conditions are sufficient and we obtain the standard synthesis of the optimal control.

2. A Dynamic Programming Derivation of the Maximum Principle

In this section we shall derive the maximum principle under very restrictive assumptions. The assumptions will be spelled out as they are needed in the course of the argument. The reader is cautioned that the assumptions made in this section are very often not fulfilled in problems of interest. Although some of the arguments can be made to hold under less restrictive assumptions, we shall not do so here. The purpose of this section is to motivate the precise statement of the maximum principle to be given in Theorem 3.1 of Section 3 below and to give some insight as to why the maximum principle is plausible.

Let $\mathscr{R}_1$ be a region of (t,x)-space and let $\mathscr{R}$ be a subregion of $\mathscr{R}_1$ such that the closure of $\mathscr{R}$ is contained in $\mathscr{R}_1$. For each point (τ,ξ) in $\mathscr{R}$ we consider the following problem. Minimize the functional

$$J(\phi,u) = g(t_1,\phi(t_1)) + \int_{\tau}^{t_1} f^0(t,\phi(t),u(t))dt \qquad (2.1)$$

subject to the state equations

$$\frac{dx}{dt} = f(t,x,u(t)), \qquad (2.2)$$

control constraints $u(t) \in \Omega(t)$, and end conditions

$$(t_0,\phi(t_0)) = (\tau,\xi) \qquad (t_1,\phi(t_1)) \in \mathscr{T}.$$

We assume that the terminal set $\mathscr{T}$ is a $C^{(1)}$ manifold of dimension q, where $0 \leq q \leq n$ and that $\mathscr{T}$ is part of the boundary of $\mathscr{R}$. See Figure 1. For simplicity we also assume that $\mathscr{T}$ can be

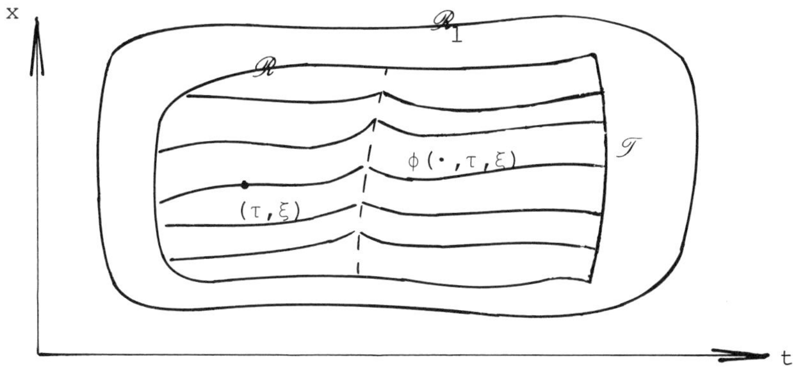

Figure 1

represented by a single coordinate patch. That is, we assume that $\mathscr{T}$ consists of all points of the form (t_1,x_1) with

$$t_1 = T(\sigma) \qquad x_1 = X(\sigma) \qquad (2.3)$$

where T and X are $C^{(1)}$ functions defined on an open parallelepiped

Σ in E^q. It is also assumed that the Jacobian matrix of the mapping (2.3),

$$\frac{\partial(T,X)}{\partial\sigma} = \begin{pmatrix} \frac{\partial T}{\partial\sigma^1} & \cdots & \frac{\partial T}{\partial\sigma^q} \\ \frac{\partial X^1}{\partial\sigma^1} & \cdots & \frac{\partial X^1}{\partial\sigma^q} \\ \vdots & & \\ \frac{\partial X^n}{\partial\sigma^1} & \cdots & \frac{\partial X^n}{\partial\sigma^q} \end{pmatrix}$$

has rank q at all points of Σ. We assume that the function g in (2.1) is defined and $C^{(1)}$ in a neighborhood of $\mathscr{T}$ and that f^0 and f are $C^{(1)}$ functions on $\mathscr{G}_1 = \mathscr{R}_1 \times \mathscr{U}$. Note that the constraint mapping Ω is assumed to be independent of x and to depend only on t.

We assume that for each (τ,ξ) in $\mathscr{R}$ the problem has a unique solution. We denote the unique optimal trajectory for the problem with initial point (τ,ξ) by $\phi(\cdot,\tau,\xi)$. The corresponding unique optimal control is denoted by $u(\cdot,\tau,\xi)$. We assume that the function $u(\cdot,\tau,\xi)$ is piecewise continuous and that at a point of discontinuity t_d the value of $u(\cdot,\tau,\xi)$ is its right hand limit; thus $u(t_d,\tau,\xi) = u(t_d+0,\tau,\xi)$. Points (t,x) on the trajectory satisfy the relation $x = \phi(t,\tau,\xi)$. In particular, note that

$$\xi = \phi(\tau,\tau,\xi).$$

The value of the optimal control at time t is $u(t) = u(t,\tau,\xi)$.

For each point (τ,ξ) in $\mathscr{R}$, let $W(\tau,\xi)$ denote the value given to the functional (2.1) by the unique optimal pair $(\phi(\cdot,\tau,\xi), u(\cdot,\tau,\xi))$. Thus, if $\mathscr{A}(\tau,\xi)$ denotes the set of admissible pairs (ϕ,u) for the problem with initial point (τ,ξ) then

$$W(\tau,\xi) = \min\{J(\phi,u): \ (\phi,u) \in \mathscr{A}(\tau,\xi)\}. \tag{2.4}$$

The function W so defined is called the value function for the problem.

Let $\tau_1 > \tau$ and let (τ_1, ξ_1) be a point on the optimal trajectory $\phi(\cdot, \tau, \xi)$. Then $\xi_1 = \phi(\tau_1, \tau, \xi)$. We assert that the optimal pair for the problem starting at (τ_1, ξ_1) is given by $(\phi(\cdot, \tau, \xi), u(\cdot, \tau, \xi))$. That is, for $t \geq \tau_1$

$$\begin{aligned} \phi(t,\tau,\xi) &= \phi(t,\tau_1,\xi_1) \\ u(t,\tau,\xi) &= u(t,\tau_1,\xi_1). \end{aligned} \tag{2.5}$$

In other words, an optimal trajectory has the property that it is optimal for the problem that starts at any point on the trajectory. To see this we write

$$W(\tau,\xi) = \int_{\tau}^{\tau_1} f^{0*}(t,\tau,\xi)\,dt + \int_{\tau_1}^{t_1} f^{0*}(t,\tau,\xi)\,dt + g(t_1, \phi(t_1,\tau,\xi)), \tag{2.6}$$

where

$$f^{0*}(t,\tau,\xi) = f^0(t, \phi(t,\tau,\xi), u(t,\tau,\xi)). \tag{2.7}$$

If $(\phi(\cdot,\tau,\xi), u(\cdot,\tau,\xi))$ were not optimal for $t \geq \tau_1$, then by (2.4) with (τ,ξ) replaced by (τ_1,ξ_1) and by our assumption of uniqueness of optimal pairs, we would have that $W(\tau_1,\xi_1)$ is strictly less than the sum of the last two terms in the right hand side of (2.6). Hence for a control u defined by

$$u(t) = \begin{cases} u(t,\tau,\xi) & \tau \leq t < \tau_1 \\ u(t,\tau_1,\xi_1) & \tau_1 \leq t \leq t_1 \end{cases}$$

the corresponding trajectory ϕ would be such that $J(\phi,u) < W(\tau,\xi)$, thus contradicting (2.4). Recall that $u(\cdot,\tau_1,\xi_1)$ is the optimal control for the problem with initial point (τ_1,ξ_1). Hence (2.5) holds.

We define a function U on $\mathscr{R}$ as follows

$$U(\tau,\xi) = u(\tau,\tau,\xi).$$

If we set $t = \tau_1$ in the second equation in (2.5) and use the definition of U we get that for all $\tau_1 \geq \tau$

$$u(\tau_1,\tau,\xi) = U(\tau_1,\xi_1) \tag{2.8}$$

where $\xi_1 = \phi(\tau_1,\tau,\xi)$. Thus at each point (τ,ξ) in $\mathscr{R}$ the value $U(\tau,\xi)$ of U is the value of the unique optimal control function associated with the unique optimal trajectory through the point. The function U is called the <u>synthesis of the optimal control</u> or <u>optimal synthesis function</u>. It is also called the <u>optimal feedback control</u>.

We now suppose that the function W is $C^{(1)}$ on $\mathscr{R}$. We shall derive a partial differential equation that W must satisfy. Consider a point (τ,ξ) in $\mathscr{R}$ and an interval $[\tau, \tau+\Delta t]$, where $\Delta t > 0$. Let v be a continuous control defined on $[\tau,\tau+\Delta t]$ satisfying $v(t) \in \Omega(t)$. We suppose that Δt is so small that the state equations (2.2) with $u(t)$ replaced by $v(t)$ have a solution ψ defined on $[\tau,\tau+\Delta t]$ and satisfying the relation $\psi(\tau) = \xi$. Let $\Delta x = \psi(\tau+\Delta t) - \psi(\tau)$. Thus, the control v transfers the system from ξ to $\xi+\Delta x$ in the time interval $[\tau,\tau+\Delta t]$. For $t \geq \tau+\Delta t$ let us use the optimal control for the problem with initial point $(\tau+\Delta t,\xi+\Delta x)$; namely $u(\cdot,\tau+\Delta t,\xi+\Delta x)$. The resulting trajectory will be $\phi(\cdot,\tau+\Delta t,\xi+\Delta x)$. Let $\tilde{u}$ denote the control obtained by using v on $[\tau,\tau+\Delta t]$ and then $u(\cdot,\tau+\Delta t,\xi+\Delta x)$. Let $\tilde{\phi}$ denote the resulting trajectory. Then $(\tilde{\phi},\tilde{u}) \in \mathscr{A}(\tau,\xi)$ and

$$W(\tau,\xi) \leq J(\tilde{\phi},\tilde{u}) = \int_{\tau}^{\tau+\Delta t} f^0(s,\psi(s),v(s))ds + \int_{\tau+\Delta t}^{t_1} f^{0*}(s,\tau+\Delta t,\xi+\Delta x)ds$$
$$+ g(t_1,\phi(t_1,\tau+\Delta t,\xi+\Delta x)),$$

where f^{0*} is defined in (2.7). The sum of the last two terms on the right is equal to $W(\tau+\Delta t,\xi+\Delta x)$. Hence

$$W(\tau+\Delta t,\xi+\Delta x)-W(\tau,\xi) \geq -\int_{\tau}^{\tau+\Delta t} f^0(s,\psi(s),v(s))ds.$$

Since W is $C^{(1)}$ on $\mathscr{R}$ we can apply Taylor's theorem to the left hand side of the preceding inequality and get

$$W_\tau(\tau,\xi)\Delta t + \langle W_\xi(\tau,\xi),\Delta x\rangle + o(|(\Delta t,\Delta x)|) \geq -\int_{\tau}^{\tau+\Delta t} f^0(s,\psi(s),v(s))ds, \tag{2.9}$$

where (W_τ,W_ξ) denotes the vector of partial derivatives of W and $o(|(\Delta t,\Delta x)|)/|(\Delta t,\Delta x)| \to 0$ as $|(\Delta t,\Delta x)| \to 0$. From the relation

$$\Delta x/\Delta t = \frac{1}{\Delta t}\int_{\tau}^{\tau+\Delta t} f(s,\psi(s),v(s))ds$$

and the continuity of f, ψ, and v it follows that

$$\lim_{\Delta t\to 0} \frac{\Delta x}{\Delta t} = f(\tau,\psi(\tau),v(\tau)) = f(\tau,\xi,v(\tau)).$$

Therefore, if we divide through by $\Delta t > 0$ in (2.9) and then let $\Delta t \to 0$, we get that

$$W_\tau(\tau,\xi) + \langle W_\xi(\tau,\xi),f(\tau,\xi,v(\tau))\rangle \geq -f^0(\tau,\xi,v(\tau)). \tag{2.10}$$

If we carry out the preceding analysis with $v(s) = u(s,\tau,\xi)$ on $[\tau,\tau+\Delta t]$, then equality holds at every step of the argument. Therefore, with the help of (2.8), we obtain the relation

$$W_\tau(\tau,\xi) = -f^0(\tau,\xi,U(\tau,\xi)) - \langle W_\xi(\tau,\xi),f(\tau,\xi,U(\tau,\xi))\rangle. \tag{2.11}$$

We now make the further assumption that the constraint mapping Ω is sufficiently smooth so that for every vector $z \in \Omega(t)$ there exists a continuous function v defined on some interval $[\tau,\tau+\Delta t]$, $\Delta t > 0$, with $v(\tau) = z$ and $v(s) \in \Omega(s)$ on $[\tau,\tau+\Delta t]$. In particular, if Ω is a constant mapping, i.e. $\Omega(t) = \mathscr{C}$ for all t, then we may take $v(s) = z$ on $[\tau, \tau+\Delta t]$. Under the assumption just made concerning Ω, we can combine (2.10) and (2.11) to get the relation

$$W_\tau(\tau,\xi) = \max_{z\varepsilon\Omega(\tau)} [-f^0(\tau,\xi,z) - \langle W_\xi(\tau,\xi), f(\tau,\xi,z)\rangle], \qquad (2.12)$$

with the maximum being attained at $z = U(\tau,\xi)$. Equation (2.12) is sometimes called Bellman's equation. Equation (2.11) is the Hamilton-Jacobi equation.

Equations (2.11) and (2.12) can be written more compactly. First define a real valued function H on $E^1 \times E^n \times E^m \times E^1 \times E^n$ by the formula

$$H(t,x,z,p^0,p) = p^0 f^0(t,x,z) + \langle p, f(t,x,z)\rangle . \qquad (2.13)$$

If we now denote a generic point in $\mathscr{R}$ by (t,x) rather than by (τ,ξ) we can write (2.11) in terms of H as follows:

$$W_t(t,x) = H(t,x,U(t,x),-1,-W_x(t,x)). \qquad (2.14)$$

Equation (2.12) can be written in the form

$$W_t(t,x) = \max_{z\varepsilon\Omega(t)} H(t,x,z,-1,-W_x(t,x)). \qquad (2.15)$$

We now suppose that the function W is of class $C^{(2)}$. Under this additional hypothesis we shall derive the Pontryagin Maximum Principle. Let (τ,ξ) again be a fixed point in $\mathscr{R}$. Consider the function F defined on $\mathscr{R}$ by the formula

$$F(x) = W_t(\tau,x) + f^0(\tau,x,U(\tau,\xi)) + \langle W_x(\tau,x), f(\tau,x,U(\tau,\xi))\rangle, \quad (2.16)$$

where W_t denotes the partial derivative of W with respect to time and W_x denotes the partial derivative vector with respect to the state variable. It follows from (2.11) that $F(\xi) = 0$. On the other hand, since $U(\tau,\xi)\varepsilon\ \Omega(\tau)$ we obtain the following inequality from (2.12) with (τ,ξ) replaced by (τ,x)

$$W_t(\tau,x) \geq -f^0(\tau,x,U(\tau,\xi)) - \langle W_x(\tau,x), f(\tau,x,U(\tau,\xi))\rangle.$$

This says that $F(x) \geq 0$. Hence the function F has a minimum at $x = \xi$. Since W is $C^{(2)}$, F is $C^{(1)}$. Therefore, since ξ is an interior point of the domain of definition of F and F attains its minimum at ξ, we have that $F_x(\xi) = 0$. If we use (2.16) to compute the partial derivatives of F with respect to the state variable and then set the partials equal to zero at $x = \xi$, we get that for $i = 1,2,\ldots,n$,

$$\frac{\partial^2 W}{\partial t\partial x^i} + \frac{\partial f^0}{\partial x^i} + \sum_{j=1}^{n} \frac{\partial^2 W}{\partial x^i \partial x^j} f^j + \sum_{j=1}^{n} \frac{\partial W}{\partial x^j} \frac{\partial f^j}{\partial x^i} = 0, \tag{2.17}$$

where the partial derivatives of W are evaluated at (τ,ξ) and the functions f^j and their partial derivatives are evaluated at $(\tau,\xi,U(\tau,\xi))$. Since (τ,ξ) is an arbitrary point in $\mathscr{R}$, it follows that (2.17) holds for the argument (t,x) and $(t,x,U(t,x))$, where (t,x) is any point in $\mathscr{R}$.

Before proceeding with our analysis we introduce some useful terminology.

DEFINITION 2.1. If $h: (t,x,z) \to h(t,x,z)$ is a function from $\mathscr{G} = \mathscr{R} \times \mathscr{U}$ to E^k, $k \geq 1$, then by the expression "the function h evaluated along the trajectory $\phi(\cdot,\tau,\xi)$" we shall mean the composite function $t \to h(t,\phi(t,\tau,\xi),u(t,\tau,\xi))$. Similarly, if w is a function defined on $\mathscr{R}$, by the expression "the function w evaluated along the trajectory $\phi(\cdot,\tau,\xi)$" we shall mean the composite function $t \to w(t,\phi(t,\tau,\xi))$.

We now let (τ,ξ) be a fixed point in $\mathscr{R}$ and consider the behavior of the partial derivative $W_x = (W_{x^1},\ldots,W_{x^n})$ along the optimal trajectory starting at (τ,ξ). We define a function $\lambda(\cdot,\tau,\xi)$: $t \to \lambda(t,\tau,\xi)$ for $[\tau,t_1]$ to E^n as follows:

$$\lambda(t,\tau,\xi) = -W_x(t,\phi(t,\tau,\xi)). \tag{2.18}$$

Since W is $C^{(2)}$ the function λ is differentiable with respect to t. Using the relation $\phi'(t,\tau,\xi) = f(t,\phi(t,\tau,\xi),u(t,\tau,\xi))$ we get

$$\frac{d\lambda^i}{dt} = -\frac{\partial^2 W}{\partial t \partial x^i} - \sum_{j=1}^{n} \frac{\partial^2 W}{\partial x^i \partial x^j} f^j \qquad i = 1,\ldots,n, \tag{2.19}$$

where the partial derivatives of W and the components of f are evaluated along the trajectory $\phi(\cdot,\tau,\xi)$. If we substitute (2.19) into (2.17) and use (2.18) we get

$$\frac{d\lambda^i}{dt} = -\left\{-\frac{\partial f^0}{\partial x^i} + \sum_{j=1}^{n} \lambda^j \frac{\partial f^j}{\partial x^i}\right\} \qquad i = 1,\ldots,n.$$

In vector-matrix notation this becomes

$$\frac{d\lambda}{dt} = -[-\frac{\partial f^0}{\partial x} + \frac{\partial f}{\partial x}\lambda], \tag{2.20}$$

where $d\lambda/dt$, $\partial f^0/\partial x$ and λ are column vectors and $\partial f/\partial x$ is the matrix of partial derivatives whose entry in the i-th row and j-th column is $\partial f^j/\partial x^i$. The partials in (2.20) are evaluated along the trajectory $\phi(\cdot,\tau,\xi)$. To summarize, we have shown that associated with the optimal trajectory $\phi(\cdot,\tau,\xi)$ there is a function $\lambda(\cdot,\tau,\xi)$ such that (2.20) holds. We point out that since $\partial f^0/\partial x$ and $\partial f/\partial x$ are evaluated along $\phi(\cdot,\tau,\xi)$ they are functions of t on the interval $[\tau,t_1]$. Hence the system (2.20) can be considered as a linear system of differential equations with time varying coefficients that the function $\lambda(\cdot,\tau,\xi)$ must satisfy. Initial conditions for this system will be discussed below.

In terms of the function H introduced in (2.13), equation (2.20) becomes

$$\lambda'(t,\tau,\xi) = -H_x(t,\phi(t,\tau,\xi),u(t,\tau,\xi),-1,\lambda(t,\tau,\xi)), \tag{2.21}$$

where the prime denotes differentiation with respect to time. From the definition of H in (2.13) it also follows that

$$\phi'(t,\tau,\xi) = H_p(t,\phi(t,\tau,\xi),u(t,\tau,\xi),-1,\lambda(t,\tau,\xi)). \qquad (2.22)$$

It follows from (2.21) and (2.22) that the functions $\phi(\cdot,\tau,\xi)$ and $\lambda(\cdot,\tau,\xi)$ satisfy the system of differential equations

$$\begin{aligned} \frac{dx}{dt} &= H_p(t,x,u(t,\tau,\xi),-1,p) \\ \frac{dp}{dt} &= -H_x(t,x,u(t,\tau,\xi),-1,p). \end{aligned} \qquad (2.23)$$

We can combine (2.14) and (2.15) and get

$$H(t,x,U(t,x),-1,W_x(t,x)) = \max_{z\varepsilon\Omega(t)} H(t,x,z,-1,W_x(t,x)). \quad (2.24)$$

Since (2.24) holds for all (t,x) in $\mathscr{R}$ it holds along the optimal trajectory $\phi(\cdot,\tau,\xi)$. For points (t,x) on the optimal trajectory $\phi(\cdot,\tau,\xi)$ the relation

$$x = \phi(t,\tau,\xi)$$

holds. From this relation and from (2.8) it follows that for such points the relation

$$u(t,\tau,\xi) = U(t,\phi(t,\tau,\xi)) \qquad (2.25)$$

holds. Relation (2.18) also holds along the optimal trajectory. Therefore, for all $\tau \leq t \leq t_1$, where t_1 is the time at which $\phi(\cdot,\tau,\xi)$ hits $\mathscr{T}$

$$\begin{aligned} &H(t,\phi(t,\tau,\xi),u(t,\tau,\xi),-1,\lambda(t,\tau,\xi)) \\ &\qquad = \max_{z\varepsilon\Omega(t)} H(t,\phi(t,\tau,\xi),z,-1,\lambda(t,\tau,\xi)). \end{aligned} \quad (2.26)$$

Equations (2.23) with boundary conditions $\phi(\tau,\tau,\xi) = \xi$ and boundary conditions on λ to be determined below together with (2.26) characterize the optimal trajectory $\phi(\cdot,\tau,\xi)$ and optimal control $u(\cdot,\tau,\xi)$ in the following way. Associated with $\phi(\cdot,\tau,\xi)$ and $u(\cdot,\tau,\xi)$ there is a function $\lambda(\cdot,\tau,\xi)$ such that $\lambda(\cdot,\tau,\xi)$ and $\phi(\cdot,\tau,\xi)$ are solutions of (2.23) with appropriate boundary conditions and such that (2.26) holds for $\tau \leq t \leq t_1$. Equations (2.23) and their appropriate boundary conditions together with relation (2.26) constitute the Pontryagin Maximum Principle under the present hypotheses. They are a set of necessary conditions that an optimal pair must satisfy. A more precise and more general formulation will be given in Theorem 3.1 below.

Equations (2.23) and (2.26) involve the optimal control $u(\cdot,\tau,\xi)$. We rewrite these equations so as to involve the synthesis U. If we substitute (2.25) into (2.21) and (2.22) we see that $\phi(\cdot,\tau,\xi)$ and $\lambda(.,\tau,\xi)$ satisfy the equations

$$\begin{aligned} \frac{dx}{dt} &= H_p(t,x,U(t,x),-1,p) \\ \frac{dp}{dt} &= -H_x(t,x,U(t,x),-1,p). \end{aligned} \tag{2.27}$$

From (2.26) we see that $\phi(\cdot,\tau,\xi)$ and $\lambda(\cdot,\tau,\xi)$ also satisfy

$$H(t,x,U(t,x),-1,p) = \max_{z\in\Omega(t)} H(t,x,z,-1,p). \tag{2.28}$$

We next derive the "transversality conditions". These are boundary conditions that the value function W and its partial derivatives must satisfy. The transversality conditions are also the boundary conditions that the function $\lambda(\cdot,\tau,\xi)$ must satisfy.

Let (τ_1,ξ_1) be the terminal point of the optimal trajectory for the problem with initial point (τ,ξ). Then there is a point σ_1 in Σ such that $\tau_1 = T(\sigma_1)$ and $\xi_1 = X(\sigma_1)$. Let Γ^i be the curve on $\mathscr{T}$ passing through (τ_1,ξ_1) defined parametrically by the equations

$$t_1 = T(\sigma_1^1,\ldots\sigma_1^{i-1},\sigma^i,\sigma_1^{i+1},\ldots\sigma_1^q)$$

$$x_1 = X(\sigma_1^1,\ldots\sigma_1^{i-1},\sigma^i,\sigma_1^{i+1},\ldots\sigma_1^q)$$

where σ^i ranges over some open interval (a^i,b^i). The curve Γ^i is obtained by holding all components of the vector σ but the i-th component fixed and letting the i-th component vary over the interval (a^i,b^i). The curve Γ^i is sometimes called the i-th coordinate curve on $\mathcal{T}$.

We now assume that $\mathcal{T}$ is n-dimensional, that each point of $\mathcal{T}$ is the terminal point of a unique trajectory and that W can be extended to a $C^{(1)}$ function in a neighborhood of $\mathcal{T}$. It follows from (2.1) and the definition of W that for (t_1,x_1) in $\mathcal{T}$

$$W(t_1,x_1) = g(t_1,x_1). \tag{2.29}$$

It therefore follows that (2.29) holds along each Γ^i, $i = 1,\ldots,n$. We may therefore differentiate (2.29) along Γ^i with respect to σ^i and get that

$$W_t \frac{\partial T}{\partial \sigma^i} + \langle W_x, \frac{\partial X}{\partial \sigma^i}\rangle = g_t \frac{\partial T}{\partial \sigma^i} + \langle g_x, \frac{\partial X}{\partial \sigma^i}\rangle$$

holds along Γ^i. We rewrite this equation as

$$\langle (W_t - g_t,\ W_x - g_x),\ (\frac{\partial T}{\partial \sigma^i},\ \frac{\partial X}{\partial \sigma^i})\rangle = 0 \qquad i = 1,\ldots,n. \tag{2.30}$$

In particular, (2.30) holds at $\sigma^i = \sigma_1^i$. We may therefore take the argument of W_t, W_x, g_t and g_x to be (τ_1,ξ_1) and the argument of $\partial T/\partial\sigma^i$ and $\partial X/\partial\sigma^i$ to be σ_1. Using (2.14) we can rewrite (2.30) as

$$\langle (H - g_t,\ W_x - g_x),\ (\frac{\partial T}{\partial \sigma^i},\ \frac{\partial X}{\partial \sigma^i})\rangle = 0 \qquad i = 1,\ldots,n, \tag{2.31}$$

where H is evaluated at $(\tau_1,\xi_1,u(\tau_1,\tau,\xi),-1,-W_x(\tau,\xi))$.

If we use (2.18), equation (2.31) can be written

$$\left\langle (H-g_t,\ -\lambda-g_x),\ \left(\frac{\partial T}{\partial \sigma^i},\ \frac{\partial X}{\partial \sigma^i}\right)\right\rangle = 0 \qquad i = 1,\ldots,n, \tag{2.32}$$

where $\lambda = \lambda(\tau_1,\tau,\xi)$. Equations (2.32) when written out become

$$(H - \frac{\partial g}{\partial t})\frac{\partial T}{\partial \sigma^i} - \sum_{j=1}^{n} (\lambda^j + \frac{\partial g}{\partial x^j}) \frac{\partial X^j}{\partial \sigma^i} = 0 \qquad i = 1,\ldots,n.$$

Since $H = -f^0 + \langle \lambda, f\rangle$, we can rewrite this system as follows:

$$-(f^0 + \frac{\partial g}{\partial t})\frac{\partial T}{\partial \sigma^i} - \sum_{j=1}^{n} \frac{\partial g}{\partial x^j}\frac{\partial X^j}{\partial \sigma^i} = \sum_{j=1}^{n} (\frac{\partial X^j}{\partial \sigma^i} - \frac{\partial T}{\partial \sigma^i} f^j)\lambda^j. \tag{2.33}$$

Here the functions f^0, f^j, $\partial g/\partial t$, and $\partial g/\partial x^j$ are evaluated at the end point (τ_1,ξ_1) of the trajectory $\phi(\cdot,\tau,\xi)$. The partial derivatives of T and X are evaluated at σ_1, the point in $\sum$ corresponding to (τ_1,ξ_1). If $\mathscr{T}$ is a q-dimensional manifold, $0 \leq q \leq n$ then (2.33) consists of q equations instead of n equations. This does not follow from our arguments here, but will be shown to hold in the next chapter.

Since the unit tangent vector to Γ^i is the unit vector in the direction of $(\partial T/\partial\sigma^i,\ \partial X/\partial\sigma^i)$, equation (2.30) states that at (τ_1,ξ_1) the vector (W_t-g_t, W_x-g_x) is either zero or is orthogonal to Γ^i. We assume that orthogonality holds. From (2.31) and (2.32) we see that that this statement is equivalent to the statement that $(H-g_t, W_x-g_x)$ is orthogonal to Γ^i at (τ_1,ξ_1) and to the statement that $(H-g_t,\ -\lambda-g_x)$ is orthogonal to Γ^i. Since this is true for each coordinate curve Γ^i, $i = 1,\ldots,n$ and since the tangent vectors to the Γ^i at (τ_1,ξ_1) generate the tangent plane to $\mathscr{T}$ at (τ_1,ξ_1), the following statement is true. <u>The vector</u> $(W_t-g_t,\ W_x-g_x)$ <u>or equivalently the vector</u> $(H-g_t,\ W_x-g_x)$ <u>or equivalently the vector</u> $(H-g_t,\ -\lambda-g_x)$ <u>evaluated at the end point of an optimal trajectory is orthogonal to</u> $\mathscr{T}$ <u>at that point</u>. This is the geometric statement of the transversality condition. The analytic statement consists of equations (2.30) or (2.31) or (2.32).

Equations (2.29) and (2.30) are the boundary conditions for the partial differential equation (2.14).

Equations (2.32), or equivalently, (2.33) specify the values of $\lambda(\cdot,\tau,\xi)$ at $t = \tau_1$. They therefore furnish the heretofore missing boundary conditions for the systems (2.23) and (2.27). Note that the equations (2.32) and (2.33) are linear in λ. Thus λ satisfies a system of linear differential equations with linear boundary conditions. We point out that the system (2.23) with boundary conditions $\phi(\tau,\tau,\xi) = \xi$ and (2.32) constitute a two-point boundary value problem in that the values of ϕ are specified at the initial time and the values of λ are specified at the terminal time.

REMARK 2.1. If one solves the partial differential equation (2.14) subject to the boundary conditions (2.29) and (2.30), one finds that the characteristic equations for the problem are the equations (2.27). We leave the verification of this to the reader.

3. Statement of Maximum Principle

In this section we shall give a precise statement of the maximum principle for Problem 1 of Chapter II. The proof will be given in Chapter 6. We shall assume that $g \equiv 0$. Problems with $g \not\equiv 0$ can be reduced to problems with $g \equiv 0$, as indicated in Section 4 of Chapter III. The statement of the maximum principle when $g \not\equiv 0$ and in other problems will be taken up in the Exercises at the end of this section.

The following assumptions will be made about the data of the problem.

ASSUMPTION 3.1. (i) The region $\mathscr{R}$ has the form $\mathscr{I}_0 \times \mathscr{X}_0$, where $\mathscr{I}_0$ is an open interval in E^1 and $\mathscr{X}_0$ is an open interval in E^n. The region $\mathscr{U}$ is an open interval in E^m. (ii) The function $\hat{f} = (f^0,f) = (f^0,f^1,\ldots,f^n)$ is of class $C^{(1)}$ on $\mathscr{X}_0$ for each (t,z) in $\mathscr{I}_0 \times \mathscr{U}$, and is Borel measurable on $\mathscr{I}_0 \times \mathscr{U}$ for each x in $\mathscr{X}_0$. (iii) For each compact interval $\mathscr{X} \subset \mathscr{X}_0$ and admissible control u, there exists a function $\mu = \mu(\cdot,\mathscr{X},u)$ defined on $\mathscr{I}_u = [t_0,t_1]$, the interval of definition of u, such that $\mu \in L_1[\mathscr{I}_u]$ and such that for almost all t in $\mathscr{I}_u$ and all x in $\mathscr{X}$

$$|\hat{f}(t,x,u(t))| \leq \mu(t)$$
$$|\hat{f}_x(t,x,u(t))| \leq \mu(t).$$

ASSUMPTION 3.2. The mapping Ω is independent of x; i.e. $\Omega(t,x) = \Omega(t,x')$ for all x and x' in $\mathscr{X}_0$. Thus, $\Omega: t \to \Omega(t)$.

ASSUMPTION 3.3. The set $\mathscr{B}$ that define the end conditions of the problem is a $C^{(1)}$ manifold of dimension q, where $0 \leq q \leq 2n+1$.

Since the statement of the maximum principle will only involve the nature of $\mathscr{B}$ in a neighborhood of the end points of an optimal trajectory there is no loss of generality in assuming that $\mathscr{B}$ can be represented by a single coordinate patch. That is, we assume that $\mathscr{B}$ is the image of an open parallelepiped Σ in E^q under a mapping

$$t_0 = T_0(\sigma) \qquad x_0 = X_0(\sigma)$$
$$t_1 = T_1(\sigma) \qquad x_1 = X_1(\sigma),$$

where the functions T_i and X_i, $i = 0,1$ are $C^{(1)}$ on Σ and the Jacobian matrix

$$(T_{0\sigma} \quad X_{0\sigma} \quad T_{1\sigma} \quad X_{1\sigma})$$

has rank q everywhere on Σ.

As in Section 2 let H be a real valued function defined on

$E^1 \times E^n \times E^m \times E^1 \times E^n$ by the formula

$$\begin{aligned} H(t,x,z,p^0,p) &= p^0 f^0(t,x,z) + \langle p, f(t,x,z)\rangle \\ &= \sum_{i=0}^{n} p^i f^i(t,x,z). \end{aligned} \tag{3.1}$$

If we set $\hat{p} = (p^0, p)$ we can write (3.1) in the equivalent form

$$H(t,x,z,\hat{p}) = \langle \hat{p}, f(t,x,z)\rangle .$$

For the remainder of this chapter we shall be concerned primarily with optimal pairs. Therefore, we shall not use special designations such as asterisks or subscripts for optimal pairs unless we feel that confusion would otherwise result.

THEOREM 3.1 (Maximum Principle in Integrated Form). Let $g \equiv 0$ and let Assumptions (3.1) to (3.3) hold. Let (ϕ,u) be an optimal pair defined on an interval $[t_0,t_1]$. Then there exists a constant $\lambda^0 \le 0$ and an absolutely continuous vector function $\lambda = (\lambda^1,\ldots,\lambda^n)$ defined on $[t_0,t_1]$ such that the following hold: (i) The vector $\hat{\lambda}(t) = (\lambda^0, \lambda(t))$ is never zero on $[t_0,t_1]$. (ii) For a.e. t in $[t_0,t_1]$

$$\begin{aligned} \phi'(t) &= H_p(t,\phi(t),u(t),\hat{\lambda}(t)) \\ \lambda'(t) &= -H_x(t,\phi(t),u(t),\hat{\lambda}(t)). \end{aligned} \tag{3.2}$$

(iii) For any admissible control v defined on the interval $[t_0,t_1]$

$$\int_{t_0}^{t_1} H(t,\phi(t),u(t),\hat{\lambda}(t))\,dt \ge \int_{t_0}^{t_1} H(t,\phi(t),v(t),\hat{\lambda}(t))\,dt. \tag{3.3}$$

(iv) If the mapping $t \to \hat{f}(t,\phi(t),u(t))$ is continuous at $t = t_i$, $i = 0,1$, then the $(2n+2)$-vector

$$(H(\pi(t_0)),-\lambda(t_0),-H(\pi(t_1)),\lambda(t_1)) \tag{3.4}$$

is orthogonal to $\mathscr{B}$ at the point $(t_0,\phi(t_0),t_1,\phi(t_1))$, where

$$\pi(t_i) = (t_i,\phi(t_i),u(t_i),\hat{\lambda}(t_i)) \qquad i = 0,1.$$

Theorem 3.1 will be proved in Chapter 6.

COROLLARY 3.1 (Pointwise Maximum Principle). If for all t, $\Omega(t) = \mathscr{C}$, where $\mathscr{C}$ is a fixed set, and $\hat{f}$ is continuous on $\mathscr{R}\times\mathscr{U}$, then

$$H(t,\phi(t),u(t),\hat{\lambda}(t)) \geq H(t,\phi(t),z,\hat{\lambda}(t)) \tag{3.5}$$

for almost all t in $[t_0,t_1]$ and all z in $\mathscr{C}$.

Corollary 3.1 will also be proved in Chapter 6. Note that if (3.5) holds then (3.3) of the theorem certainly holds.

COROLLARY 3.2. Let $\Omega(t) = \mathscr{C}$ for all t, where $\mathscr{C}$ is a fixed set. Let $\hat{f}$ be continuous on $\mathscr{R}\times\mathscr{U}$ and continuously differentiable on $\mathscr{R}$. Let u be bounded on $[t_0,t_1]$. Then there exists an absolutely continuous function h defined on $[t_0,t_1]$ such that

$$h(t) = H(t,\phi(t),u(t),\hat{\lambda}(t)) \qquad \text{a.e.}$$

and

$$h'(t) = H_t(t,\phi(t),u(t),\hat{\lambda}(t)) \qquad \text{a.e.}$$

Moreover, if the problem is autonomous; i.e. $\hat{f}$ is independent of t and $\mathscr{B}$ is independent of t_0 and t_1, then

$$H(t,\phi(t),u(t),\hat{\lambda}(t)) = c \qquad \text{a.e.},$$

where c is a constant.

If u is piecewise continuous and $\mathscr{C}$ is closed then $t \to H(t,\phi(t),u(t),\hat{\lambda}(t))$ is absolutely continuous and the derivative of this mapping is $t \to H_t(t,\phi(t),u(t),\hat{\lambda}(t))$.

Corollary 3.2 will also be proved in Chapter 6.

<u>DEFINITION 3.1.</u> An admissible pair (ϕ,u) that satisfies the

conclusions of Theorem 3.1 is called an <u>extremal pair</u>. The function ϕ is called an <u>extremal trajectory</u> and the control u is called an <u>extremal control</u>.

Equations (3.2) state that the functions $\hat{\lambda}$ and ϕ are solutions of the system

$$\frac{dx}{dt} = H_p(t,x,u(t),p^0,p)$$
$$\frac{dp}{dt} = -H_x(t,x,u(t),p^0,p). \qquad (3.6)$$

If (ϕ,u) is an extremal pair then we shall call the vector $(t,\phi(t),u(t),\hat{\lambda}(t))$, where $\hat{\lambda}$ is as in Theorem 3.1, an <u>extremal element at</u> t. We shall denote extremal elements as follows:

$$\pi(t) = (t,\phi(t),u(t),\hat{\lambda}(t)). \qquad (3.7)$$

Thus, equations (3.2) can be written as

$$\phi'(t) = H_p(\pi(t)) \qquad \lambda'(t) = -H_x(\pi(t)).$$

We have already written the transversality conditions using extremal elements.

Since $H_p = f$, the first equation in (3.2) reduces to $\phi'(t) = f(t,\phi(t),u(t))$ a.e. This merely states that ϕ is a trajectory corresponding to the control u. The second equation in (3.2) can be written

$$\lambda'(t) = -[\hat{f}_x(t,\phi(t),u(t))]\hat{\lambda}(t), \qquad (3.8)$$

where $\hat{f}_x$ is the matrix of n rows and $(n+1)$ columns whose entry is the i-th row, $i = 1,\dots,n$ and j-th column, $j = 0,1,\dots,n$ is $\partial f^j/\partial x^i$ evaluated along the trajectory. Thus λ satisfies a system of linear differential equations, which when written out in component form becomes

$$\frac{dp^i}{dt} = -\lambda^0 \frac{\partial f^0}{\partial x^i}(t,\phi(t),u(t)) - \sum_{j=1}^{n} \frac{\partial f^j}{\partial x^i}(t,\phi(t),u(t))p^j,$$

$i = 1,\ldots,n.$

The statement that (3.4) is orthogonal to $\mathscr{B}$ at the end point of the trajectory ϕ is usually referred to as the "transversality condition". The analytic formulation of the transversality condition is taken up in Exercise 3.4 below.

The maximum principle is a necessary condition for optimality. It is not a sufficient condition, as the following example shows.

EXAMPLE 3.1. Let

$$f^0(t,x,z) = az^2 - 4bxz^3 + 2btz^4 \qquad a > 0, \quad b > 0.$$

Let $\Omega(t,x) = E^1$, let the state equation be $dx/dt = u(t)$, and let $\mathscr{B}$ consist of the single point $(t_0,x_0,t_1,x_1) = (0,0,1,0)$. The problem is to minimize

$$J(\phi,u) = \int_0^1 f^0(t,\phi(t),u(t))dt.$$

The function H in this problem is given by

$$H(t,x,z,\hat{p}) = p^0(az^2 - 4bxz^3 + 2btz^4) + pz.$$

Equations (3.6) become

$$\begin{aligned} \frac{dx}{dt} &= H_p = u(t) \\ \frac{dp}{dt} &= -H_x = 4p^0 b(u(t))^3. \end{aligned} \tag{3.9}$$

Since $\mathscr{B}$ is zero dimensional the transversality condition gives no information about λ^0, $\lambda(0)$, or $\lambda(1)$.

Let ϕ, u and λ be identically zero on $[0,1]$ and let $\lambda^0 = -1$. Then (3.9) is satisfied, and

$$H(t,\phi(t),z,\hat{\lambda}(t)) = -z^2(a+2btz^2).$$

Since $a > 0$, $b > 0$ and $t \geq 0$, the right hand side of the last equation is non-positive for all t in $[0,1]$. On the other hand $H(t,\phi(t),u(t),\hat{\lambda}(t)) = 0$, so that (3.5) holds. Hence (ϕ,u) is an extremal pair. Note that in this example strict inequality holds in (3.5) whenever $z \neq u(t)$.

To see that (ϕ,u) is not optimal we first note that $J(\phi,u) = 0$. Let k be a fixed number satisfying $0 < k < 1$ and let h be a small positive number. Define a control u_h as follows:

$$u_h(t) = \begin{cases} k/h & 0 \leq t < h \\ -k/(1-h) & h \leq t < 1. \end{cases}$$

Let ϕ_h be the corresponding trajectory. It is then a straightforward calculation, which we leave to the reader, to show that $\lim_{h\to 0} J(\phi_h,u_h) = -\infty$. Therefore (ϕ,u) cannot be optimal.

EXERCISE 3.1. Show that $J(\phi_h,u_h) \to -\infty$ as $h \to 0$.

In the exercises which follow and in the remainder of these notes we shall denote the end point of a trajectory ϕ by $e(\phi)$. Thus

$$e(\phi) = (t_0,\phi(t_0),t_1,\phi(t_1)). \tag{3.10}$$

EXERCISE 3.2. Let $\mathcal{T}_0$ be a point and let $\mathcal{T}_1$ be an n-dimensional manifold of class $C^{(1)}$. Show that if an extremal trajectory is not tangent to $\mathcal{T}_1$ then $\lambda^0 \neq 0$ and $\lambda(t_1)$ is unique.

EXERCISE 3.3. Show that if $\lambda^0 \neq 0$, then we may assume that $\lambda^0 = -1$.

EXERCISE 3.4. Show that the statement, "The (2n+2)-vector (3.4) is orthogonal to $\mathcal{B}$ at the end point $e(\phi)$" is equivalent to the following statement. The vector $(\lambda^0, \lambda(t_0), \lambda(t_1))$ satisfies the following system of equations:

$$\lambda^0 \left(f^0(P_0)\frac{\partial T_0}{\partial \sigma^i} - f^0(P_1)\frac{\partial T_1}{\partial \sigma^i}\right) + \sum_{j=1}^{n} \lambda^j(t_0)\left(f^j(P_0)\frac{\partial T_0}{\partial \sigma^i} - \frac{\partial X_0^j}{\partial \sigma^i}\right)$$

$$- \sum_{j=1}^{n} \lambda^j(t_1)\left(f^j(P_1)\frac{\partial T_1}{\partial \sigma^i} - \frac{\partial X_1^j}{\partial \sigma^i}\right) = 0 \qquad i = 1,\ldots,q,$$

where

$$P_i = (t_i, \phi(t_i), u(t_i)) \qquad i = 0,1$$

and where the partial derivatives of T_i and X_i, $i = 0,1$ are evaluated at the value of σ that maps into the end point $e(\phi)$.

EXERCISE 3.5. If we assume that $g \not\equiv 0$ and that g is a $C^{(1)}$ function in a neighborhood of $\mathscr{B}$ in E^{2n+2}, then the statement of the maximum principle is as in Theorem 3.1, except for the transversality condition, which now reads as follows. The vector

$$(H(\pi(t_0)) - \lambda^0 g_{t_0},\ -\lambda(t_0) - \lambda^0 g_{x_0},\ -H(\pi(t_0)) - \lambda^0 g_{t_1},\ \lambda(t_1) - \lambda^0 g_{x_1}),$$

where the partial derivative of g are evaluated at $e(\phi)$, is orthogonal to $\mathscr{B}$ at $e(\phi)$.

EXERCISE 3.6 (Isoperimetric Constraints). Suppose that we impose the following additional constraints on admissible pairs (ϕ,u):

$$\int_{t_0}^{t_1} h^i(t,\phi(t),u(t))\,dt = c^i \qquad i = 1,\ldots,r,$$

where $h = (h^1,\ldots,h^r)$ has the same properties as f in Theorem 3.1 and $c = (c^1,\ldots,c^r)$ is a given vector. Let

$$\tilde{H}(t,x,z,\hat{p},\rho) = H(t,x,z,\hat{p}) + \langle \rho, h(t,x,z)\rangle .$$

Show that if $g \equiv 0$ and if (ϕ,u) is an optimal pair with interval of definition $[t_0,t_1]$ for the isoperimetric problem just formulated, then the following maximum principle holds.

There exist a constant $\lambda^0 \le 0$, an absolutely continuous vector function $\lambda = (\lambda^1,\ldots,\lambda^n)$, and a constant vector $\gamma = (\gamma^1,\ldots,\gamma^r)$ such that the following hold: (i) The vector $(\hat{\lambda}(t),\gamma) = (\lambda^0,\lambda(t),\gamma)$ is never zero on $[t_0,t_1]$. (ii) For a.e. t in $[t_0,t_1]$

$$\phi'(t) = \tilde{H}_p(t,\phi(t),u(t),\hat{\lambda}(t),\gamma)$$

$$\lambda'(t) = -\tilde{H}_x(t,\phi(t),u(t),\hat{\lambda}(t),\gamma).$$

(iii) For any admissible control v defined on the interval $[t_0,t_1]$

$$\int_{t_0}^{t_1}\tilde{H}(t,\phi(t),u(t),\hat{\lambda}(t),\gamma)\,dt \ge \int_{t_0}^{t_1}\tilde{H}(t,\phi(t),v(t),\hat{\lambda}(t),\gamma)\,dt.$$

(iv) If the mapping $t \to (\hat{f}(t,\phi(t),u(t)),\ h(t,\phi(t),u(t)))$ is continuous at $t = t_i$, $i = 0,1$, then the $(2n+2)$ vector

$$(\tilde{H}(\tilde{\pi}(t_0)),\ -\hat{\lambda}(t_0),\ -\tilde{H}(\tilde{\pi}(t_1)),\ \hat{\lambda}(t_1))$$

is orthogonal to $\mathscr{B}$ at the end point $e(\phi)$, where

$$\tilde{\pi}(t_i) = (t_i,\phi(t_i),u(t_i),\hat{\lambda}(t_i),\gamma) \qquad i = 0,1.$$

EXERCISE 3.7 (Parameter Optimization). Consider the problem of control and parameter optimization described in III.5. Let $\hat{f} = (f^0,f)$ be defined on $\mathscr{I}_0 \times \mathscr{X}_0 \times \mathscr{W}_0 \times \mathscr{U}$, where $\mathscr{I}_0$, $\mathscr{X}_0$ and $\mathscr{U}$ are as in Assumption 3.1 and $\mathscr{W}_0$ is an open interval in E^k. For each (t,z) in $\mathscr{I}_0 \times \mathscr{U}$ let $\hat{f}$ be of class $C^{(1)}$ on $\mathscr{X}_0 \times \mathscr{W}_0$ and for each (x,w) in $\mathscr{X}_0 \times \mathscr{W}_0$ let $\hat{f}$ be Borel measurable on $\mathscr{I}_0 \times \mathscr{U}$. Let μ be as in Assumption 3.1 and let

$$|\hat{f}(t,x,w,u(t))| \le \mu(t) \qquad |\hat{f}_x(t,x,w,u(t))| \le \mu(t)$$

$$|\hat{f}_w(t,x,w,u(t))| \le \mu(t).$$

Let W be an open set in $\mathscr{W}_0$ and let the parameter values w be constrained to lie in W. Let (ϕ_0, u_0, w_0) be a solution to the problem and let (ϕ_0, u_0) be defined in an interval $[t_0, t_1]$. Let

$$\tilde{H}(t,x,w,z,\hat{p}) = \langle \hat{p}, \hat{f}(t,x,w,z) \rangle.$$

Show that there exists a constant $\lambda^0 \leq 0$ and an absolutely continuous vector $\lambda = (\lambda^1, \ldots, \lambda^n)$ such that the following hold: (i) The vector $\hat{\lambda}(t) = (\lambda^0, \lambda(t))$ is never zero on $[t_0, t_1]$. (ii) For a.e. t in $[t_0, t_1]$

$$\phi_0'(t) = \tilde{H}_p(t, \phi_0(t), w_0, u_0(t), \hat{\lambda}(t))$$

$$\lambda'(t) = -\tilde{H}_x(t, \phi_0(t), w_0, u_0(t), \hat{\lambda}(t)).$$

(iii) For any admissible control v defined on the interval $[t_0, t_1]$,

$$\int_{t_0}^{t} \tilde{H}(t, \phi_0(t), w_0, u_0(t), \hat{\lambda}(t))\,dt \geq \int_{t_0}^{t} \tilde{H}(t, \phi_0(t), w_0, v(t), \hat{\lambda}(t))\,dt.$$

(iv) If the mapping $t \to \hat{f}(t, \phi_0(t), w_0, u_0(t))$ is continuous at $t = t_i$, $i = 0,1$, then the $(2n+2)$ vector

$$(\tilde{H}(\tilde{\pi}(t_0)),\ -\lambda(t_0),\ -\tilde{H}(\tilde{\pi}(t_1)),\ \lambda(t_1))$$

is orthogonal to $\mathscr{B}$ at $e(\phi_0)$, where

$$\tilde{\pi}(t_i) = (t_i, \phi_0(t_i), w_0, u_0(t_i), \hat{\lambda}(t_i)) \qquad i = 0,1.$$

Moreover,

$$\int_{t_0}^{t_1} \langle \hat{\lambda}(t), \frac{\partial \hat{f}}{\partial w^j}(t, \phi_0(t), w_0, u_0(t)) \rangle\,dt = 0 \qquad j = 1, \ldots, k.$$

EXERCISE 3.8. Let (ϕ, u) be an extremal pair. Show that if u is piecewise continuous, then the mapping H^* defined by $H^*(t) = H(\pi(t))$ is absolutely continuous and

$$\frac{dH^*}{dt} = H_t,$$

where both sides of the equation are evaluated at $\pi(t)$.

4. An Example

In this section we shall illustrate how the maximum principle and the existence theorems are used to find an optimal control. The example is also useful in pointing out the difficulties to be encountered in dealing with more complicated systems. The reader will note that all of the information in the maximum principle is used to solve the problem.

We consider the following resource allocation problem. Let $x_P(t)$ denote the rate of production at time t of a commodity such as steel. Let $x_I(t)$ denote the rate of investment of this commodity. It is assumed that the investment is used to increase productive capacity. In the case of steel investment can be thought of as using steel to produce new steel mills, mining equipment, and transport facilities. All of these uses are lumped together under the heading of investment. Let $x_C(t)$ denote the rate of consumption of the commodity at time t. In the case of steel consumption can be thought of as the production of consumer goods such as automobiles. We assume that

$$x_P(t) = x_I(t) + x_C(t) \qquad x_I(t) \geq 0 \quad x_C(t) \geq 0 \tag{4.1}$$

and

$$x_P(t) = C + \int_0^t x_I(s)ds, \tag{4.2}$$

where C is the initial rate of production. Equation (4.1) reflects the assumption that all of the commodity produced in a given period must either be allocated to investment or consumption. Equation (4.2) reflects the assumption that the commodity allocated to investment is used to increase productive capacity. Let $T > 0$ be given. The total consumption in time T is then

$$P = \int_0^T x_C(s)ds.$$

At each instant of time t in $[0,T]$ the production planner is to choose $x_I(t)$ and $x_C(t)$ so that P is maximized.

We now formulate this problem as a control problem. Let $x(t)$ denote the rate of production at time t. Since the commodity is not destroyed, $x(t) \geq 0$. Let $u(t)$ denote the fraction of the commodity produced at time t that is allocated to investment. Thus $0 \leq u(t) \leq 1$ and $1-u(t)$ represents the fraction allocated to consumption. Hence

$$x_I(t) = u(t)x(t)$$
$$x_C(t) = (1-u(t))x(t).$$

The problem posed in the preceding paragraph is therefore equivalent to the following control problem. Minimize

$$J(\phi,u) = -\int_0^T (1-u(s))\phi(s)ds \tag{4.3}$$

subject to

$$\frac{dx}{dt} = u(t)x \qquad x_0 = C \tag{4.4}$$

$$0 \leq u(t) \leq 1 \qquad x \geq 0 , \tag{4.5}$$

where $C > 0$, T is fixed, and the terminal state x_1 is non-negative, but otherwise arbitrary.

In the control formulation $\Omega(t) = [0,1]$ for all t and

$$\mathscr{B} = \{(t_0,x_0,t_1,x_1): t_0 = 0,\ x_0 = C,\ t_1 = T,\ x_1 \geq 0\}. \tag{4.6}$$

Hence $\mathscr{B}$ and Ω satisfy the hypotheses of Theorem III 5.1. Also

$f^0(t,x,z) = -(1-z)x$ and $f(t,x,z) = zx$, so that f^0 and f satisfy the hypotheses of Theorem III 5.1. It follows from the state equations and the initial condition (4.4) that for any control u satisfying (4.5), the corresponding trajectory will satisfy

$$C \leq \phi(t) \leq Ce^t. \tag{4.7}$$

Hence the constraint $x \geq 0$ is always satisfied and so can be omitted from further consideration. Since $t_1 = T$, (4.7) shows that all trajectories lie in a compact set. The sets $\mathscr{Q}^+(t,x)$ in this problem are given by

$$\begin{aligned}\mathscr{Q}^+(t,x) &= \{(y^0,y): y^0 \geq (z-1)x,\ y = zx,\ 0 \leq z \leq 1\}\\ &= \{(y^0,y): y^0 \geq y-x,\ 0 \leq y \leq x\}.\end{aligned}$$

For each (t,x) the set $\mathscr{Q}^+(t,x)$ is closed and convex. Thus all of the hypotheses of Theorem III 5.1 are satisfied and an optimal pair exists.

Let (ϕ,u) now denote an optimal pair. We shall use Theorem 3.1 and Corollary 3.1 to determine (ϕ,u) and in the process we shall show that (ϕ,u) is unique.

In the present example the function H defined in (3.1) becomes

$$H(t,x,z,\hat{p}) = -p^0(1-z)x + pzx.$$

Thus

$$H_p = zx \qquad -H_x = p^0(1-z) - pz.$$

The function $\hat{\lambda}$ of Theorem 3.1 and the optimal pair (ϕ,u) satisfy equations (3.2), which in the present example become

$$\begin{aligned}\phi'(t) &= u(t)\phi(t)\\ \lambda'(t) &= \lambda^0(1-u(t)) - \lambda(t)u(t).\end{aligned} \tag{4.8}$$

Moreover, since $\hat{f}$ is continuous and $\Omega(t) = [0,1]$ for all t, Corollary 3.1 is applicable and (3.5) holds. In the present example (3.5) becomes

$$(\lambda^0+\lambda(t))\phi(t)u(t) \geq (\lambda^0+\lambda(t))\phi(t)z \tag{4.9}$$

for all z in $[0,1]$ and almost all t in $[0,T]$.

Since $\mathscr{B}$ is given by (4.6), any tangent vector to $\mathscr{B}$ at any point of $\mathscr{B}$ is a scalar multiple of $(0,0,0,1)$. Therefore, the transversality condition, which states that (3.4) is orthogonal to $\mathscr{B}$ at $e(\phi)$, the end point of ϕ, reduces to the condition

$$\lambda(T) = 0 \tag{4.10}$$

in the present case. Since $\hat{\lambda}(t) = (\lambda^0,\lambda(t))$ is never zero, it follows that $\lambda^0 \neq 0$. Since $\lambda^0 \leq 0$, it follows that $\lambda^0 < 0$. Therefore we may divide both sides of the second equation in (4.8) and both sides of the inequality (4.9) by $(-\lambda^0) > 0$. If we set $\tilde{\lambda} = \lambda/(-\lambda_0)$ and then relabel $\tilde{\lambda}$ as λ, we get that ϕ, u and $\hat{\lambda}$ satisfy

$$\begin{aligned} \phi'(t) &= u(t)\phi(t) \\ \lambda'(t) &= -(1-u(t)) - \lambda(t)u(t) \end{aligned} \tag{4.11}$$

and the inequality

$$(-1+\lambda(t))\phi(t)u(t) \geq (-1+\lambda(t))\phi(t)z \tag{4.12}$$

for all $0 \leq z \leq 1$ and a.e. t in $[0,T]$. Comparing (4.8) and (4.9) with (4.11) and (4.12), we see that we have shown that we can take $\lambda^0 = -1$ in (4.8) and (4.9). In general, whenever it can be shown that $\lambda^0 \neq 0$, it may be assumed that $\lambda^0 = -1$. (See Exercise 3.3.)

It follows from (4.11) that ϕ and λ are solutions of the system of equations

$$\frac{dx}{dt} = u(t)x$$
$$\frac{dp}{dt} = -(1-u(t)) - u(t)p \,. \qquad (4.13)$$

The boundary conditions are given by (4.10) and $\phi(0) = C$. Thus we know the initial value ϕ and the terminal value of λ: Not knowing the values of both ϕ and λ at the same point makes matters difficult, as we shall see.

Since $C > 0$, it follows from (4.7) that $\phi(t) > 0$ for all t. Thus, although we do not know $\phi(T)$, the value of ϕ at T, we do know that $\phi(T) > 0$.

From (4.12) we see that for a.e. t, $z = u(t)$ maximizes the expression

$$(-1+\lambda(t))\phi(t)z \qquad (4.14)$$

subject to $0 \leq z \leq 1$. Therefore the sign of the coefficient of z in (4.14) is most important. If this coefficient is > 0, then $z = 1$ maximizes; if this coefficient is < 0 then $z = 0$ maximizes. Since $\phi(t) > 0$ for all t, the determining factor is $(-1+\lambda(t))$.

Let $\phi(T) = \xi$. Since $\xi > 0$ and since $\lambda(T) = 0$ the coefficient of z at $t = T$ in (4.14) is negative. Moreover, by the continuity of ϕ and λ there exists a maximal interval of the form $(T-\delta, T]$ such that the coefficient of z on this interval is negative. Hence $z = 0$ maximizes (4.14) for all t in this interval. Hence $u(t) = 0$ for all $T-\delta < t \leq T$.

At the initial point $t = 0$ no such analysis can be made since $\lambda(0)$ and therefore the sign of the coefficient of z are unknown. This suggests that we attempt to work backward from an arbitrary point $\xi > 0$ on the terminal manifold $t = T$, as was done in the preceding paragraph.

We have already noted that there exists a maximal interval $(T-\delta,T]$ on which $u(t) = 0$. This is the maximal interval with right hand end point T on which $-1+\lambda(t) < 0$. We now determine δ. From the first equation in (4.13) with $u(t) = 0$ and from the assumption $\phi(T) = \xi$ we see that $\phi(t) = \xi$ on this interval. From the second equation in (4.13) with $u(t) = 0$ and from (4.10) we see that $\lambda(t) = (T-t)$ on this interval. Therefore, if $T > 1$ it follows that $-1+\lambda(t) < 0$ for $T-1 < t \leq T$ and that $-1+\lambda(T-1) = 0$. Thus, $\delta = 1$. In summary, we have established that if $T > 1$ then on the interval $(T-1,T]$ the following hold

$$u(t) = 0 \qquad \phi(t) = \xi \qquad \lambda(t) = T-t.$$

If we define $u(T-1) = 0$, then the preceding hold on the interval $[T-1,T]$.

If $T \leq 1$, then $u(t) = 0$, $\phi(t) = \xi$, where $\xi = C$, is the optimal pair. It is clear from the construction that the optimal pair is unique. Thus if the "planning horizon" (the value of T) is too short, the optimal policy is to consume.

We now return to the case in which $T > 1$ and determine (ϕ,u) to the left of $T-1$. The reader is advised to graph the functions ϕ, u and λ on the interval $[T-1,T]$ and to complete the graph as the functions are being determined to the left of $T-1$. We rewrite the second equation in (4.13) as

$$\frac{dp}{dt} = -1 + (1-p)u(t). \qquad (4.15)$$

We consider this differential equation for λ on the interval $[0,T-1]$. Since λ is continuous on $[0,T]$ we have the terminal condition $\lambda(T-1) = 1$ from the discussion in the next to the last paragraph.

Since $\lambda(T-1) = 1$ and $0 \leq u(t) \leq 1$, it follows from (4.15) and the continuity of λ that on an interval $(T-1-\delta_1,T-1]$ we have

$\lambda'(t) < 0$. Hence λ is increasing as we go <u>backwards</u> in time on this interval. Hence $\lambda(t) > 1$ for t in some interval $(T-1-\delta_2, T-1)$. Since $\phi(t) > 0$ for all t it follows from (4.14) that $z = 1$ maximizes (4.14) on $(T-1-\delta_2, T-1)$. Hence $u(t) = 1$ on this interval and equations (4.13) become

$$\frac{dx}{dt} = x \qquad \phi(T-1) = \xi$$

$$\frac{dp}{dt} = -p \qquad \lambda(T-1) = 1.$$

Hence

$$\lambda(t) = \exp(T-1-t) \qquad \phi(t) = \xi \exp(t-T+1) \tag{4.16}$$

on the interval $(T-1-\delta_2, T-1)$. But then $\lambda(t) > 1$ for all $0 \leq t < T-1$ so that $(T-1-\delta_2, T-1) \equiv (0, T-1)$. Therefore, on $[0, T-1)$, λ and ϕ are given by (4.16) and $u(t) = 1$.

At $t = 0$ we have $\phi(0) = \xi \exp(-T+1)$. We require that $\phi(0) = C$. Hence $\xi = C \exp(T-1)$, and so

$$\phi(t) = Ce^t.$$

on the interval $0 \leq t \leq T-1$.

The pair (ϕ, u) that we have determined is an extremal pair. From the procedure used to determine (ϕ, u) it is clear that (ϕ, u) is unique. Therefore, since we know that an optimal pair exists and must be extremal, it follows that (ϕ, u) is indeed optimal. Moreover, it is unique. We point out that although the existence theorem guaranteed the existence of a measurable optimal control, the application of the maximum principle yielded a control that was piecewise continuous.

The procedure used in the preceding example is one that can often be applied. It illustrates the difficulties arising because the value of ϕ is specified at the initial point and the value of λ

is specified at the terminal point. In small scale problems that can be attacked analytically one can proceed backwards from an arbitrary terminal point and adjust the constants of integration so as to obtain the desired initial point. In large scale problems, or problems that must be solved with a computer, this is not so easy. We shall not pursue these matters here.

5. Relationship with the Calculus of Variations

In this section we investigate the relationship between the maximum principle and the first order necessary conditions in the calculus of variations. We show, in detail, how the classical first order necessary conditions for the simple problem in the calculus of variations can be obtained from the maximum principle. In Exercise 5.1 we ask the reader to derive the first order necessary conditions for the problem of Bolza from the maximum principle. In Exercise 5.2 we ask the reader to derive the maximum principle for a certain class of problems from the results stated in Exercise 5.1.

In II.6 we showed that the simple problem in the calculus of variations can be formulated as a control problem as follows. Minimize

$$J(\phi) = g(e(\phi)) + \int_{t_0}^{t_1} f^0(t,\phi(t),u(t))dt$$

subject to

$$\frac{dx}{dt} = u(t) \qquad\qquad e(\phi) \in \mathscr{B}$$

$$\Omega(t,x) = \mathscr{U},$$

where $\mathscr{B}$ is a given set in E^{2n+2}, $e(\phi)$ denotes the end point $(t_0,\phi(t_0),t_1,\phi(t_1))$ of ϕ and $\mathscr{U}$ is an open set in E^n. We shall assume that f^0 is of class $C^{(1)}$ on $\mathscr{G} = \mathscr{R} \times \mathscr{U}$, that $\mathscr{B}$ is a q dimensional manifold of class $C^{(1)}$ in E^{2n+2}, $0 \leq q \leq 2n+1$, and that g is identically zero. We shall show that Theorem 3.1 reduces to the usual first order necessary conditions that a minimizing curve must satisfy.

The function H in the present case is given by the formula

$$H(t,x,z,\hat{p}) = p^0 f^0(t,x,z) + \langle p,z \rangle.$$

Let ϕ be a solution of the variational problem and let ϕ be defined on an interval $[t_0,t_1]$. Then $(\phi,u) = (\phi, \phi')$ is a solution of the corresponding control problem. The pair (ϕ,u) therefore satisfies the conditions of Theorem 3.1. Moreover, since the constraint set is fixed, Corollary 3.1 holds. Thus, there exists a scalar $\lambda^0 \le 0$ and an absolutely continuous vector function $\lambda = (\lambda^1,\dots,\lambda^n)$ defined on $[t_0,t_1]$ such that $(\lambda^0,\lambda(t)) \ne 0$ for all t in $[t_0,t_1]$ and such that for a.e. t in $[t_0,t_1]$

$$\phi'(t) = H_p(\pi(t)) = u(t) \tag{5.1}$$

$$\lambda'(t) = -H_x(\pi(t)) = -\lambda^0 f_x^0(t,\phi(t),u(t)) \tag{5.2}$$

and

$$H(t,\phi(t),z,\hat{\lambda}(t)) = \lambda^0 f^0(t,\phi(t),z) + \langle \lambda(t),z\rangle \tag{5.3}$$

is maximized over $\mathscr{U}$ at $z = u(t)$. Moreover, the vector

$$(H(\pi(t_0)), -\lambda(t_0), -H(\pi(t_1)), \lambda(t_1)) \tag{5.4}$$

is orthogonal to $\mathscr{B}$ at $e(\phi)$. (Recall that $g \equiv 0$.)

We assert that $\lambda^0 \ne 0$. For if $\lambda^0 = 0$, then from (5.3) we get that for a.e. t in $[t_0,t_1]$

$$\langle \lambda(t),u(t)\rangle \ge \langle \lambda(t),z\rangle$$

for all z in $\mathscr{U}$. This says that for a.e. t the linear function $z \to \langle \lambda(t),z\rangle$ is maximized at some point $u(t)$ of the open set $\mathscr{U}$. This can only happen if $\lambda(t) = 0$. But then $(\lambda^0,\lambda(t)) = 0$, which cannot be.

Since $\lambda^0 \neq 0$ we may henceforth take $\lambda^0 = -1$ in (5.1) to (5.4). (See Exercise 3.3.) From (5.2) we get

$$\lambda(t) = \int_{t_0}^{t} f_x^0(s,\phi(s),u(s))ds + C \tag{5.5}$$

for some constant vector C. From (5.3) we get that the mapping $z \to H(t,\phi(t),z,-1,\lambda(t))$ is maximized at $z = u(t)$. Since $\mathscr{U}$ is open, $u(t)$ is an interior point of $\mathscr{U}$. Since the mapping $z \to H(t,\phi(t),z,-1,\lambda(t))$ is differentiable, the derivative is zero at $z = u(t)$. Thus

$$H_z(t,\phi(t),u(t),-1,\lambda(t)) = 0$$

and therefore

$$\lambda(t) = f_z^0(t,\phi(t),u(t)) \tag{5.6}$$

for a.e. t in $[t_0,t_1]$.

From (5.1), (5.5) and (5.6) we get that

$$f_z^0(t,\phi(t),\phi'(t)) = \int_{t_0}^{t} f_x^0(s,\phi(s),\phi'(s))ds + C \quad \text{a.e.} \tag{5.7}$$

Equation (5.7) is sometimes called the Euler equation in integrated form or the du-Bois Reymond equation. In the elementary theory it is assumed that ϕ' is piecewise continuous. Equation (5.7) then holds everywhere and the function f_z^{0*} defined by the formula

$$f_z^{0*}(t) = f_z^0(t,\phi(t),\phi'(t)) \tag{5.8}$$

is continuous even at corners of ϕ. By a corner of ϕ we mean a point at which ϕ' has a jump discontinuity. This result is known as the Weierstrass-Erdmann corner condition.

From (5.7) we get that

$$\frac{d}{dt}(f_z^{0*}) = f_x^{0*} \quad \text{a.e.,} \tag{5.9}$$

where f_z^{0*} is defined by (5.8) and f_x^{0*} denotes the mapping

$t \to f_x^{0*}(t,\phi(t),\phi'(t))$. Equation (5.9) is the Euler equation. If we assume that ϕ' is piecewise continuous then (5.9) holds between corners of ϕ.

We next discuss the transversality condition. To simplify matters we assume that u is continuous at $t = t_i$, $i = 0,1$ and that (5.1) holds at these points. If in (5.4) we take $\lambda^0 = -1$ and use (5.1) and (5.6), then (5.4) becomes

$$(-f^{0*}(t_0)+\langle f_z^{0*}(t_0),\phi'(t_0)\rangle,\ -f_z^{0*}(t_0),\ f^{0*}(t_1)-\langle f_z^{0*}(t_1),\phi'(t_1)\rangle,\ f_z^{0*}(t_1)), \tag{5.10}$$

where $f^{0*}(t) = f^0(t,\phi(t),\phi'(t))$ and f_z^{0*} is given by (5.8). The transversality condition now states that (5.10) is orthogonal to $\mathscr{B}$ at $e(\phi)$.

If we take $\lambda^0 = -1$ and use (5.1) and (5.6), then the statement that $z = u(t)$ maximizes (5.3) over $\mathscr{U}$ becomes the following statement. For almost all t in $[t_0,t_1]$ and all z in $\mathscr{U}$

$$\begin{aligned} -f^0(t,\phi(t),\phi'(t)) + \langle f_z^0(t,\phi(t),\phi'(t)),\phi'(t)\rangle &\geq \\ -f^0(t,\phi(t),z) + \langle f_z^0(t,\phi(t),\phi'(t)),z\rangle&. \end{aligned} \tag{5.11}$$

If we introduce the function $\mathscr{E}$ defined by

$$\mathscr{E}(t,x,z,y) = f^0(t,x,z) - f^0(t,x,y) - \langle f^0(t,x,y),z-y\rangle$$

then (5.11) is equivalent to the statement that

$$\mathscr{E}(t,\phi(t),z,\phi'(t)) \geq 0 \tag{5.12}$$

for almost all t and all z in $\mathscr{U}$. The inequality (5.12) is known as the Weierstrass condition. Note that the left hand side of (5.12) consists of the first order terms in the Taylor expansion of the mapping $z \to f^0(t,\phi(t),z)$ about the point $z = \phi'(t)$.

If we assume that ϕ' is piecewise continuous, then (5.12) will certainly hold between corners. If $t = \tau$ is a corner, then by letting $t \to \tau+0$ and $t \to \tau-0$ we get that (5.12) holds for the one-sided limits obtained by letting $t \to \tau \pm 0$.

We now suppose that for fixed (t,x) in $\mathscr{R}$ the function f^0 is $C^{(2)}$ on $\mathscr{U}$. Hence for each t the mapping $z \to H(t,\phi(t),z,\hat{\lambda}(t))$ is $C^{(2)}$ on $\mathscr{U}$. For a.e. t this function is maximized at $z = u(t)$, which since $\mathscr{U}$ is open must be an interior point of $\mathscr{U}$. Therefore the quadratic form determined by the matrix $H_{zz}(t,\phi(t),u(t),\hat{\lambda}(t))$ is negative semi-definite for almost all t. But

$$H_{zz}(t,\phi(t),u(t),\hat{\lambda}(t)) = -f^0_{zz}(t,\phi(t),\phi'(t))$$

by virtue of (5.1), and the fact that we can take $\lambda^0 = -1$. Hence the quadratic form determined by $f^0_{zz}(t,\phi(t),\phi'(t))$ is positive semi-definite for almost all t. Thus for all $\eta = (\eta^1,\ldots,\eta^n) \neq 0$ and a.e. t

$$\langle \eta, f^0_{zz}(t,\phi(t),\phi'(t))\eta \rangle \geq 0. \tag{5.13}$$

This is known as Legendres condition.

We continue to assume that for fixed (t,x) in $\mathscr{R}$ the function f^0 is $C^{(2)}$ on $\mathscr{U}$. Let ϕ' be piecewise continuous. We say that ϕ is non-singular if for all t in $[t_0,t_1]$ such that t is not a corner, the matrix $f^0_{zz}(t,\phi(t),\phi'(t))$ is non-singular. We shall show that if ϕ is non-singular then ϕ is $C^{(2)}$ between corners. This result is known as Hilbert's differentiability theorem.

Let τ be a point in $[t_0,t_1]$ that is not a corner. Consider the following equation for the n-vector w

$$f^0_z(t,\phi(t),w) - \int_{t_0}^{t} f^0_x(s,\phi(s),\phi'(s))ds - C = 0 \tag{5.14}$$

where C is as in (5.7). From (5.7) we see that at $t = \tau$, $w = \phi'(\tau)$

is a solution. Moreover, by hypothesis the matrix $f^0_{zz}(\tau,\phi(\tau),\phi'(\tau))$ is non-singular. This matrix is the jacobian matrix for the system of equations (5.14). Therefore by the implicit function theorem there exists a unique $C^{(1)}$ solution, say ω, of the system (5.14) on an interval $(\tau-\delta,\tau+\delta)$ that does not include any corners of ϕ. On the other hand, from (5.7) we see that ϕ' is a solution on $(\tau-\delta,\tau+\delta)$. Therefore $\phi' = \omega$. Hence ϕ' is $C^{(1)}$ and ϕ is $C^{(2)}$ on $(\tau-\delta,\tau+\delta)$. Since τ is an arbitrary point between corners we get that ϕ is $C^{(2)}$ between corners.

If we now assume further that f^0 is of class $C^{(2)}$ on $\mathscr{G} = \mathscr{R} \times \mathscr{U}$ and that ϕ is of class $C^{(2)}$ between corners, we may use the chain rule and from (5.9) get the following equation between corners

$$f^{0*}_{zz}\phi'' + f^{0*}_{zx}\phi' + f^{0*}_{zt} - f^{0*}_{x} = 0,$$

where the asterisk indicates that the functions are evaluated at $(t,\phi(t),\phi'(t))$. The functions ϕ' and ϕ'' are evaluated at t.

EXERCISE 5.1. Consider the problem of Bolza as formulated in II.6. We now assume that the function $F = (F^1,\ldots,F^\mu)$ that defines the differential equation side conditions II(6.2) is given by

$$F^i(t,x,x') = x'^i - G^i(t,x,\tilde{x}') \qquad i = 1,\ldots,\mu, \qquad (5.15)$$

where $\tilde{x}' = (x'^{\mu+1},\ldots,x'^n)$. We assume the functions f^0 and F are of class $C^{(1)}$ on $\mathscr{G} = \mathscr{R} \times \mathscr{U}$. Note that because of (5.15) this amounts to assuming that the function $G = (G^1,\ldots,G^\mu)$ is of class $C^{(1)}$ on an appropriate region of $(t,x,\tilde{x}')$-space. The set $\mathscr{B}$ is assumed to be a $C^{(1)}$ manifold of dimension q, where $0 \leq q \leq 2n+1$ and the function g is assumed to be $C^{(1)}$ in a neighborhood of $\mathscr{B}$. Let $\hat{\rho} = (\rho^0,\rho) = (\rho^0,\rho^1,\ldots,\rho^\mu)$ and let

$$L(t,x,x',\hat{\rho}) = -\rho^0 f^0(t,x,x') + \sum_{i=1}^{\mu} \rho^i F^i(t,x,x')$$
$$= -\rho^0 f^0(t,x,x') + \langle \rho, F(t,x,x')\rangle.$$

Under these assumptions show that if ϕ is a minimizing function for the problem of Bolza then the following conditions hold.

(a) (Lagrange Multiplier Rule). There exists a constant $\psi^0 \leq 0$ and a measurable function $\psi = (\psi^1,\ldots,\psi^\mu)$ defined on the interval $[t_0,t_1]$ such that for all t, $(\psi^0,\psi(t)) \neq 0$ and

$$L_{x'}(t,\phi(t),\phi'(t),\hat{\psi}(t)) = \int_{t_0}^{t} L_x(s,\phi(s),\phi'(s),\hat{\psi}(s))ds + C,$$

where C is an appropriate constant vector and $\hat{\psi}(t) = (\psi^0,\psi(t))$. Moreover, if

$$P(t) = (t,\phi(t),\phi'(t),\hat{\psi}(t)) \quad \text{and} \quad A = L - \langle x', L_{x'}\rangle$$

then the $(2n+2)$-vector

$$(-A(P(t_0))-\psi^0 g_{t_1}(e(\phi)), \quad -L_{x'}(P(t_0))-\psi^0 g_{x_0}(e(\phi)),$$
$$A(P(t_1))-\psi^0 g_{t_1}(e(\phi)), \quad L_{x'}(P(t_1))-\psi^0 g_{x_1}(e(\phi)))$$

is orthogonal to $\mathscr{B}$ at $e(\phi)$.

The last statement is usually called the transversality condition.

(b) (Weierstrass Condition) If

$$\mathscr{E}(t,x,x',X',\hat{\rho}) = L(t,x,X',\hat{\rho}) - L(t,x,x',\hat{\rho})$$
$$-\langle F_{x'}(t,x,x'), X'-x'\rangle,$$

then for all X' and almost all t

$$\mathscr{E}(t,\phi(t),\phi'(t),X',\hat{\psi}(t)) \geq 0,$$

where ϕ and $\hat{\psi}$ are as in (a).

(c) (Clebsch Condition) The inequality

$$\langle \eta, L_{x'x'}(t,\phi(t),\phi'(t),\hat{\psi}(t))\eta \rangle \geq 0$$

holds for almost all t and all $\eta = (\eta^1,\ldots,\eta^n) \neq 0$ such that

$$L_{x'}(t,\phi(t),\phi'(t),\hat{\psi}(t))\eta = 0,$$

where ϕ and $\hat{\psi}$ are as in (a).

EXERCISE 5.2. Consider the control problem with control constraints as in II.6 and assume that the constraint qualification II(6.6) holds. Assume that the terminal set $\mathcal{B}$ is a $C^{(1)}$ manifold of dimension q, where $0 \leq q \leq 2n+1$. Assume that the functions g, f^0, f and R are of class $C^{(1)}$ on their domains of definition. Assuming that the results of Exercise 5.1 hold for the problem of Bolza in the calculus of variations, prove the following theorem.

Let (ϕ,u) be an optimal control defined on an interval $[t_0,t_1]$. Then there exists a constant $\lambda_0 \leq 0$, an absolutely continuous function $\lambda = (\lambda^1,\ldots,\lambda^n)$ defined on $[t_0,t_1]$, and a measurable function $\nu = (\nu^1,\ldots,\nu^r)$ defined on $[t_0,t_1]$ such that the following hold. (i) The vector $\hat{\lambda}(t) = (\lambda^0,\lambda(t))$ never vanishes and $\nu(t) \leq 0$ a.e. (ii) For a.e. t in $[t_0,t_1]$

$$\phi'(t) = H_p(t,\phi(t),u(t),\hat{\lambda}(t))$$

$$\lambda'(t) = -H_x(t,\phi(t),u(t),\hat{\lambda}(t)) - R_x(t,\phi(t),u(t))\nu(t)$$

$$H_z(t,\phi(t),u(t),\hat{\lambda}(t)) + R_z(t,\phi(t),u(t))\nu(t) = 0$$

$$\nu^i(t)R^i(t,\phi(t),u(t)) = 0 \qquad i = 1,\ldots,r.$$

(iii) For almost all t in $[t_0,t_1]$

$$H(t,\phi(t),u(t),\hat{\lambda}(t)) \geq H(t,\phi(t),z,\hat{\lambda}(t))$$

for all z satisfying $R(t,\phi(t),z) \geq 0$. (iv) The transversality condition as given in Exercise 3.5 holds. (v) At each t in $[t_0,t_1]$ let $\hat{R}$ denote the vector formed from R by taking those components of R that vanish at $(t,\phi(t),u(t))$. Then for almost all t

$$\langle e,(H(\pi(t)) + \langle \nu(t),R(t,\phi(t),u(t))\rangle)_{zz}e\rangle \geq 0$$

for all vectors $e = (e^1,\ldots,e^m)$ such that $\hat{R}_z(t,\phi(t),u(t))e = 0$.

6. Systems Linear in the State Variable

In this section we apply the maximum principle to the problem of minimizing

$$J(\phi,u) = g(e(\phi)) + \int_{t_0}^{t_1}(\langle a_0(s),\phi(s)\rangle + h_0(s,\phi(s)))ds$$

subject to

$$\frac{dx}{dt} = A(t)x + h(t,u(t)), \tag{6.1}$$

control constraints Ω and terminal constraints $\mathscr{B}$. The following assumption will be made in this section.

ASSUMPTION 6.1. (i) The constraint mapping Ω is a constant map; i.e. $\Omega(t) = \mathscr{C}$ for all t, where $\mathscr{C}$ is a fixed set in E^m. (ii) The set $\mathscr{B}$ is a $C^{(1)}$ manifold of dimension q, $0 \leq q \leq 2n+1$. (iii) The vector function a_0 and the matrix function A are bounded and measurable. (iv) The functions $h_0\colon (t,z) \to h_0(t,z)$ and $h\colon (t,z) \to h(t,z)$ are measurable in t and continuous in z. (v) The function g is $C^{(1)}$ in a neighborhood of $\mathscr{B}$ in E^{2n+2}.

We assume that an optimal pair (ϕ,u) exists for this problem and that $[t_0,t_1]$ is the interval of definition of (ϕ,u). We make the following assumption about (ϕ,u).

ASSUMPTION 6.2. (i) At $t = t_0$ and $t = t_1$ the mappings

$$t \to A(t)\phi(t) + h(t,u(t))$$
$$t \to \langle a^0(t),\phi(t)\rangle + h^0(t,u(t))$$

are continuous. This, of course, is to enable us to use the transversality condition. (ii) At the end point $e(\phi)$ the vector

$$(g_{t_0}, g_{x_0}, g_{t_1}, g_{x_1}) \tag{6.2}$$

is neither zero nor orthogonal to $\mathscr{B}$.

Theorem IV.6.3 is an existence theorem for systems linear in the state. In this theorem the matrix A and the functions h, h^0 and a^0 are assumed to be continuous. It is easy to verify that Theorem IV.6.3 is valid if A, a^0, h^0 and h satisfy the conditions in Assumption 6.1.

As noted in the first sentence in the proof of Theorem IV.6.3, we can assume without loss of generality that $f^0 \equiv 0$. The functional J is then given by

$$J(\phi,u) = g(e(\phi)). \tag{6.3}$$

We henceforth assume that J is given by (6.3).

We now investigate the form that the maximum principle takes in this problem. The function H becomes

$$H(t,x,z,\hat{p}) = \langle p,A(t)x\rangle + \langle p,h(t,z)\rangle. \tag{6.4}$$

Therefore, if a^{ij} denotes the element in the i-th row and j-th column of A, we get that

$$H_{x^j}(t,x,z,\hat{p}) = \sum_{i=1}^{n} p^i a^{ij}(t).$$

The second equation in (3.6) therefore becomes

$$\frac{dp}{dt} = -A(t)^*p, \tag{6.5}$$

where the asterisk denotes "transpose".

Let λ be a solution of (6.5) satisfying the initial condition

$$\lambda(t_0) = \eta. \tag{6.6}$$

Let Ψ denote the fundamental matrix solution of (6.5) which satisfies $\Psi(t_0) = I$. Thus

$$\Psi'(t) = -A(t)^*\Psi(t) \qquad \Psi(t_0) = I. \tag{6.7}$$

Then

$$\lambda(t) = \Psi(t)\eta. \tag{6.8}$$

The system (6.5) is adjoint to the system

$$\frac{dx}{dt} = A(t)x. \tag{6.9}$$

Let Φ denote the fundamental matrix solution of (6.9) satisfying $\Phi(t_0) = I$. Then by the rule for the differentiation of a product and by (6.9) we get

$$\begin{aligned}(\Phi^*\Psi)' &= (\Phi^{*\prime})\Psi + \Phi^*\Psi' \\ &= (\Phi^*A^*)\Psi + \Phi^*(-A^*\Psi) = 0.\end{aligned}$$

Hence $\Phi^*(t)\Psi(t) = C$ for all t, where C is a constant matrix. But

$$\Phi^*(t_0) = \Psi(t_0) = I,$$

so that $C = I$ and we have that

$$\Psi(t) = \Phi^{-1*}(t) \tag{6.10}$$

for all t. Equation (6.8) can therefore also be written as

$$\lambda(t) = \Phi^{-1*}(t)\eta.$$

The transversality condition for problems with g not identically zero was given in Exercise 3.5. If we set

$$f^*(t_i) = A(t_i)\phi(t_i) + h(t_i,u(t_i)) \qquad i = 0,1,$$

then the transversality condition given in Exercise 3.5 takes the following form in the present situation. The vector

$$(\langle\lambda(t_0),f^*(t_0)\rangle-\lambda^0 g_{t_0},\ -\lambda(t_0)-\lambda^0 g_{x_0},\ -\langle\lambda(t_1),f^*(t_1)\rangle-\lambda^0 g_{t_1},\ \lambda(t_1)-\lambda^0 g_{x_1})$$

is orthogonal to $\mathscr{B}$ at $e(\phi)$, where all the partial derivatives of g are evaluated at $e(\phi)$. By virtue of (6.6) and (6.8) this vector can also be written as

$$(\langle\eta,f^*(t_0)\rangle-\lambda^0 g_{t_0},\ -\eta-\lambda^0 g_{x_0},\ -\langle\Psi(t_1)\eta,f^*(t_1)\rangle-\lambda^0 g_{t_1},\ \Psi(t_1)\eta-\lambda^0 g_{x_1}). \tag{6.11}$$

We now show that $\eta \neq 0$. For if $\eta = 0$, then by (6.8), $\lambda(t) = 0$ on $[t_0,t_1]$. Hence by the maximum principle $\lambda^0 \neq 0$. Also, if $\eta = 0$, (6.11) becomes

$$\lambda^0(g_{t_0},g_{x_0},g_{t_1},g_{x_1}).$$

Since $\lambda^0 \neq 0$, the transversality condition implies that the vector (6.2) is either zero or is orthogonal to $\mathscr{B}$ at $e(\phi)$. This, however, was ruled out at the outset.

EXERCISE 6.1. Obtain an analytic formulation of the transversality condition as a system of q linear equations in the unknowns $(\lambda^0,\eta^1,\ldots,\eta^n)$ by setting the inner product of (6.11) with each of q linearly independent tangent vectors to $\mathscr{B}$ at $e(\phi)$ equal to zero.

In the present discussion $\Omega(t) = \mathscr{C}$ for all t so that

Corollary 3.1 is applicable. From (6.4) we see that (3.5) is equivalent to the following inequality in the present discussion:

$$\langle \lambda(t), h(t,u(t))\rangle \geq \langle \lambda(t), h(t,z)\rangle \tag{6.12}$$

for a.e. t in $[t_0, t_1]$ and all z in $\mathscr{C}$. From (6.8) and (6.10) we get that

$$\langle \lambda(t), h(t,z)\rangle = \langle \Psi(t)\eta, h(t,z)\rangle$$

$$= \langle \eta, \Psi^*(t)h(t,z)\rangle = \langle \eta, \Phi^{-1}(t)h(t,z)\rangle.$$

Thus, (6.12) is equivalent to the inequality

$$\langle \eta, \Phi^{-1}(t)h(t,u(t))\rangle \geq \langle \eta, \Phi^{-1}(t)h(t,z)\rangle. \tag{6.13}$$

We summarize the preceding discussion with the following theorem, which gives the maximum principle for systems linear in the state variable.

THEOREM 6.1. Let (ϕ, u) minimize the functional (6.3) subject to (6.1), control constraints Ω and terminal set $\mathscr{B}$. Let Assumptions 6.1 and 6.2 hold. Let Φ denote the fundamental matrix solution of (6.9) such that $\Phi(t_0) = I$. Then there exists a non zero vector η in E^n and a scalar $\lambda^0 \leq 0$ such that the vector (6.11) is orthogonal to $\mathscr{B}$ at $e(\phi)$ and such that for a.e. t in $[t_0, t_1]$

$$\max\{\langle \eta, \Phi^{-1}(t)h(t,z)\rangle : \quad z \in \mathscr{C}\}$$

occurs at $z = u(t)$.

REMARK 6.1. By virtue of (6.10) the quantity to be maximized can also be written as $\langle \eta, \Psi^*(t)h(t,z)\rangle$.

REMARK 6.2. Note that, in principle, for systems linear in the state variable we only need the initial value η of the function

λ in order to determine an extremal trajectory. Of course, the terminal value $\lambda(t_1)$ will also do, since by virtue of (6.8), $\eta = \Psi^{-1}(t_1)\lambda(t_1)$. To determine η we use the transversality condition. This involves knowing $e(\phi)$ and $u(t_i)$, $i = 0,1$. Once u is known the variation of parameters formula gives the extremal trajectory.

7. Linear Systems

A linear system is one in which the function h is given by

$$h(t,z) = B(t)z + d(t), \tag{7.1}$$

where B is an $n \times m$ matrix and d is an n-vector. The system (6.1) becomes

$$\frac{dx}{dt} = A(t)x + B(t)z + d(t). \tag{7.2}$$

For (iii) and (iv) of Assumption 6.1 to hold, B and d must be bounded measurable functions.

The maximum principle for linear systems is an immediate consequence of Theorem 6.1 and (7.1).

THEOREM 7.1. Let (ϕ,u) minimize the functional (6.3) subject to (7.2), control constraints Ω and terminal constraint $\mathscr{B}$. Let Assumptions 6.1 and 6.2 hold. Let Φ be the fundamental solution matrix of (6.9) satisfying $\Phi(t_0) = I$. Then there exists a non zero vector η in E^n and a scalar $\lambda^0 \leq 0$ such that the vector (6.11) is orthogonal to $\mathscr{B}$ at $e(\phi)$ and such that for a.e. t in $[t_0,t_1]$

$$\max\{\langle \eta\Phi^{-1}(t)B(t),z\rangle : \ z \in \mathscr{C}\} \tag{7.3}$$

occurs at $z = u(t)$.

REMARK 7.1. By virtue of (6.10) we can write

$$\max\{\langle \eta\Psi^*(t)B(t), z\rangle : \quad z \in \mathscr{C}\}$$

in place of (7.3). In either case we have that $z = u(t)$ maximizes a linear form in z over a set $\mathscr{C}$. To emphasize this we shall let

$$L(t,\eta,z) = \langle \eta\Phi^{-1}(t)B(t), z\rangle = \langle \eta\Psi^*(t)B(t), z\rangle. \quad (7.4)$$

Let ψ_j denote the function comprising the j-th column of Ψ, let $\eta = (\eta^1,\ldots,\eta^n)$, let $B^j(t)$ denote the j-th column of $B(t)$, and let b^{ik} denote the entry in the i-th row and k-th column of B. Then the coefficient of z^j in (7.4) is

$$\sum_{i,k=1}^{n} \eta^i \psi_i^k(t) b^{kj}(t) = \eta\Psi^*(t)B^j(t).$$

In many problems the set $\mathscr{C}$ is the cube given by

$$\mathscr{C} = \{z: |z^i| \le 1, \ i = 1,\ldots,m\}. \quad (7.5)$$

In this situation an optimal control often can be characterized very simply.

COROLLARY 7.1. Let $\mathscr{C}$ be given by (7.5). For each $j = 1,\ldots,m$ let

$$E_j(\eta) = \{t: \eta\Psi^*(t)B^j(t) \ne 0\}$$

have measure zero. Then for almost all t in $[t_0,t_1]$

$$u^j(t) = \operatorname{signum} \eta\Psi^*(t)B^j(t). \quad (7.6)$$

Since for a.e. t, $u(t)$ maximizes $z \to L(t,\eta,z)$ over $\mathscr{C}$ and since $\eta\Psi^*(t)B^j(t)$ is the coefficient of z^j, the result is immediate.

<u>DEFINITION 7.1.</u> A linear system (7.2) is said to be <u>normal with respect to</u> $\mathscr{C}$ on an interval $[t_0,t_1]$ if for every non-zero n-vector μ and for a.e. t in $[t_0,t_1]$,

$$\max\{L(t,\mu,z): z \in \mathscr{C}\}$$

occurs at a unique point $z^*(t)$ in $\mathscr{C}$.

Note that whether a system is or is not normal on a given interval is determined by the matrices A and B and by the constraint set $\mathscr{C}$. At the end of this section we shall develop criteria for normality that involve conditions on A, B, and $\mathscr{C}$ that are relatively easy to verify.

If $\mathscr{C}$ is given by (7.5) a system is normal if and only if the set $E_j(\mu)$ has measure zero for each μ in E^m and each $j = 1,\ldots,m$. Thus, Corollary 7.1 states that if a system is normal with respect to $\mathscr{C}$ and $\mathscr{C}$ is given by (7.5), then the optimal control is given by (7.6).

We now investigate the structure of an optimal control when $\mathscr{C}$ is a compact convex set. If $\mathscr{C}$ is compact and convex then by the Krein-Milman Theorem (Lemma IV.6.2) the set of extreme points $\mathscr{C}_e$ of $\mathscr{C}$ is non-empty. The following Corollary of Theorem 7.1 holds.

COROLLARY 7.2. Let $\mathscr{C}$ be compact and convex set and let the system be normal. If u is an optimal control then $u(t) \in \mathscr{C}_e$ for almost all t.

Proof. If the conclusion were false then $u(t) \notin \mathscr{C}_e$ for t in a set E of positive measure. Hence for $t \in E$, there exist points $z_1(t)$ and $z_2(t)$ in $\mathscr{C}$ and real numbers $\alpha(t) > 0$, $\beta(t) > 0$, with $\alpha(t)+\beta(t) = 1$ such that $u(t) = \alpha(t)z_1(t) + \beta(t)z_2(t)$. Since the system is normal, the linear function $L(t,\eta,\cdot)$ achieves its maximum at a unique point $z^*(t)$ for a.e. t in E. By the maximum principle, the maximum is achieved at $u(t)$, so that $z^*(t) = u(t)$. Hence $L(t,\eta,u(t)) > L(t,\eta,z_1(t))$ and $L(t,\eta,u(t)) > L(t,\eta,z_2(t))$ a.e. in E. Hence

$$\begin{aligned} L(t,\eta,u(t)) &= \alpha(t)L(t,\eta,u(t)) + \beta(t)L(t,\eta,u(t)) \\ &> \alpha(t)L(t,\eta,z_1(t)) + \beta(t)L(t,\eta,z_2(t)) \\ &= L(t,\eta,\alpha(t)z_1(t)+\beta(t)z_2(t)) = L(t,\eta,u(t)), \end{aligned}$$

which is a contradiction.

DEFINITION 7.2. Let $\mathscr{C}$ be a compact polyhedron $\mathscr{P}$ with vertices $e_1,\ldots,e_k$. A control u is said to be bang-bang on an interval $[t_0,t_1]$ if for a.e. t in $[t_0,t_1]$, $u(t)$ is equal to one of the vertices.

If $\mathscr{C} = \mathscr{P}$, Corollary 7.2 can be restated as follows.

COROLLARY 7.3. Let the system be normal and let the constraint set be a compact polyhedron $\mathscr{P}$. Then any optimal control is bang-bang.

REMARK 7.2. The bang-bang principle (Theorem IV.6.4) tells us that if u is an optimal control, then there is another optimal control u^* that is bang-bang. The system is not assumed to be normal. Corollary 7.3, on the other hand, tells us that if a system is normal, then any optimal control must be bang-bang. Similarly, for an arbitrary compact convex constraint set $\mathscr{C}$, the bang-bang principle tells us that if u is an optimal control then there is an optimal control u^* such that $u^*(t) \in \mathscr{C}_e$. If the system is normal then Corollary 7.2 says that if u is an optimal control, u itself must have the property that $u(t) \in \mathscr{C}_e$.

The preceding results do not guarantee uniqueness of the optimal control for normal systems. The next theorem gives reasonable conditions under which an optimal control is unique.

THEOREM 7.2. Let $\mathscr{C}$ be compact and convex, let the system be normal, let $\mathscr{B}$ be a relatively open convex subset of a linear variety in E^{2n+2}, and let g be given by

$$g(t_0,x_0,t_1,x_1) = g_1(x_0,x_1) + g_2(t_0,t_1), \qquad (7.7)$$

where g_1 is convex. Let u_1 and u_2 be two optimal controls defined on the same interval $[t_0,t_1]$. Then $u_1 = u_2$ a.e. on $[t_0,t_1]$.

Proof. Let ϕ_1 be the trajectory corresponding to u_1 and let ϕ_2 be the trajectory corresponding to u_2. Define $u_3 = (u_1+u_2)/2$. Since $\mathscr{C}$ is convex, $u_3(t) \in \mathscr{C}$. Let ϕ_3 be the trajectory corresponding to u_3 that satisfies the initial condition $\phi_3(t_0) = (\phi_1(t_0) + \phi_2(t_0))/2$. Then

$$\begin{aligned}\phi_3(t) &= \Phi(t)\{\phi_3(t_0) + \tfrac{1}{2}\int_{t_0}^{t} \Phi^{-1}(s)[B(s)(u_1(s)+u_2(s)) + 2d(s)]ds\} \\ &= (\phi_1(t) + \phi_2(t))/2.\end{aligned}$$

and

$$e(\phi_3) = (e(\phi_1) + e(\phi_2))/2.$$

Since $\mathscr{B}$ is a convex subset of a linear variety it follows that $e(\phi_3) \in \mathscr{B}$. Hence (ϕ_3,u_3) is an admissible pair.

Let $\mu = \inf\{J(\phi,u): (\phi,u) \text{ admissible}\}$. Then

$$\mu = J(\phi_1,u_1) = J(\phi_2,u_2).$$

From the definition of μ, from (6.6), from (7.7), from the convexity of g_1 and the assumption that ϕ_1 and ϕ_2 have the same initial and terminal times we get

$$\begin{aligned}\mu \le J(\phi_3,u_3) &= g(e(\phi_3)) = g((e(\phi_1) + e(\phi_2))/2) \\ &\le \tfrac{1}{2}g(e(\phi_1)) + \tfrac{1}{2}g(e(\phi_2)) = \mu.\end{aligned}$$

Thus $J(\phi_3,u_3) = \mu$, and the pair (ϕ_3,u_3) is optimal. By Corollary 7.2, $u_3(t) \in \mathscr{C}_e$ a.e. This contradicts the definition of u_3 unless $u_1 = u_2$ a.e.

REMARK 7.3. For problems with t_0 and t_1 fixed g automatically has the form (7.7) with $g_2 \equiv 0$. If we assume that g is a convex function of (t_0, x_0, t_1, x_1) then the assumption that g has the form (7.7) can be dropped.

DEFINITION 7.3. The linear system (7.2) is said to be strongly normal on an interval $[t_0, t_1]$ with respect to a constraint set $\mathscr{C}$ if for every non-zero vector μ in E^n, $\max\{L(t,\mu,z): z \in \mathscr{C}\}$ is attained at a unique $z^*(t)$ in $\mathscr{C}$ at all but a finite set of points in $[t_0, t_1]$.

DEFINITION 7.4. A control u is said to be piecewise constant on an interval $[t_0, t_1]$ if there exist a finite number of disjoint open subintervals (τ_j, τ_{j+1}) such that the union of the closed subintervals $[\tau_j, \tau_{j+1}]$ is $[t_0, t_1]$ and such that u is constant on each of the open subintervals (τ_j, τ_{j+1}).

The next theorem gives a characterization of the optimal control in strongly normal systems that is of practical significance. Simple criteria for strong normality will be given in Theorem 7.4 and its corollaries.

THEOREM 7.3. Let (ϕ, u) be an optimal pair and let Assumption 6.1 and 6.2 hold. Let the matrix B be continuous and let the constraint set $\mathscr{C}$ be a compact polyhedron $\mathscr{P}$. Let the system (7.2) be strongly normal on $[t_0, t_1]$, the interval of definition of (ϕ, u). Then u is piecewise constant on $[t_0, t_1]$ with values in the set of vertices of $\mathscr{P}$.

Proof. If we remove the points t_0, t_1 and the finite set of points at which the maximum of $L(t,\eta,z)$ is not achieved at a unique $z^*(t)$, we obtain a finite collection of disjoint open intervals (τ_j, τ_{j+1}) such that the union of the closed intervals $[\tau_j, \tau_{j+1}]$ is the interval $[t_0, t_1]$. Let J denote one of the intervals (τ_j, τ_{j+1}).

From the proof of Corollary 7.2 it is seen that for each t in J, $u(t)$ is equal to one of the vertices e_i, $i = 1,\ldots,k$, of $\mathscr{P}$. Let M_i denote the set of points t in J at which $u(t) = e_i$. Then not all of the M_i, $i = 1,\ldots,k$ are empty, the sets M_i are pairwise disjoint and $J = \cup M_i$. We now show that if M_i is not empty then it is open. For let $\tau \in M_i$. Then

$$L(\tau,\eta,e_i) > L(\tau,\eta,e_j) \quad \text{for all} \quad j \neq i. \tag{7.8}$$

Since for fixed η, e_i the mapping $t \to L(t,\eta,e_i)$ is continuous, (7.8) holds in a neighborhood of τ. Hence all points of this neighborhood are in M_i and hence M_i is open. Since J is connected and since $J = \cup M_j$, where the M_j are open and pairwise disjoint, it follows that for $j \neq i$ the set M_j must be empty. Thus $u(t) = e_i$ in J, and the theorem is proved.

The conclusion of Theorem 7.3 is much stronger than that of Corollary 7.3. Here we assert that the optimal control is piecewise constant with values at the vertices $e_1,\ldots,e_k$ of $\mathscr{P}$, while in Corollary 7.3 we merely assert that the optimal control is measurable with values at the vertices of $\mathscr{P}$. Of course, the assumptions are more stringent here.

We conclude this section with a presentation of criteria for strong normality.

THEOREM 7.4. Let the state equations be given by (7.2). Let A be of class $C^{(n-2)}$ on a compact interval $\mathscr{I}$ and let B be of class $C^{(n-1)}$ on $\mathscr{I}$. Let the constraint set be a compact polyhedron $\mathscr{P}$. Let

$$\begin{aligned} B_1(t) &= B(t) \\ B_j(t) &= -A(t)B_{j-1}(t) + B'_{j-1}(t) \qquad j = 2,\ldots,n. \end{aligned} \tag{7.9}$$

If for every vector w in E^m that is parallel to an edge of $\mathscr{P}$

the vectors

$$B_1(t)w,\ B_2(t)w,\dots,B_n(t)w \tag{7.10}$$

are linearly independent for all t in $\mathscr{I}$, then the system (7.2) is strongly normal with respect to $\mathscr{P}$ on $\mathscr{I}$.

Proof. Suppose the conclusion is false. Then there exists a non-zero vector η in E^n and an infinite set of points E in $\mathscr{I}$ such that for t in E, the maximum over $\mathscr{P}$ of $L(t,\eta,z)$ is not achieved at a unique $z^*(t)$ in $\mathscr{P}$. Since for fixed (t,η) the mapping $z \to L(t,\eta,z)$ is linear and since $\mathscr{P}$ is a compact polyhedron, the maximum over $\mathscr{P}$ of $L(t,\eta,z)$ is attained on some face of $\mathscr{P}$. Since there are only a finite number of faces of $\mathscr{P}$, there exists an infinite set $E_1 \subset E$ and a face $\mathscr{P}_F$ of $\mathscr{P}$ such that for t in E_1, the maximum over $\mathscr{P}$ is attained on $\mathscr{P}_F$. Hence if e_1 and e_2 are two distinct vertices in $\mathscr{P}_F$, $L(t,\eta,e_1) = L(t,\eta,e_2)$ for all t in E_1. Hence if $w = e_1-e_2$,

$$L(t,\eta,w) = \langle \eta\Psi^*(t),B(t)w\rangle = 0$$

for all t in E_1. From the first equation in (7.9) we get

$$L(t,\eta,w) = \langle \eta\Psi^*(t),B_1(t)w\rangle = 0 \tag{7.11}$$

for all t in E_1.

Since E_1 is an infinite set and $\mathscr{I}$ is compact, E_1 has a limit point τ in $\mathscr{I}$. From (7.11) and the continuity of B_1 and Ψ^* we get

$$L(\tau,\eta,w) = \langle \eta\Psi^*(\tau),B_1(\tau)w\rangle = 0. \tag{7.12}$$

By hypothesis, the matrix A is of class $C^{(n-2)}$. Hence the fundamental matrix Ψ of the system adjoint to (6.9) is of class

$C^{(n-1)}$. Since $B_1 = B$ and B is assumed to be of class $C^{(n-1)}$, it follows from the first equality in (7.11) that the mapping $t \to L(t,\eta,w)$ is of class $C^{(n-1)}$ on $\mathscr{I}$. Also,

$$L'(t,\eta,w) = \langle \eta\Psi^{*\prime}(t), B_1(t)w\rangle + \langle \eta\Psi^*(t), B_1'(t)w\rangle$$

From (6.7) we get

$$\Psi^{*\prime}(t) = -\Psi^*(t)A(t).$$

If we substitute this into the preceding equation we get

$$L'(t,\eta,w) = \langle \eta\Psi^*(t), (-A(t)B_1(t)+B_1'(t))w\rangle.$$

From the second equation in (7.9) we get

$$L'(t,\eta,w) = \langle \eta\Psi^*(t), B_2(t)w\rangle. \tag{7.13}$$

The derivative of a function has a zero between any two zeros of the function. Therefore $L'(t,\eta,w) = 0$ for all t in an infinite set E_2 having τ as a limit point. From (7.13) and the continuity of Ψ^* and B_2 it follows that

$$\langle \eta\Psi^*(\tau), B_2(\tau)w\rangle = 0.$$

We can proceed inductively in this manner and get

$$\langle \eta\Psi^*(\tau), B_j(\tau)w\rangle = 0 \qquad j = 1,\ldots,n.$$

Since the n vectors $B_1(t)w,\ldots,B_n(t)w$ are assumed to be linearly independent, we must have $\eta\Psi^*(\tau) = 0$. This, however, is impossible since $\eta \neq 0$ and $\Psi^*(\tau)$ is non-singular. This contradiction proves the theorem.

COROLLARY 7.4.1. Let A and B be constant matrices. If for every vector w in E^m that is parallel to an edge of $\mathscr{P}$, the vectors

$$Bw, ABw, A^2Bw, \ldots, A^{n-1}Bw$$

are linearly independent, then the system (7.2) is strongly normal with respect to $\mathscr{P}$ on $\mathscr{I}$.

The corollary follows from the observation that if A and B are constant matrices then

$$B_j = (-A)^{j-1}B \qquad j = 1,\ldots,n.$$

If the set $\mathscr{P}$ is a parallelepiped with axes parallel to the coordinate axes, then the only vectors w that we need consider are the standard basis vectors $w_1,\ldots,w_m$ in E^m. Here, w_i is the m-vector whose i-th component is equal to one and all of whose other components are zero. Let b^j denote the j-th column of the matrix B. Then $b^j = Bw_j$, and Corollary 7.4.1 yields the following

COROLLARY 7.4.2. Let A and B be constant matrices and let $\mathscr{P}$ be a parallelepiped with axes parallel to the coordinate axes. Let b^j denote the j-th column of B. For each $j = 1,\ldots,m$, let

$$b^j, Ab^j, A^2b^j, \ldots, A^{n-1}b^j$$

be linearly independent. Then the system (7.2) is strongly normal with respect to $\mathscr{P}$ on $\mathscr{I}$.

8. The Linear Time Optimal Problem

In the linear time optimal problem it is required to transfer a given point x_0 to another given point x_1 in minimum time by means of a linear system. More precisely, in the linear time optimal problem it is required to minimize

$$J(\phi,u) = t_1$$

subject to the state equation (7.2), constraint condition Ω, and end condition $\mathscr{B}$, where

$$\mathscr{B} = \{(t_0,x_0,t_1,x_1): t_0 = t_0', \ x_0 = x_0', \ x_1 = x_1'\},$$

with t_0', x_0' and x_1' given. The function g is now $g(t_1) = t_1$.

If $\Omega(t) = \mathscr{C}$, where $\mathscr{C}$ is a fixed compact convex set, and if the system (7.2) is normal with respect to $\mathscr{C}$, then by Corollary 7.4 any optimal control u has the form $u(t) \in \mathscr{C}_e$ a.e. If $\mathscr{C}$ is a compact polyhedron $\mathscr{P}$ then any optimal control u is bang-bang. We assert that it is also unique. If u_1 and u_2 are two optimal controls, then since the problem is one of minimizing t_1, it follows that u_1 and u_2 are both defined on the same interval $[t_0,t_1^*]$, where t_1^* is the minimum time. Since the function g is now given by $g(t_1) = t_1$, it is of the form (7.7) with $g_1 \equiv 0$. Note that g_1 is therefore convex. Also, for all t_1

$$\nabla g(t_1) = (0,0_n,1,0_n), \tag{8.1}$$

where 0_n is the n-dimensional zero vector. The vector on the right in (8.1) is also the unit tangent vector to $\mathscr{B}$ in this case. Therefore $\nabla g(t_1) \neq 0$ and $\nabla g(t_1)$ is never orthogonal to $\mathscr{B}$. Finally, $\mathscr{B}$ is an n-dimensional linear variety. Thus all of the hypotheses of Theorem 7.2 are satisfied and the two controls must be equal. We summarize our results in the following theorem.

THEOREM 8.1. In the linear time optimal problem if the constraint set is a fixed compact convex set $\mathscr{C}$ and the system is normal with respect to $\mathscr{C}$, then the optimal control u is unique and $u(t) \in \mathscr{C}_e$ a.e.

A large class of linear time optimal problems has the property that extremal controls are unique. This property in the presence of an existence theorem guarantees that an extremal control is optimal.

Actually, for the class of systems in question, the arguments used to show uniqueness of extremal controls prove directly, without reference to existence theorems, that an extremal control is unique and is optimal.

THEOREM 8.2. Let the system equations be given by (7.2) with $d \equiv 0$. Let $\mathscr{C}$ be a compact convex set with the origin of E^m an interior point of $\mathscr{C}$. Let the system be normal with respect to $\mathscr{C}$. Let (ϕ_1, u_1) be an extremal pair for the time optimal problem with terminal state $x_1 = 0$. Let the terminal time at which ϕ_1 reaches the origin be t_1. Let (ϕ_2, u_2) be an admissible pair which transfers x_0 to the origin in time $t_2 - t_0$. Then $t_2 \geq t_1$ with equality holding if and only if $u_1(t) = u_2(t)$ a.e.

Suppose there exists a pair (ϕ_2, u_2) for which $t_2 \leq t_1$. From the variation of parameters formula we get

$$0 = \Phi(t_1)\{x_0 + \int_{t_0}^{t_1} \Phi^{-1}(s)B(s)u_1(s)ds\}$$

$$0 = \Phi(t_2)\{x_0 + \int_{t_0}^{t_2} \Phi^{-1}(s)B(s)u_2(s)ds\},$$

where Φ is the fundamental matrix for the system (6.9) satisfying $\Phi(t_0) = I$. If we multiply the first equation by $\Phi(t_1)^{-1}$ on the left and multiply the second equation by $\Phi(t_2)^{-1}$ on the left we get

$$-x_0 = \int_{t_0}^{t_1} \Phi^{-1}(s)B(s)u_1(s)ds = \int_{t_0}^{t_2} \Phi^{-1}(s)B(s)u_2(s)ds. \qquad (8.2)$$

Since u_1 is an extremal control there exists a non-zero vector η in E^n such that for a.e. t in $[t_0, t_1]$, $u_1(t)$ maximizes $L(t,\eta,z)$ over $\mathscr{C}$. If we compute $\langle \eta, -x_0 \rangle$ in (8.2) we get

$$\int_{t_0}^{t_2} + \int_{t_2}^{t_1} \langle \eta\Phi^{-1}(s)B(s), u_1(s)\rangle ds = \int_{t_0}^{t_2} \langle \eta\Phi^{-1}(s)B(s), u_2(s)\rangle ds.$$

Therefore

$$\int_{t_0}^{t_2}\{L(s,\eta,u_1(s))-L(s,\eta,u_2(s))\}ds = -\int_{t_2}^{t_1} L(s,\eta,u_1(s))ds. \tag{8.3}$$

Since u_1 is extremal and the system is normal with respect to $\mathscr{C}$, $u_1(t) \in \mathscr{C}_e$ a.e. Since 0 is an interior point of $\mathscr{C}$, $u_1(t) \neq 0$ a.e. and

$$L(t,\eta(t),u_1(t)) > L(t,\eta,0) = 0.$$

Hence the right hand side of (8.3) is ≤ 0, with equality holding if and only if $t_1 = t_2$. On the other hand, since the system is normal

$$L(t,\eta,u_1(t)) \geq L(t,\eta,u_2(t))$$

for a.e. t, with equality holding if and only if $u_1(t) = u_2(t)$ a.e. Hence the integral on the left in (8.3) is ≥ 0 with equality holding if and only if $u_1(t) = u_2(t)$ a.e. Therefore, each side of (8.2) is equal zero. This implies that $t_2 = t_1$ and $u_2 = u_1$ a.e., and the theorem is proved.

9. Linear Plant-Quadratic Criterion Problem

In the class of problems to be studied in this section the state equations are

$$\frac{dx}{dt} = A(t)x + B(t)z + d(t) \tag{9.1}$$

and the function f^0 is given by

$$f^0(t,x,z) = \langle x,X(t)x\rangle + \langle z,R(t)z\rangle. \tag{9.2}$$

Existence theorems for such problems are given in Chapter III. Problems with compact constraint sets are considered in Corollary III.5.1. Problems with unbounded controls are considered in Exercise III.6.5.

Unless stated otherwise, the following assumptions will be in effect throughout this section.

ASSUMPTION 9.1. (i) The matrices A, B, X, amd R in (9.1) and (9.2) are continuous on an interval $[a,b]$, as is the function d in (9.1). (ii) For each t in $[a,b]$ the matrix $X(t)$ is symmetric, positive semi-definite and the matrix $R(t)$ is symmetric, positive definite. (iii) For each t in $[a,b]$, $\Omega(t) = \mathcal{O}$, where $\mathcal{O}$ is a fixed open set in E^m. (iv) The set $\mathcal{B}$ has the following form: (t_0,x_0) fixed; $(t_1,x_1) \in \mathcal{T}_1$, where $\mathcal{T}_1$ is a $C^{(1)}$ manifold of dimension n; $t_0,t_1 \in [a,b]$. (v) The function $g: (t_1,x_1) \to g(t_1,x_1)$ is $C^{(1)}$ on E^{n+1}.

The problem to be studied is that of minimizing

$$J(\phi,u) = g(t_1,\phi(t_1)) + \frac{1}{2}\int_{t_0}^{t_1}\{\langle\phi(s),X(s)\phi(s)\rangle + \langle u(s),R(s)u(s)\rangle\}ds \tag{9.3}$$

subject to the state equations (9.1), the control constraints Ω, and the terminal condition $\mathcal{B}$, where the data of the problem satisfy Assumption 9.1.

We now characterize optimal pairs by means of the maximum principle. The function H is given by

$$\begin{aligned} H(t,x,z,\hat{p}) = p^0/2\{\langle x,X(t)x\rangle + \langle z,R(t)z\rangle\} + \langle p,A(t)x\rangle \\ + \langle p,B(t)z\rangle + \langle p,d(t)\rangle . \end{aligned} \tag{9.4}$$

Thus,

$$H_x(t,x,z,\hat{p}) = p^0X(t)x + A(t)^*p,$$

where the asterisk denotes transpose.

We now consider an optimal pair (ϕ,u). We make the following assumption.

ASSUMPTION 9.2. The trajectory ϕ is not tangent to $\mathcal{T}_1$ at $(t_1,\phi(t_1))$.

From the form of the right hand sides of (9.1) and (9.2) and from Assumption 9.1-(ii) it follows that in order for us to be able to apply the transversality conditions to this problem it suffices to assume that the mapping $t \to u(t)$ is continuous at t_0 and at t_1. We shall see later, by arguments that do not involve the transversality condition that the optimal control u must be continuous. Let us assume for the moment that we have already shown this.

By virtue of (iv) of Assumption 9.1, the set $\mathscr{B}$ is the n-dimensional manifold consisting of all points of the form (t_0,x_0,t_1,x_1) with (t_0,x_0) fixed and (t_1,x_1) in a specified n-dimensional $C^{(1)}$ manifold $\mathscr{T}_1$. Thus the transversality condition given in Exercise 3.5 takes the following form in the present case. The vector

$$(-H(\pi(t_1))-\lambda^0 g_{t_1},\ \lambda(t_1)-\lambda^0 g_{x_1}),$$

where the partial derivatives of g are evaluated at $(t_1,\phi(t_1))$, is orthogonal to $\mathscr{T}_1$ at $(t_1,\phi(t_1))$.

If we assume that the trajectory ϕ is not tangent to $\mathscr{T}_1$ at its terminal point $(t_1,\phi(t_1))$ then $\lambda^0 \neq 0$. For if $\lambda^0 = 0$, the transversality condition would state that

$$(-\langle\lambda(t_1),f(t_1)\rangle,\lambda(t_1)) \tag{9.5}$$

is orthogonal to $\mathscr{T}_1$ at $(t_1,\phi(t_1))$, where $f(t_1)$ denotes the right hand side of (9.1) evaluated at $(t_1,\phi(t_1),u(t_1))$. The arguments used to establish the result of Exercise 3.2 are applicable here and show that (9.5) cannot be orthogonal to $\mathscr{T}_1$ at $(t_1,\phi(t_1))$ if the trajectory ϕ is not tangent to $\mathscr{T}_1$ at this opoint. Since $\lambda^0 \neq 0$, it follows from Exercise 3.3 that we may take $\lambda^0 = -1$. The transversality condition now states that the vector

$$(g_{t_1}+f_1^0-\langle\lambda(t_1),f_1\rangle,\ g_{x_1}+\lambda(t_1)) \tag{9.6}$$

is orthogonal to $\mathcal{T}_1$ at $(t_1,\phi(t_1))$, where the partial derivatives of g are evaluated at $(t_1,\phi(t_1))$, f_1^0 denotes (9.2) evaluated at $(t_1,\phi(t_1),u(t_1))$ and f_1 denotes the right hand side of (9.1) evaluated at $(t_1,\phi(t_1),u(t_1))$. We remark that the arguments used in Exercise 3.2 now show that $\lambda(t_1)$ is unique.

Equations (3.2) now become

$$\begin{aligned} \frac{d\phi}{dt} &= A(t)\phi(t) + B(t)u(t) + d(t) \\ \frac{d\lambda}{dt} &= X(t)\phi(t) - A^*(t)\lambda(t). \end{aligned} \tag{9.7}$$

Since the constraint set is fixed, (3.5) holds. From (9.4) we see that in the present context, (3.5) is equivalent to the inequality

$$- \frac{1}{2}\langle u(t), R(t)u(t)\rangle + \langle \lambda(t), B(t)u(t)\rangle \geq$$
$$- \frac{1}{2}\langle z, R(t)z\rangle + \langle \lambda(t), B(t)z\rangle$$

for all z in $\mathcal{O}$ and almost all t in $[t_0,t_1]$. Thus, for almost every t in $[t_0,t_1]$ the mapping

$$z \to - \frac{1}{2}\langle z, R(t)z\rangle + \langle \lambda(t), B(t)z\rangle \tag{9.8}$$

attains its maximum over $\mathcal{O}$ at $z = u(t)$. But $\mathcal{O}$ is open, so that the derivative of the mapping (9.8) is zero at $z = u(t)$. Hence

$$-R(t)u(t) + B^*(t)\lambda(t) = 0.$$

Since $R(t)$ is non-singular for all t, we get that

$$u(t) = R^{-1}(t)B^*(t)\lambda(t) \quad \text{a.e.} \tag{9.9}$$

Note that since B, R and λ are continuous, the optimal control is also continuous.

If we now substitute (9.9) into the first equation in (9.7) we get the following theorem from the maximum principle.

THEOREM 9.1. Let (ϕ,u) be an optimal pair with interval of definition $[t_0,t_1]$. Let Assumption 9.2 hold. Then there exists an absolutely continuous function $\lambda = (\lambda^1,\ldots,\lambda^n)$ defined on $[t_0,t_1]$ such that (ϕ,λ) is a solution of the linear system

$$\begin{aligned} \frac{dx}{dt} &= A(t)x + B(t)R^{-1}(t)B^*(t)p + d(t) \\ \frac{dp}{dt} &= X(t)x - A^*(t)p \end{aligned} \tag{9.10}$$

and such that the vector (9.6) is orthogonal to $\mathscr{T}_1$ at $(t_1,\phi(t_1))$. The optimal control is given by (9.9).

We now specialize the problem by taking $\mathscr{T}_1$ to be the hyperplane $t_1 = T$; i.e.

$$\mathscr{T}_1 = \{(t_1,x_1): t_1 = T,\ x_1 \text{ free}\}, \tag{9.11}$$

and by taking g to be given by

$$g(x_1) = \frac{1}{2}\langle x_1, Gx_1\rangle, \tag{9.12}$$

where G is a positive semi-definite symmetric matrix.

REMARK 9.1. If (9.11) holds then every tangent vector to $\mathscr{T}_1$ has its first component equal to zero. On the other hand, a tangent vector to the trajectory ϕ has its first component always different from zero. Moreover, it follows from (9.1) and the continuity of an optimal control u that the trajectory has a tangent vector at all points. Hence, if (9.11) holds then Assumption 9.2 is automatically satisfied.

COROLLARY 9.1. If (9.11) and (9.12) hold, then ϕ and λ satisfy the system (9.10) subject to the boundary conditions

$$\phi(t_0) = x_0 \qquad \lambda(t_1) = -G\phi(t_1). \tag{9.13}$$

The first condition is a restatement of the initial condition already imposed. The second follows from the orthogonality of (9.6) to $\mathscr{T}_1$ at the terminal point of the trajectory and from (9.12).

An admissible pair (ϕ,u) that satisfies the conditions of Theorem 9.1 will be called an <u>extremal pair</u>. If (9.11) and (9.12) hold, then an extremal pair satisfies (9.13).

In the next theorem we show that if (9.11) and (9.12) hold, then an extremal pair is unique and must be optimal. This will be done without reference to any existence theorems previously established.

THEOREM 9.2. Let (9.11) and (9.12) hold. Let (ϕ,u) be an extremal pair and let (ϕ_1,u_1) be any other admissible pair. Then $J(\phi_1,u_1) \geq J(\phi,u)$, with equality holding if and only if $u = u_1$. In that event $\phi = \phi_1$.

<u>Proof</u>. First note that because the system (9.1) is linear and (t_0,x_0) is fixed, if $u = u_1$ then $\phi = \phi_1$. Let

$$\phi_f = \phi(T) \qquad \phi_{1f} = \phi_1(T).$$

Since $X(t)$ is positive semi-definite and $R(t)$ is positive definite for all t and since G is positive semi-definite, we get

$$\begin{aligned} 0 \leq\ & \langle(\phi_{1f}-\phi_f),G(\phi_{1f}-\phi_f)\rangle \\ & + \int_{t_0}^{T} \{\langle(\phi_1-\phi),X(\phi_1-\phi)\rangle + \langle(u_1-u),R(u_1-u)\rangle\}dt, \end{aligned}$$

with equality holding if and only if $u_1 = u$. Hence

$$\begin{aligned} 0 \leq\ & 2J(\phi_1,u_1) + 2J(\phi,u) - 2\langle\phi_{1f},G\phi_f\rangle \\ & - 2\int_{t_0}^{T} \{\langle\phi_1,X\phi\rangle + \langle u_1,Ru\rangle\}dt, \end{aligned}$$

which we rewrite as

$$J(\phi_1,u_1)+J(\phi,u) \geq \langle\phi_{1f},G\phi_f\rangle + \int_{t_0}^{T} \{\langle\phi_1,X\phi\rangle + \langle u_1,Ru\rangle\}dt. \tag{9.14}$$

Since (ϕ,u) is an extremal pair, there is an absolutely continuous vector λ such that λ and ϕ are solutions of (9.10) that satisfy (9.13) and such that (9.9) holds. We now substitute for $X\phi$ in the right hand side of (9.14) from the second equation in (9.10) and substitute for u in the right hand side of (9.14) from (9.9). We get

$$J(\phi_1,u_1) + J(\phi,u) \geq \langle \phi_{1f}, G\phi_f \rangle + \int_{t_0}^{T} \{\langle \phi_1, \lambda' + A^*\lambda \rangle + \langle u_1, B^*\lambda \rangle\} dt. \tag{9.15}$$

The integral on the right in (9.15) can be written as

$$\int_{t_0}^{T} \{\langle \phi_1, \lambda' \rangle + \langle A\phi_1 + Bu_1, \lambda \rangle\} dt.$$

Since (ϕ_1,u_1) is admissible we have from (9.1) that

$$A\phi_1 + Bu_1 = \phi_1' - d.$$

Substituting this into the last integral gives

$$\int_{t_0}^{T} \{\langle \phi_1, \lambda' \rangle + \langle \phi_1', \lambda \rangle - \langle d, \lambda \rangle\} dt.$$

Therefore, we can rewrite (9.15) as follows

$$J(\phi_1,u_1) + J(\phi,u) \geq \langle \phi_{1f}, G\phi_f \rangle + \langle \phi_{1f}, \lambda(t_1) \rangle - \langle x_0, \lambda(t_0) \rangle - \int_{t_0}^{T} \langle d, \lambda \rangle dt.$$

If we now use (9.13) we get

$$J(\phi_1,u_1) + J(\phi,u) \geq -\langle x_0, \lambda(t_0) \rangle - \int_{t_0}^{T} \langle d, \lambda \rangle dt.$$

Recall that equality holds if and only if $u_1 = u$, in which case $\phi_1 = \phi$. Therefore if we take $u_1 = u$ in the preceding inequality we get

$$2J(\phi,u) = -\langle x_0, \lambda(t_0) \rangle - \int_{t_0}^{T} \langle d, \lambda \rangle dt. \tag{9.16}$$

Substituting (9.16) into the preceding inequality gives

$$J(\phi_1,u_1) \geq J(\phi,u),$$

with equality holding if and only if $u_1 = u$.

EXERCISE 9.1. Consider the linear quadratic problem with t_0 and t_1 fixed, $\mathscr{T}_1$ a linear variety of dimension n, and $g: x_1 \to g(x_1)$ a convex function. Suppose also that the constraint set $\mathscr{O}$ is convex. Show, without appealing to the maximum principle of Theorem 9.1, that if (ϕ,u) is an optimal pair then (ϕ,u) is unique.

The linear plant quadratic criterion problem posed in this section, with $\mathscr{T}_1$ as in (9.11) and g as in (9.12), admits a very elegant and relatively simple synthesis of the optimal control. The determination of this synthesis will take up the remainder of this section.

For the problem with fixed initial point (τ,ξ), with $a \leq \tau < T$ and $\xi \in R^n$, it follows from Exercise III.6.5 that there exists an optimal pair $(\phi(\cdot,\tau,\xi),u(\cdot,\tau,\xi))$. By Theorem 9.1 this pair is extremal and for $\tau \leq t \leq T$

$$u(t,\tau,\xi) = R^{-1}(t)B^*(t)\lambda(t,\tau,\xi).$$

It then follows from Theorem 9.2 that the optimal pair for the problem with initial point (τ,ξ) is unique. Therefore, as in Section 2, we obtain a field $\mathscr{F}$ of optimal trajectories. We obtain a synthesis of the optimal control, or feedback control, U as follows:

$$U(\tau,\xi) = u(\tau,\tau,\xi) = R^{-1}(\tau)B^*(\tau)\lambda(\tau,\tau,\xi). \tag{9.17}$$

This holds for all $a \leq \tau < T$ and for all ξ, since we may choose any such (τ,ξ) to be the initial point for the problem.

The feedback law in (9.17) is not satisfactory since it requires knowing the value of the adjoint variable, of multiplier, λ at the initial point. If the formalism of Section 2 is valid, then by (2.18) we have $W_x(\tau,\xi) = -\lambda(\tau,\tau,\xi)$ and we can write

$$U(\tau,\xi) = -R^{-1}(\tau)B^*(\tau)W_x(\tau,\xi). \tag{9.18}$$

This leads us to investigate the value function W for the present problem. We shall proceed formally, as in Section 2, assuming that all functions have the required number of derivatives existing and continuous. In this way we shall obtain insights and conjectures as to the structure of the feedback control. We shall then show rigorously, by other methods, that these conjectures are valid.

We henceforth suppose that $d = 0$ in (9.1).

The function W satisfies the Hamilton-Jacobi equation (2.14), which in the present case becomes

$$W_t = -\frac{1}{2}\langle x,Xx\rangle - \frac{1}{2}\langle U,RU\rangle - \langle W_x,Ax\rangle - \langle W_x,BU\rangle.$$

In this relation and in what follows we shall omit the arguments of the functions involved. Using (9.18) we can rewrite this equation as follows:

$$W_t = -\frac{1}{2}\langle x,Xx\rangle - \frac{1}{2}\langle R^{-1}B^*W_x,B^*W_x\rangle - \langle W_x,Ax\rangle + \langle W_x,BR^{-1}B^*W_x\rangle.$$

Hence

$$W_t = -\frac{1}{2}\langle x,Xx\rangle + \frac{1}{2}\langle W_x,BR^{-1}B^*W_x\rangle - \langle W_x,Ax\rangle. \tag{9.19}$$

The form of equation (9.19) leads to the conjecture that there exists a solution of the Hamilton-Jacobi equation (2.14) of the form

$$W(t,x) = \frac{1}{2}\langle x,P(t)x\rangle \tag{9.20}$$

where for each t, $P(t)$ is a symmetric matrix. For then

$$W_x = P(t)x \qquad W_t = \frac{1}{2}\langle x,P'(t)x\rangle, \tag{9.21}$$

and for proper choice of $P(t)$ we would have a quadratic form in the left equal to a quadratic form on the right.

If we assume a solution of the form (9.20), substitute (9.21) into (9.19), and recall that $P^* = P$, we get

$$\frac{1}{2}\langle x, P'x\rangle = -\frac{1}{2}\langle x, Xx\rangle + \frac{1}{2}\langle x, PBR^{-1}B^*Px\rangle - \langle x, PAx\rangle. \tag{9.22}$$

For any matrix M, we can write

$$M = (M+M^*)/2 + (M-M^*)/2.$$

Hence

$$\langle x, Mx\rangle = \frac{1}{2}\langle x, (M+M^*)x\rangle \quad \text{for all} \quad x.$$

If we apply this observation to the matrix PA in (9.22) we get

$$\frac{1}{2}\langle x, P'x\rangle = -\frac{1}{2}\langle x, Xx\rangle + \frac{1}{2}\langle x, PBR^{-1}B^*Px\rangle - \frac{1}{2}\langle x, (PA+A^*P)x\rangle.$$

Therefore, if a solution to the Hamilton-Jacobi equation of the form (9.20) exists, the matrix P must satisfy the following differential equation

$$P' = -X + PBR^{-1}B^*P - (PA+A^*P). \tag{9.23}$$

Moreover, since

$$W(T, x_1) = g(x_1) = \frac{1}{2}\langle x_1, Gx_1\rangle,$$

it follows from (9.20) that the solution of (9.23) must satisfy the initial condition

$$P(T) = G. \tag{9.24}$$

Equation (9.23) is sometimes called the matrix Riccati equation. If a solution of (9.23) satisfying (9.24) exists, then from the first relation in (9.21) and from (9.18) we would expect the optimal synthesis or feedback control law to be given by

$$U(t,x) = -R^{-1}(t)B^*(t)P(t)x. \tag{9.25}$$

Note that the control law is linear in x, and its determination merely requires the solution of a differential equation.

We now show that the state of affairs suggested by the analysis in the last few paragraphs is indeed true.

THEOREM 9.3. Let $\mathcal{T}_1$ be as in (9.11), let g be as in (9.12), let $d = 0$ in (9.1), and let the constraint set $\mathcal{O}$ contain the origin. Then the problem of minimizing (9.3) subject to (9.1), control constraint Ω and terminal set $\mathcal{B}$, where the data of the problem satisfy Assumption 9.1 has an optimal synthesis. This synthesis is given by (9.25) and holds for all $a \leq t < T$ and all x in R^n. The matrix $P(t)$ is symmetric for each t and the function P is a solution, defined for all $a \leq t \leq T$ of the matrix Riccati equation (9.23) with initial condition (9.24).

Proof. It follows from standard existence and uniqueness theorems for ordinary differential equations that (9.23) has a unique solution satisfying (9.24) on some interval $(T-\delta,T]$. Note that if P is a solution of (9.23) satisfying (9.24), then so is P^*. By the uniqueness of solutions we then get that $P = P^*$, so that P is symmetric.

Let τ be any point on $(T-\delta,T)$ and let ξ be any point in E^n. We shall use the solution P obtained in the previous paragraph to construct an extremal for the problem with initial point (τ,ξ). By Theorem 9.2 this extremal will be unique and will furnish the minimum for the problem with initial point (τ,ξ).

Consider the linear system

$$\frac{dx}{dt} = A(t)x - B(t)R^{-1}(t)B^*(t)P(t)x \tag{9.26}$$

subject to initial conditions $x(\tau) = \xi$. We denote the solution of this system by $\phi(\ ,\tau,\xi)$. This solution is defined on the interval of definition of P and is unique.

Let

$$\lambda(t,\tau,\xi) = -P(t)\phi(t,\tau,\xi), \quad \tau \leq t \leq T \tag{9.27}$$

Note that

$$\lambda(T,\tau,\xi) = -P(T)\phi(T,\tau,\xi) = -G\phi(T,\tau,\xi),$$

where the last equality follows from (9.24). Thus λ satisfies (9.13).

If we differentiate (9.27) and then use (9.23) and (9.26), we get

$$\begin{aligned}\frac{d\lambda}{dt} = -\frac{dP}{dt}\phi - P\frac{d\phi}{dt} &= (X - PBR^{-1}B^*P + PA + A^*P) \\ &\quad - P(A-BR^{-1}B^*P)\phi = X\phi + A^*P\phi.\end{aligned}$$

If we now use (9.27) we get

$$\frac{d\lambda}{dt} = X\phi - A^*\lambda.$$

Hence by Theorem 9.1 and Corollary 9.1, $\phi(\cdot,\tau,\xi)$, and $\lambda(\cdot,\tau,\xi)$ determine an extremal pair $(\phi(\cdot,\tau,\xi), u(\cdot,\tau,\xi))$ with

$$u(t,\tau,\xi) = R^{-1}(t)B^*(t)\lambda(t,\tau,\xi).$$

It now follows from Theorem 9.2 that this extremal pair is the unique optimal pair for the problem.

From the definition of λ in (9.27), and the last equation it follows that

$$u(t,\tau,\xi) = -R^{-1}(t)B^*(t)P(t)\phi(t,\tau,\xi).$$

Therefore, since $\phi(\tau,\tau,\xi) = \xi$,

$$u(\tau,\tau,\xi) = -R^{-1}(\tau)B^*(\tau)P(\tau)\xi.$$

Since (τ,ξ) is an arbitrary point in $(T-\delta,T) \times E^n$ and since the optimal pair from (τ,ξ) is unique we obtain a synthesis of the optimal control by setting

$$U(\tau,\xi) = u(\tau,\tau,\xi).$$

Hence the optimal synthesis in $(T-\delta,T) \times E^n$ can be written as

$$U(t,x) = -R^{-1}(t)B^*(t)P(t)x,$$

where we have written a generic point as (t,x) instead of (τ,ξ). This, however, is precisely the relation (9.25).

We now show that the solution P of (9.23) with initial condition (9.24) is defined on the entire interval $[a,T]$. It will then follow that (9.25) holds for all $a \leq t \leq T$ and all x in E^n.

Let us now suppose that $\delta > 0$ is such that $(T-\delta,T]$ is the maximal interval with T as right hand end point on which the solution P is defined. From the standard existence theorem in the theory of ordinary differential equations and from the form of equation (9.23) it follows that $P(t)$ must be unbounded as $t \to T-\delta$ from the right. We shall show that if $T-\delta \geq a$, then P is bounded as $t \to T-\delta$. From this it will, of course, follow that P is defined on $[a,T]$ and (9.25) holds for all $a \leq t \leq T$ and x in E^n.

From the existence theorem for linear quadratic problems (Exercise III.6.5) and from Theorem 9.2 it follows that for all (τ,ξ) in $[a,T] \times E^n$ the function

$$W(\tau,\xi) = J(\phi(\cdot,\tau,\xi),u(\cdot,\tau,\xi))$$

is defined, where $(\phi(\cdot,\tau,\xi),u(\cdot,\tau,\xi))$ is the unique optimal pair for the problem with initial point (τ,ξ). The function W so defined is called the value function or value. Let $\tilde{\phi}(\cdot,\tau,\xi)$ denote the trajectory for the problem corresponding to the control $\tilde{u}$, where $\tilde{u}(t) = 0$ on $[\tau,T]$. Then

$$0 \le W(\tau,\xi) \le J(\tilde{\phi}(\cdot,\tau,\xi),\tilde{u}), \tag{9.28}$$

where the leftmost inequality follows from (ii) of Assumption 9.1 and from (9.12). From (9.1) with $d = 0$ we see that

$$\tilde{\phi}(t,\tau,\xi) = \Phi(t,\tau)\xi,$$

where $\Phi(\cdot,\tau)$ is the fundamental matrix for the system $dx/dt = A(t)x$ satisfying $\Phi(\tau,\tau) = I$. Therefore, from (9.3) and (9.12), we get

$$J(\tilde{\phi}(\cdot,\tau,\xi),\tilde{u}) = \langle \xi\Phi^*(T,\tau), G\Phi(T,\tau)\xi\rangle$$
$$+ \int_\tau^T \langle \xi\Phi^*(s,\tau), X(s)\Phi(s,\tau)\rangle\, ds.$$

From this it follows that given a compact set $\mathscr{X}$ in E^n, there exists a constant M, depending on $\mathscr{X}$ such that for all $a \le \tau \le T$ and all ξ in $\mathscr{X}$

$$J(\tilde{\phi}(\cdot,\tau,\xi),\tilde{u}) \le M.$$

Combining this inequality with (9.28) shows that given a compact set $\mathscr{X}$ in E^n, there exists a constant M, depending on $\mathscr{X}$ such that for all $a \le \tau \le T$ and all ξ in $\mathscr{X}$

$$0 \le W(\tau,\xi) \le M. \tag{9.29}$$

In (9.16), which was derived in the course of proving Theorem 9.2, we have an expression for $J(\phi(\tau,\xi),u(\tau,\xi))$, and hence for $W(\tau,\xi)$. Since $d = 0$ in the present discussion we have from (9.16),

$$W(\tau,\xi) = -\tfrac{1}{2}\langle \xi, \lambda(\tau,\tau,\xi)\rangle. \tag{9.30}$$

Here λ is the adjoint function, or multiplier function, for the problem with initial point (τ,ξ). It is not assumed here that λ is given by (9.27).

We now consider points (τ,ξ) such that $T-\delta < \tau \leq T$. For such points (9.27) holds. Thus

$$\lambda(\tau,\tau,\xi) = -P(\tau)\phi(\tau,\tau,\xi) = -P(\tau)\xi.$$

Substituting this into (9.30) gives

$$W(\tau,\xi) = \frac{1}{2}\langle\xi,P(\tau)\xi\rangle, \tag{9.31}$$

which is valid for $T-\delta < \tau \leq T$ and all ξ. From this and from (9.29) we get that for all $T-\delta \leq \tau \leq T$ and all ξ in a compact set $\mathscr{X}$,

$$0 \leq \frac{1}{2}\langle\xi,P(\tau)\xi\rangle \leq M.$$

Hence $P(\tau)$ must be bounded on $T-\delta \leq \tau \leq T$, and the theorem is proved.

In the course of proving Theorem 9.3 we also proved the following.

COROLLARY 9.3. The value function W is given by (9.31) for all $a \leq \tau \leq T$ and ξ in E^n.

CHAPTER VI

PROOF OF THE MAXIMUM PRINCIPLE

1. Introduction

This chapter is devoted to the proof of the maximum principle, Theorem V.3.1. We shall actually prove a theorem that is more general than the maximum principle and shall obtain the maximum principle as a special case of this theorem. An essential property of an optimal trajectory is used to motivate the introduction of a concept called $\mathscr{F}$-$\mathscr{N}$ extremality. A necessary condition for $\mathscr{F}$-$\mathscr{N}$ extremality is then stated (Theorem 3.1 of this chapter), and it is shown how this implies Theorem V.3.1.

To prove Theorem 3.1 below we proceed as follows. First we establish the effects of perturbations of a control on the end points of the corresponding trajectory. Certain convex sets of perturbations are then considered and it is shown that to first order, the end points of the perturbed trajectories generate a cone in a euclidean space. It is next shown that as a consequence of the optimality assumption, this cone is separated from a certain half-line. The analytic consequences of this separation constitute Theorem 3.1 and hence Theorem V.3.1.

In the last section of this chapter we prove Corollaries V.3.1 and V.3.2 which give sharpened versions of the maximum principle that are valid in important cases.

2. $\mathscr{F}$-$\mathscr{N}$ Extremals

In Theorem V.3.1 it is assumed that $g \equiv 0$. In proving the theorem it is more convenient to use the Mayer formulation. The equivalence of the two formulations was shown in II.4.

Let $\hat{f} = (f^0, f) = (f^0, f^1, \ldots, f^n)$, let $\hat{x} = (x^0, x) =$

$(x^0, x^1, \ldots, x^n)$ and let $\hat{\phi} = (\phi^0, \phi) = (\phi^0, \phi^1, \ldots, \phi^n)$. The problem we now consider is the following.

PROBLEM 2.1. Minimize

$$J(\hat{\phi}, u) = \phi^0(t_1) \tag{2.1}$$

subject to the system equations

$$\frac{d\hat{x}}{dt} = \hat{f}(t, x, u(t)), \tag{2.2}$$

control constraints $u(t) \in \Omega(t)$ and end conditions

$$\{(t_0, \phi(t_0), t_1, \phi(t_1)) \in \mathscr{B}, \quad \phi^0(t_0) = 0, \quad \phi^0(t_1) \text{ free}\}. \tag{2.3}$$

Note that the right hand side of (2.2) is an $(n+1)$-vector and that $\hat{f}$ depends only on $x = (x^1, \ldots, x^n)$ and not on x^0. Also note that a pair $(\hat{\phi}, u)$ is admissible for Problem 2.1 if and only if (ϕ, u) is admissible for the original Lagrange problem.

Let the notation be as in V.3. For every control u for which there exists a trajectory ϕ defined on an interval $[t_0, t_1] \subset \mathscr{I}_0$ such that (ϕ, u) is admissible, we obtain a function F_u defined on $[t_0, t_1] \times \mathscr{X}_0$ as follows:

$$\hat{F}_u(t, x) = \hat{f}(t, x, u(t)). \tag{2.4}$$

It follows from Assumption V.3.1-(ii) that for each x in $\mathscr{X}_0$ the function $t \to \hat{F}_u(t, x)$ is measurable on $[t_0, t_1]$ and for each t in $[t_0, t_1]$ the function $x \to \hat{F}_u(t, x)$ is $C^{(1)}$ on $\mathscr{X}_0$. From Assumption V.3.1-(iii) we get that for each compact set $\mathscr{X} \subset \mathscr{X}_0$ and each control u there exists a non-negative function $\mu = \mu(\cdot, \mathscr{X}, u)$ in $L_1[t_0, t_1]$ such that

$$|\hat{F}_u(t, x)| \leq \mu(t) \qquad \left|\frac{\partial \hat{F}_u}{\partial x}(t, x)\right| \leq \mu(t) \tag{2.5}$$

for almost all t in $[t_0,t_1]$.

If we set $\hat{F}_u(t,x) = 0$ for $t \notin [t_0,t_1]$, then $\hat{F}_u$ is defined on $\mathscr{I}_0 \times \mathscr{X}_0$ and the statements in the preceding paragraph hold with $[t_0,t_1]$ replaced by $\mathscr{I}_0$. We henceforth assume this to be the case.

Let $\mathscr{F}'$ denote the class of functions $\hat{F}_u$ obtained as in the preceding paragraph. The set $\mathscr{F}'$ is the set of "right hand sides of (2.2)" obtained by letting u vary over all admissible controls. Thus, Problem 2.1 can be rephrased as follows.

PROBLEM 2.2. Minimize (2.1) subject to

$$\frac{d\hat{x}}{dt} = \hat{F}_u(t,x) \qquad \hat{F}_u \in \mathscr{F}' \tag{2.6}$$

and end condition (2.3).

From the standard existence and uniqueness theorem in the theory of ordinary differential equations and from the second inequality in (2.5) it follows that once $\hat{F}_u$ is chosen and $(t_0,\phi(t_0))$ is specified, then the trajectory ϕ is uniquely determined. Thus in Problem 2.2 we have at our disposal the function $\hat{F}_u$ and the point $(t_0,\phi(t_0))$. Recall that $\phi^0(t_0) = 0$ is already specified. The function $\hat{F}_u$ and the point $(t_0,\phi(t_0))$ are to be chosen in such a way that (2.3) holds and $\phi^0(t_1)$ is minimized.

Let $(\hat{\phi}^*,u^*)$ be an optimal pair for Problem 2.1. Then corresponding to u^* we obtain a function $\hat{F}^* = \hat{F}_{u^*}$ which is optimal for Problem 2.2.

We now present the property of an optimal pair $(\hat{\phi}^*,u^*)$ that will be used in the proof of Theorem V.3.1.

Let

$$\mathscr{N}' = \{(t_0,\hat{x}_0,t_1,\hat{x}_1): (t_0,x_0,t_1,x_1) \in \mathscr{B},\ x_0^0 = 0,\ x_1 \leq \phi^{*0}(t_1)\} \tag{2.7}$$

$$\mathscr{M}' = \{(t_0,\hat{x}_0,t_1,\hat{x}_1): (t_0,x_0,t_1,x_1) \in \mathscr{B},\ x_0^0 = 0,\ x_1^0 = \phi^{*0}(t_1)\} \tag{2.8}$$

$$\mathcal{E}' = \{e(\hat{\phi}): \hat{\phi} \in \mathcal{A}_T\},$$

where $\mathcal{A}_T$ now denotes the set of admissible trajectories for Problem 2.1.

If we let $\mathcal{N}_1'$ consist of those points $(t_0,\hat{x}_0,t_1,\hat{x}_1)$ with $(t_0,x_0,t_1,x_1) \in \mathcal{B}$, $x_0^0 = 0$ and $\phi^{*0}(t_1)-\varepsilon < x_1^0 < \phi^*(t_1)+\varepsilon$ for some $\varepsilon > 0$ then since $\mathcal{B}$ is a $C^{(1)}$ manifold of dimension q, the following is true. The set $\mathcal{N}_1'$ is a $C^{(1)}$ manifold of dimension $q+1$. The set $\mathcal{M}'$ is a $C^{(1)}$ manifold of dimension q which is contained in $\mathcal{N}_1'$ and divides $\mathcal{N}_1'$ into two sets, one of which is $\mathcal{N}'$ and the other is $\mathcal{N}_1'-\mathcal{N}'$. Note that $\mathcal{M}'$ is contained in the boundary of $\mathcal{N}'$ and that $\mathcal{N}'$ can be considered as that part of $\mathcal{N}_1'$ that lies on one side of $\mathcal{M}'$.

We shall say that <u>$\mathcal{N}$ is a manifold with upper boundary $\mathcal{M}$</u> if there exists a manifold $\mathcal{N}_1$ such that $\mathcal{N}_1$, $\mathcal{N}$ and $\mathcal{M}$ are related as $\mathcal{N}_1'$, $\mathcal{N}'$ and $\mathcal{M}'$ are. We leave the formulation of a formal definition to the reader.

The statement that $(\hat{\phi}^*,u^*)$ is an optimal pair for Problem 2.1 implies that in a neighborhood of

$$e(\hat{\phi}^*) = (t_0,\hat{\phi}^*(t_0),t_1,\hat{\phi}^*(t_1))$$

no other admissible trajectory with end point in the neighborhood can have its zero-th component less than $\phi^{*0}(t_1)$. Actually, this property characterizes a local minimum. Thus, if $(\hat{\phi}^*,u^*)$ is an optimal pair, then $e(\hat{\phi})$ has the following property:

(i) $e(\hat{\phi}^*) \in \mathcal{M}'$

(ii) There exists a neighborhood $\mathcal{O}'$ of $e(\hat{\phi}^*)$ such that

$$(\mathcal{O}' \cap \mathcal{E}' \cap \mathcal{N}') \subset \mathcal{M}'.$$

Figure 2.1 illustrates the situation when $\hat{x} = (x^0,x^1)$, the initial point is fixed, and the terminal set is a one dimensional manifold $\mathscr{T}$.

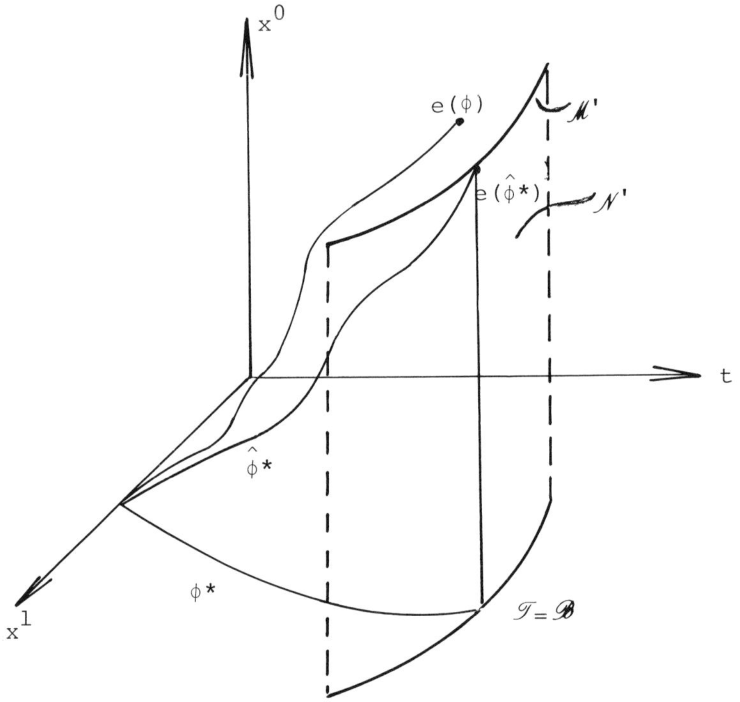

Figure 1

The preceding discussion leads to the following considerations. Let $\mathscr{Y}_0$ be an open interval in E^{n+1} and let points in E^{n+1} be denoted by $y = (y^0, y^1, \ldots, y^n)$. Let $\mathscr{F}$ be a family of functions defined on $\mathscr{I}_0 \times \mathscr{Y}_0$ with range in E^{n+1}. We denote the elements in $\mathscr{F}$ by F and assume the following.

ASSUMPTION 2.1. (i) Each F in $\mathscr{F}$ is measurable on $\mathscr{I}_0$ for each fixed y in $\mathscr{Y}_0$ and is $C^{(1)}$ on $\mathscr{Y}_0$ for each fixed t in $\mathscr{I}_0$. (ii) For each F in $\mathscr{F}$ and compact subset $\mathscr{Y} \subset \mathscr{Y}_0$ there exists a non-negative function $\mu = \mu(\cdot, \mathscr{Y}, F)$ in $L_1[\mathscr{I}_0]$ such that

$$|F(t,y)| \leq \mu(t) \qquad |F_y(t,y)| \leq \mu(t) \tag{2.9}$$

a.e. in $\mathscr{I}_0$.

REMARK 2.1. The functions $\hat{F}$ in $\mathscr{F}'$ are defined on $\mathscr{I}_0 \times \mathscr{X}_0$. Hence if we set $\mathscr{Y}_0 = E^1 \times \mathscr{X}_0$ and take

$$y = (y^0, x) \qquad y^0 \in E^1, \qquad x \in \mathscr{X}^0, \tag{2.10}$$

it follows from (2.5) and the continuity and measurability properties of the functions $\hat{F}_u$ noted on page 241, that the family $\mathscr{F}'$ satisfies Assumption 2.1.

Consider the family of differential equations

$$y' = F(t,y) \qquad F \in \mathscr{F} \qquad y(t_0) = y_0. \tag{2.11}$$

Let ψ be a solution of (2.11) defined on an interval $[t_0,t_1] \subset \mathscr{I}_0$. Let $\mathscr{S}$ denote the set of solutions of (2.11) and their intervals of definition. Thus

$$\mathscr{S} = \{[\psi,t_0,t_1]: \psi'(t) = F(t,\psi(t)), F \in \mathscr{F}, t \in [t_0,t_1] .$$

Let $e(\psi) = (t_0,\psi(t_0),t_1,\psi(t_1))$ and let

$$\mathscr{E} = \{e(\psi): \psi \in \mathscr{S}\}.$$

Let $\mathscr{N}$ be a differentiable manifold with a subset $\mathscr{M}$ of its boundary, where $\mathscr{M}$ is a differentiable manifold. The sets $\mathscr{N}$ and $\mathscr{M}$ are in $E^{2(n+1)+2}$.

<u>DEFINITION 2.1</u>. A solution $\bar{\psi}$ of (2.11) for some $\bar{F}$ in $\mathscr{F}$ is said to be an $\mathscr{F}$-$\mathscr{N}$ extremal if

(i) $e(\bar{\psi}) \in \mathscr{M}$

(ii) There exists a neighborhood $\mathscr{O}$ of $e(\bar{\psi})$ such that

$$(\mathscr{O} \cap \mathscr{E} \cap \mathscr{N}) \subset \mathscr{M}.$$

REMARK 2.2. If $(\hat{\phi}^*, u^*)$ is an optimal pair for Problem 2.1, then $\hat{\phi}^*$ is optimal for Problem 2.2 and $\hat{\phi}^*$ is an $\mathscr{F}'$-$\mathscr{N}'$ extremal. We shall deduce necessary conditions that an $\mathscr{F}$-$\mathscr{N}$ extremal must satisfy. These will give the desired necessary conditions for $(\hat{\phi}^*, u^*)$.

We now make further assumptions about $\mathscr{F}$. Let

$$P^r = \{\alpha: \alpha \in E^r, \quad \alpha^i \geq 0, \quad \sum_{i=1}^{r} \alpha^i = 1\}. \tag{2.12}$$

Let

$$\text{co } \mathscr{F} = \{h: h = \sum_{i=1}^{r} \alpha^i F_i, \ F_i \in \mathscr{F}, \ \alpha \in P^r \ \text{ for some } \ r\}.$$

The set co $\mathscr{F}$ will be called the convex hull of $\mathscr{F}$.

DEFINITION 2.2. A family $\mathscr{F}$ of functions F is said to be quasiconvex if $\mathscr{F}$ satisfies Assumption 2.1 and has the following property.

Let there be given an $\varepsilon > 0$, a compact set $\mathscr{Y} \subset \mathscr{Y}_0$, a finite collection of elements $F_1, \ldots, F_r$ in $\mathscr{F}$ and an α in P^r. Then there exists a function $g(\cdot,\cdot,\alpha,\varepsilon)$ defined on $\mathscr{I}_0 \times \mathscr{Y}$, depending on α, ε, $\mathscr{Y}$, and $F_1, \ldots, F_r$ with the following properties:

(i) The function g is measurable in t for fixed y in $\mathscr{Y}$ and is $C^{(1)}$ on $\mathscr{Y}$ for fixed t in $\mathscr{I}_0$.

(ii) The function F_α defined by

$$F_\alpha(t,y) = \sum_{i=1}^{r} \alpha^i F_i(t,y) - g(t,y,\alpha,\varepsilon)$$

is in $\mathscr{F}$.

(iii) There exists a function μ in $L_1[\mathscr{I}_0]$ depending on $\mathscr{Y}$ and $F_1, \ldots, F_r$ but not depending on α or ε such that

$$|g(t,y,\alpha,\varepsilon)| \leq \mu(t) \qquad |g_y(t,y,\alpha,\varepsilon)| \leq \mu(t) \tag{2.13}$$

for all y in $\mathscr{Y}$, all $\alpha \in P^r$, and almost all t in $\mathscr{I}_0$.

(iv) If $\mathscr{Z}$ is a family of equicontinuous functions defined on $\mathscr{I}_0$ with range in $\mathscr{Y}$, then for every z in $\mathscr{Z}$ and every t and t' in $\mathscr{I}_0$,

$$\left|\int_t^{t'} g(s,z(s),\alpha,\varepsilon)ds\right| < \varepsilon^2. \tag{2.14}$$

(v) If $\{\alpha_n\}$ is a sequence of points in P^r such that $\alpha_n \to \alpha$, then for every y in $\mathscr{Y}$

$$g(\cdot,y,\alpha_n,\varepsilon) \to g(\cdot,y,\alpha,\varepsilon)$$

in measure on $\mathscr{I}_0$.

Note that since P^r is closed the point α in (v) is in P^r.

LEMMA 2.1. The family $\mathscr{F}'$ is quasiconvex.

The elements of $\mathscr{F}'$ are the functions $\hat{F}_u$ defined by (2.4). Hence a finite collection of elements in $\mathscr{F}'$ can be written as a set of functions $\hat{f}_1,\ldots,\hat{f}_r$ defined by

$$\hat{f}_i(t,x) = \hat{f}(t,x,u_i(t)) \qquad i = 1,\ldots,r,$$

where the function $\hat{f}$ is the function in the right hand side of (2.2) and $u_1,\ldots,u_r$ are controls for Problem 2.1.

In Remark 2.1 we noted that the functions $\hat{F}_u$ in $\mathscr{F}'$ fulfill Assumption 2.1. The existence of a function g satisfying (i)-(iv) of Definition 2.1 is a consequence of Theorem IV.4.2. In that theorem replace n by $n+1$, take the functions f_i to be the present $\hat{f}_i$ and take

$$p^i(t) = \alpha \qquad q = r.$$

Then the required function g is given by

$$g(t,y,\alpha,\varepsilon) = \lambda(t,x),$$

where λ is defined in IV. (4.14) and y is as in (2.10).

To show that (v) holds we note that since

$$g(t,y,\alpha_n,\varepsilon) = \sum_{i=1}^{r} \alpha_n^i \hat{f}_i(t,x) - \hat{f}_{\alpha_n}(t,x)$$

and

$$g(t,y,\alpha,\varepsilon) = \sum_{i=1}^{r} \alpha^i \hat{f}_i(t,x) - \hat{f}_\alpha(t,x)$$

it suffices to show that

$$\hat{f}_{\alpha_n}(\cdot,x) \to f_\alpha(\cdot,x) \tag{2.15}$$

in measure on $\mathscr{I}_0$.

To see that this is so we first recall the definition of the functions $\hat{f}_\alpha$ and $\hat{f}_{\alpha_n}$ in Theorem IV.4.1. The interval $\mathscr{I}_0$ is first written as the union of an appropriate finite collection of subintervals

$$\mathscr{I}_0 = \bigcup_{j=1}^{k} I_j,$$

where $I_j = [t_j, t_{j+1}]$ with $\ldots < t_{j-1} < t_j < t_{j+1} < \ldots$, and where k depends on ε. To define $\hat{f}_\alpha$ each I_j is then partitioned into a finite number of non-overlapping subintervals designated from left to right as $E_{j1},\ldots,E_{jr}$. The length of E_{ji} is $\alpha^i|I_j|$. For t interior to E_{ji} we set $\hat{f}_\alpha(t,x) = \hat{f}_i(t,x)$, $i = 1,\ldots,r$. At the end points $\hat{f}_\alpha$ is defined in an arbitrary manner. Similarly, to define $\hat{f}_{\alpha_n}$ each interval I_j (the I_j's are the same for all α_n and α) is partitioned into a finite number of non-overlapping subintervals, designated from left to right as $E^n_{j1}, E^n_{j2},\ldots,E^n_{jr}$. The length of E^n_{ji} is $\alpha^i_n|I_j|$. For t interior to E^n_{ji}, set $\hat{f}_{\alpha_n}(t,x) = f_i(t,x)$, $i = 1,\ldots,r$. At the end points $\hat{f}_{\alpha_n}$ is defined in an arbitrary manner.

Since there are a finite number of intervals I_j it suffices to show that (2.15) holds in measure on each I_j. To simplify notation let us consider I_1 and let us suppose that I_1 has length one.

Then $E_{1i} = [\alpha^{i-1}, \alpha^i]$ and $E^n_{1i} = [\alpha^{i-1}_n, \alpha^i_n]$, where $\alpha^0_n = \alpha^0$ is the left hand end point of I_1. Since $\alpha_n \to \alpha$, it follows that for every $\eta > 0$ there exists a positive integer n_0 such that for $n > n_0$ and for all $j = 1,\ldots,r$, the point α^{j-1}_n either belongs to $[\alpha^{j-1}, \alpha^j]$ or to $[\alpha^{j-2}, \alpha^{j-1}]$; the point α^j_n either belongs to $[\alpha^{j-1}, \alpha^j]$ or to $[\alpha^j, \alpha^{j+1}]$; and

$$\sum_{i=1}^{r} |\alpha^i - \alpha^i_n| < \eta.$$

Here α with a negative or zero superscript denotes the left hand end point of I_1. It now follows from the definition of $\hat{f}_{\alpha_n}$ and of $\hat{f}_\alpha$ that $f_{\alpha_n}(t,x) = f_\alpha(t,x)$, except for t in a set of measure less than η. Thus (2.15) holds in measure on I_1.

3. A Necessary Condition for $\mathcal{F}$-$\mathcal{N}$ Extremality

Theorem 3.1 of this section gives necessary conditions that must be satisfied by an $\mathcal{F}$-$\mathcal{N}$ extremal. The maximum principle, Theorem V.3.1, is then deduced as a corollary of this theorem. The proof of Theorem 3.1 will be given in Sections 4-7.

For each function $F = (F^0, F^1, \ldots, F^n)$ in $\mathcal{F}$ we define a function H_F as follows:

$$H_F(t,y,\ell) = \sum_{i=0}^{n} \ell^i F^i(t,y) = \langle \ell, F(t,y) \rangle. \tag{3.1}$$

Note that ℓ is an $(n+1)$-vector. Each function H_F is real valued, is linear in ℓ, is $C^{(1)}$ in y and is measurable in t.

THEOREM 3.1. Let $\mathcal{F}$ be a quasiconvex family. Let $\bar{\psi}$ be an $\mathcal{F}$-$\mathcal{N}$ extremal defined on $[t_0, t_1]$ and let

$$\bar{\psi}'(t) = \bar{F}(t, \bar{\psi}(t)) \quad \text{a.e.} \tag{3.2}$$

Let

$$\bar{H}(t,y,\ell) = H_{\bar{F}}(t,y,\ell).$$

Then there exists an absolutely continuous function $\eta = (\eta^0, \eta^1, \ldots, \eta^n)$ defined on $[t_0,t_1]$ such that: (i) $\eta(t)$ is never zero on $[t_0,t_1]$; (ii) for almost all t in $[t_0,t_1]$

$$\bar{\psi}'(t) = \bar{H}_\ell(t,\bar{\psi}(t),\eta(t)) = \bar{F}(t,\bar{\psi}(t)) \tag{3.3}$$

$$\eta'(t) = -\bar{H}_y(t,\bar{\psi}(t),\eta(t)) = -\langle \eta(t), \bar{F}_y(t,\bar{\psi}(t))\rangle; \tag{3.4}$$

(iii) the inequality

$$\int_{t_0}^{t_1} \bar{H}(s,\bar{\psi}(s),\eta(s))ds \geq \int_{t_0}^{t_1} H_F(s,\bar{\psi}(s),\eta(s))ds \tag{3.5}$$

holds for all F in $\mathscr{F}$. Moreover, if the mapping $t \to \bar{F}(t,\bar{\psi}(t))$ is continuous at the end points t_0 and t_1, then the $2(n+1)+2$ dimensional vector

$$(\bar{H}(t_0,\bar{\psi}(t_0),\eta(t_0)),\ -\eta(t_0),\ -\bar{H}(t_1,\bar{\psi}(t_1),\eta(t_1)),\ \eta(t_1)) \tag{3.6}$$

is orthogonal to $\mathscr{M}$ at $e(\bar{\psi})$.

We now deduce the maximum principle, Theorem V.3.1, from this theorem.

We noted in Remark 2.2 that the solution $(\hat{\phi}^*, u^*)$ of Problem 2.1 has the property that $\hat{\phi}^*$ is an $\mathscr{F}'$-$\mathscr{N}'$ extremal. Hence Theorem 3.1 is applicable with $\mathscr{F}$ replaced by $\mathscr{F}'$, with $\bar{\psi}$ replaced by $\hat{\phi}^*$ and with y as in (2.10). The functions $\hat{F}$ in $\mathscr{F}'$ are given by (2.4) and $\bar{F}$ is replaced by $\hat{F}^* = \hat{F}^*_u$, which is optimal for Problem 2.2. The function H_F is replaced by $H_{\hat{F}}$. From (3.1) and (2.4) we get that $H_{\hat{F}}$ is given by the formula

$$H_{\hat{F}}(t,y,\ell) = \sum_{i=0}^{n} \ell^i \hat{F}^i_u(t,x) = \sum_{i=0}^{n} \ell^i f^i(t,x,u(t)), \tag{3.7}$$

where $\hat{f} = (f^0, f) = (f^0, f^1, \ldots, f^n)$ is the right hand side of (2.2) and u is an admissible control. Similarly we get

$$H_{\hat{F}*}(t,y,\ell) = \sum_{i=0}^{n} \ell^i f^i(t,x,u*(t)). \qquad (3.8)$$

From (3.8) we see that $H_{\hat{F}*}$ is independent of y^0. Hence the first component of the vector equation (3.4) is

$$\frac{d\eta^0}{dt} = 0.$$

Thus η^0 is a constant, which we designate as λ^0. Let

$$\lambda = (\eta^1,\ldots,\eta^n) \quad \text{and} \quad \hat{\lambda} = (\lambda^0,\lambda) = (\lambda^0,\lambda^1,\ldots,\lambda^n).$$

If we now observe that the functions $\hat{F}_u$ in $\mathcal{F}'$ do not depend on y^0, then the last n component of (3.4) give

$$\lambda'(t) = -\lambda^0 f_x^0(t,\phi*(t),u*(t)) - \langle \lambda(t), f_x(t,\phi*(t),u*(t))\rangle.$$

The last n components of (3.3) give

$$\phi*'(t) = f(t,\phi*(t),u*(t)).$$

The last two equations above are precisely equations V. (3.2) of Theorem V.3.1 with the optimal pair designated as $(\phi*,u*)$ instead of (ϕ,u).

From (3.5), (3.7) and (3.8) we get (3.3) of Theorem V.3.1.

From the orthogonality of (3.6) to $\mathcal{M} = \mathcal{M}'$ at $e(\overline{\psi})$ and from (2.8) it follows that V (3.4) is orthogonal to $\mathcal{B}$ at $e(\phi*)$.

Let ν be any vector in $E^{2(n+1)+2}$ of the form

$$(\tau-t_0,\ \hat{\xi}_0-\hat{\phi}*(t_0),\ \tau_1-t_1,\ \hat{\xi}_1-\hat{\phi}*(t_1)),$$

with

$$\xi_1^0 - \phi^{*0}(t_1) < 0, \qquad (3.9)$$

and in the tangent plane to $\mathcal{N}$ at $e(\hat{\phi}*(t_1))$. In the course of proving Theorem 3.1 it will be shown that the inner product of (3.6) with any such ν is always ≥ 0. One such ν is the vector with all entries zero except for $\xi_1^0-\phi^{*0}(t_1)$. The inner product of this particular ν and (3.6) is

$$\eta^0(t_1)(\xi_1^0 - \phi^{*0}(t_1)) \geq 0.$$

This and (3.9) imply that $\eta^0(t_1) \leq 0$. But $\eta^0(t_1) = \lambda^0$, so that $\lambda^0 \leq 0$. All statements in Theorem V.3.1 are now satisfied.

4. Perturbations of the Extremal Trajectory

We begin this section with statements of results that we shall need concerning solutions of the system of differential equations

$$\frac{dy}{dt} = G(t,y,\beta) \qquad y(\tau) = \eta \tag{4.1}$$

where $t \in \mathscr{I}_0$, $y \in \mathscr{Y}_0$ and β is in some open region B of euclidean space. We shall denote solutions of (4.1) by $\theta = \theta(\cdot,\tau,\eta,\beta)$. By virtue of the initial condition $y(\tau) = \eta$ we have $\eta = \theta(\tau,\tau,\eta,\beta)$.

Although the statements in Lemmas 4.1 and 4.2 below are well known, not all of the statements appear in the form we give in any single standard reference. For the reader's convenience we shall refer him to a single source and we shall indicate how one gets the statements in the lemmas that are not found in the given reference.

LEMMA 4.1. For each (y,β) in $\mathscr{Y}_0 \times B$ let $G(\cdot,y,\beta)$ be measurable on $\mathscr{I}_0$ and for each t in $\mathscr{I}_0$ and $\beta \in B$ let $G(t,\cdot,\beta)$ be $C^{(1)}$ on $\mathscr{Y}_0$. Let there exist a compact interval $\mathscr{I} \subset \mathscr{I}_0$, a compact interval $\mathscr{Y}$ contained in $\mathscr{Y}_0$, and a function μ in $L_1[\mathscr{I}]$ such that for all (t,y,β) in $\mathscr{I} \times \mathscr{Y} \times B$

$$|G(t,y,\beta)| \leq \mu(t) \qquad |G_y(t,y,\beta)| \leq \mu(t). \tag{4.2}$$

Then for each (τ,η,β) in the interior of $\mathscr{I} \times \mathscr{Y} \times B$ there exists a unique solution $\theta(\cdot,\tau,\eta,\beta)$ of (4.1) defined on a maximal interval $(\omega_-(\tau,\eta,\beta),\ \omega_+(\tau,\eta,\beta))$. Moreover, the one-sided limits

$$\lim_{\tau \to \omega_-} \theta(t,\tau,\eta,\beta) = \theta_-$$

$$\lim_{\tau \to \omega_+} \theta(t,\tau,\eta,\beta) = \theta_+$$

exist and the points (ω_-,θ_-) and (ω_+,θ_+) are boundary points of $\mathcal{I} \times \mathcal{Y}$.

For each fixed β in B the existence of a solution θ of (4.1) on some closed interval $[\tau-\varepsilon(\eta,\beta),\ \tau+\varepsilon(\eta,\beta)]$, where $\varepsilon(\eta,\beta) > 0$, follows from Theorem 68.4 of [38]. We now show that θ is unique on any interval on which it is defined. Let ψ be another solution of (4.1) and for $\theta(t,\tau,\eta,\beta)$ let us simply write $\theta(t)$. It then follows from the relation $\psi(\tau) = \theta(\tau) = \eta$, from the mean value theorem, and from (4.2) that for any t in the common interval of definition of θ and ψ,

$$|\theta(t)-\psi(t)| \leq \int_\tau^t |G(s,\theta(s),\beta)-G(s,\psi(s),\beta)|ds$$

$$\leq \int_\tau^t \mu(s)|\theta(s)-\psi(s)|ds.$$

The uniqueness now follows from Gronwall's Lemma. (Lemma IV.4.2).

Having established the existence of a closed interval with τ in its interior on which a unique solution of (4.1) exists, it follows that there is a maximal open interval $(\omega_-(\tau,\eta,\beta),\ \omega_+(\tau,\eta,\beta))$ contained in $\mathcal{I}$ on which a unique solution of (4.1) exists. We now show that the one-sided limits θ_- and θ_+ exist. Let t_1 and t_2 be any two points satisfying $\tau < t_1 < t_2 < \omega_+(\tau,\eta,\beta)$. Then by virtue of (4.2)

$$|\theta(t_2)-\theta(t_1)| \leq \int_{t_1}^{t_2} |G(s,\theta(s),\beta)|ds \leq \int_{t_1}^{t_2} \mu(s)ds.$$

If we let $t_1 \to \omega_+$ and $t_2 \to \omega_+$, then it follows from the integrability of μ and from the Cauchy criterion that θ_+ exists. A similar argument shows that θ_- exists.

If the point $(\omega_+(\tau,\eta,\beta),\theta_+)$ were an interior point of $\mathscr{I}\times\mathscr{Y}$ then by the existence and uniqueness results given above we would get that there exists a unique solution of

$$\frac{dy}{dt} = G(t,y,\beta) \qquad y(\omega_+) = \theta_+$$

defined on some interval with ω_+ in its interior. To the left of ω_+ this solution would agree with θ. Thus, we would contradict the maximality of $(\omega_-(\tau,\eta,\beta),\ \omega_+(\tau,\eta,\beta))$. Hence $(\omega_+(\tau,\eta,\beta),\theta_+)$ is a boundary point of $\mathscr{I}\times\mathscr{Y}$. A similar argument shows that $(\omega_-(\tau,\eta,\beta),\theta_-)$ is a boundary point of $\mathscr{I}\times\mathscr{Y}$.

LEMMA 4.2. Let G be as in Lemma 4.1, except that for each t in $\mathscr{I}_0$ we assume that $G(t,\cdot,\cdot)$ is $C^{(1)}$ on $\mathscr{Y}_0\times B$. Let $(\tau,\bar{\eta},\bar{\beta})$ be a point in the interior of $\mathscr{I}\times\mathscr{Y}\times B$ and let t' and t'' be two points in $\mathscr{I}$ such that $\omega_-(\tau,\bar{\eta},\bar{\beta}) < t' < t'' < \omega_+(\tau,\bar{\eta},\bar{\beta})$. Then the following statements are true: (i) there exist positive numbers ε_1 and δ such that for each (η,β) in $\mathscr{Y}\times B$ satisfying

$$|\eta-\bar{\eta}| < \delta \qquad |\beta-\bar{\beta}| < \delta \tag{4.4}$$

there is a unique solution $\theta(\cdot,\tau,\eta,\beta)$ of (4.1) defined on the interval

$$t'-\varepsilon_1 \le t \le t''+\varepsilon_1. \tag{4.5}$$

(ii) The function $\theta(t,\tau,\cdot,\cdot)$ is differentiable with respect to η and β and the partial derivatives θ_η and θ_β are continuous on the set defined by (4.4) and (4.5). (iii) For each (η,β) satisfying (4.4) the matrix function $\theta_\eta(\cdot,\tau,\eta,\beta)$ is a fundamental matrix of solutions for the system

$$\nu' = G_y(t,\theta(t,\tau,\eta,\beta),\beta)\nu$$

on the interval (4.5) and $\theta_\eta(\cdot,\tau,\eta,\beta)$ satisfies the initial condition

$$\theta_\eta(\tau,\tau,\eta,\beta) = I,$$

where I is the $(n+1) \times (n+1)$ identity matrix.

Statements (i) and (ii) of Lemma 4.2 follow from parts (C_1) and (C_2) of Theorem 69.4 of [38]. Statement (iii) is established in the course of proving Theorem 69.4 of [38]; see equation (P) and related text on page 362 of [38].

COROLLARY 4.1. Let

$$B = \{\beta\colon \beta = (\varepsilon\alpha,\varepsilon) = \varepsilon(\alpha^1,\ldots,\alpha^r,1),\ \alpha \in P^r,\ -\tfrac{1}{2} < \varepsilon < \tfrac{1}{2}\}, \tag{4.6}$$

where P^r is as in (2.12), and let

$$\theta_\varepsilon = \theta_\beta \frac{\partial\beta}{\partial\varepsilon} \qquad G_\varepsilon = G_\beta \frac{\partial\beta}{\partial\varepsilon}. \tag{4.7}$$

Then for each (η,β) satisfying (4.4) the function $\theta_\varepsilon(\cdot,\tau,\eta,\beta)$ is a solution of

$$\begin{aligned} \nu' &= G_y(t,\theta(t,\tau,\eta,\beta),\beta)\nu + G_\varepsilon(t,\theta(t,\tau,\eta,\beta),\beta) \\ \nu(\tau) &= 0 \end{aligned} \tag{4.8}$$

on the interval (4.5).

<u>Proof</u>. Consider the system

$$\begin{aligned} \frac{dy}{dt} &= G(t,y,(\varepsilon\alpha,\varepsilon)) \qquad y(\tau) = \eta \\ \frac{d\varepsilon}{dt} &= 0 \qquad \varepsilon(\tau) = \varepsilon, \end{aligned} \tag{4.9}$$

where α is a fixed element in P^r. Since we consider a fixed α, the system (4.9) does not involve a parameter. Lemma 4.2 is valid, with obvious modifications, if we consider β as fixed. Hence Lemma

4.2, with statements involving the parameter β deleted, is applicable to the system (4.9). It follows from Lemma 4.2 that $\theta(\cdot,\tau,\eta,\beta)$ and $\varepsilon(\cdot,\tau,\eta,\beta) = \varepsilon$, where $\beta = (\alpha\varepsilon,\varepsilon)$, is the unique solution of (4.9) on an interval $t'-\varepsilon_1 \le t \le t"+\varepsilon_1$ for all η satisfying (4.4) and ε satisfying $|\varepsilon| \le \varepsilon_2$ for some $\varepsilon_2 > 0$. It follows from (iii) of Lemma 4.2 that for $t'-\varepsilon_1 \le t \le t"+\varepsilon_1$,

$$\begin{pmatrix} \theta_\eta & \theta_\varepsilon \\ \varepsilon_\eta & \varepsilon_\varepsilon \end{pmatrix}' = \begin{pmatrix} G_y & G_\varepsilon \\ 0 & 0 \end{pmatrix} \begin{pmatrix} \theta_\eta & \theta_\varepsilon \\ \varepsilon_\eta & \varepsilon_\varepsilon \end{pmatrix},$$

where the prime in the matrix equation denotes differentiation with respect to t, the functions θ_η, θ_ε, ε_η, ε_ε and their derivatives are evaluated at (t,τ,η,β), and the functions G_y and G_ε are evaluated at $(t,\theta(t,\tau,\eta,\beta),\beta)$. Here θ_η and G_y are $(n+1) \times (n+1)$ matrices, θ_ε and G_ε are $(n+1)$-dimensional column vectors, ε_η is an $(n+1)$-dimensional row vector and ε_ε is a scalar. From the relation $\varepsilon(t,\tau,\eta,\beta) = \varepsilon$ it follows that $\varepsilon_\varepsilon = 1$. From this and from the matrix equation above it follows that $\theta_\varepsilon(\cdot,\tau,\eta,\beta)$ satisfies the differential equation (4.8). From the relation

$$\begin{pmatrix} \theta_\eta(\tau,\tau,\eta,\beta) & \theta_\varepsilon(\tau,\tau,\eta,\beta) \\ \varepsilon_\eta(\tau,\tau,\eta,\beta) & \varepsilon_\varepsilon(\tau,\tau,\eta,\beta) \end{pmatrix} = \begin{pmatrix} I & 0 \\ 0 & 1 \end{pmatrix},$$

which follows from (iii) of Lemma (4.2), we see that $\theta_\varepsilon(\tau,\tau,\eta,\beta) = 0$. Thus, all statements in (4.8) are established.

REMARK 4.1. Let B be as in (4.6). Let δt_0 and δt_1 be scalars, and let δw be an $(n+1)$-vector such that $(\delta t_0,\delta t_1,\delta w)$ is in a fixed compact set K in E^{n+3}. Let τ and $\bar{\eta}$ be fixed and let $\bar{\beta} = 0$. Then $\bar{\beta}$ corresponds to $\varepsilon = 0$. Let $[t_0,t_1]$ be a closed interval contained in the interior of the open interval

$(\omega_-(\tau,\bar{\eta},0),\ \omega_+(\tau,\bar{\eta},0))$. Recall that $(\omega_-(\tau,\bar{\eta},0),\ \omega_+(\tau,\bar{\eta},0))$ is the maximal interval on which the solution $\theta(\cdot,\tau,\bar{\eta},0)$ of (4.1) is defined. Let $\eta = \bar{\eta}+\varepsilon\delta w$. It then follows from Lemma 4.2 that there exists an $\varepsilon_0 > 0$ such that the function

$$\theta(\cdot,\tau,\bar{\eta}+\varepsilon\delta w,\beta) \qquad \beta = (\varepsilon\alpha,\varepsilon)$$

is a solution of (4.1) on the interval $[t_0+\varepsilon\delta t_0, t_1+\varepsilon\delta t_1]$ for all $(\delta t_0,\delta t_1,\delta w)$ in K, all α in P^r, and all $0 \leq \varepsilon \leq \varepsilon_0$. Since (4.8) is a linear system for ν it follows that θ_ε is a solution of (4.8) on $[t_0+\varepsilon\delta t_0,\ t_1+\varepsilon\delta t_1]$ for all α in P^r, all $(\delta t_0,\delta t_1,\delta w)$ in K and all $0 \leq \varepsilon \leq \varepsilon_0$.

Let $\bar{\psi}$ be an $\mathscr{F}$-$\mathscr{N}$ extremal defined on an interval $[t_0,t_1]$ that is interior to $\mathscr{I}_0$. Then if τ is a point in (t_0,t_1), the function $\bar{\psi}$ is the unique solution on $[t_0,t_1]$ of the differential equation

$$y' = \bar{F}(t,y) \qquad y(\tau) = \bar{\psi}(\tau), \tag{4.10}$$

for some $\bar{F}$ in $\mathscr{F}$. Let $\mathscr{I}$ be a compact interval contained in $\mathscr{I}_0$ and let $\mathscr{Y}$ be a compact interval contained in $\mathscr{Y}_0$ such that the graph of $\bar{\psi}$ is in the interior of $\mathscr{I} \times \mathscr{Y}$.

We wish to perturb the right hand side of (4.10) by an element δF in co $\mathscr{F} - \bar{F}$. Unfortunately, this may not be possible since the resulting element $\bar{F} + \varepsilon\delta F$ need not belong to $\mathscr{F}$. Since $\mathscr{F}$ is quasiconvex we can add a "small" element g to $\bar{F} + \varepsilon\delta F$ so that $\bar{F} + \varepsilon\delta F+g$ is in $\mathscr{F}$. We now make this precise.

Let $F_1,\ldots,F_r$ be r fixed functions in $\mathscr{F}$ and let α be an element in P^r. Define

$$\delta F = \sum_{i=1}^{r} \alpha^i F_i - \bar{F}.$$

Note that δF depends on $F_1,\ldots,F_r$ and on α. We shall not exhibit this dependence in the notation, but it is important for the reader to keep this dependence in mind. Let $0 \le \varepsilon \le 1$. Then

$$\bar{F} + \varepsilon\delta F = (1-\varepsilon)\bar{F} + \varepsilon \sum_{i=1}^{r} \alpha^i F_i . \tag{4.11}$$

The right hand side of (4.11) is a convex combination of $(r+1)$ functions in $\mathcal{F}$. Since $\mathcal{F}$ is quasiconvex there exists a function $g = g(\cdot,\cdot,\alpha,\varepsilon)$ defined on $\mathcal{J}_0 \times \mathcal{Y}_0$, depending on α, ε, $\bar{F}$, $F_1,\ldots,F_r$ and $\mathcal{Y}$, possessing the properties listed in Definition 2.2, and such that

$$\bar{F} + \varepsilon\delta F + g \in \mathcal{F}.$$

It follows from (2.9) in Assumption 2.1 that we may suppose the function μ in (iii) of Definition 2.2 to be such that (2.13) holds and that

$$\begin{aligned} |\bar{F}(t,y)| &\le \mu(t) \qquad & |F_i(t,y)| &\le \mu(t) \\ |\bar{F}_y(t,y)| &\le \mu(t) \qquad & |F_{iy}(t,y)| &\le \mu(t) \end{aligned} \tag{4.12}$$

for all $i = 1,\ldots,r$ and (t,y) in $\mathcal{J} \times \mathcal{Y}$.

We shall consider the family of perturbations of (4.10) given by

$$\frac{dy}{dt} = \bar{F}(t,y) + \varepsilon\delta F(t,y) + g(t,y,\alpha,\varepsilon), \tag{4.13}$$

where α is in P^r for some fixed r and $0 \le \varepsilon \le 1$. We shall also perturb the interval of definition of $\bar{\psi}$ by considering intervals $[t_0+\varepsilon\delta t_0,\ t_1+\varepsilon\delta t_1]$, where δt_0 and δt_1 range over a compact set. Finally, we shall perturb the initial condition $y(\tau) = \bar{\psi}(\tau)$ by considering initial conditions $\bar{\psi}(\tau) + \varepsilon\delta w$, where δw ranges over a compact set in E^{n+1}. The effects of such perturbations on the trajectory $\bar{\psi}$ are summarized in the following lemma.

LEMMA 4.3. Let $F_1,\ldots,F_r$ be r given functions in $\mathcal{F}$. Let

δt_0 and δt_1 be scalars, let δw be a vector in E^{n+1}, and let α be an element in P^r. Then there exists an $\varepsilon_0 > 0$ such that for every $0 < \varepsilon \leq \varepsilon_0$ there exists a solution $\psi = \psi(\cdot, \delta w, \alpha, \varepsilon)$ of (4.13) defined on the interval $[t_0+\varepsilon\delta t_0, t_1+\varepsilon\delta t_1]$ and satisfying the initial condition

$$\psi(\tau) = \overline{\psi}(\tau) + \varepsilon\delta w. \tag{4.14}$$

The solution ψ has the form

$$\psi(t,\delta w,\alpha,\varepsilon) = \overline{\psi}(t) + \varepsilon\delta\overline{\psi}(t) + E(t,\delta w,\alpha,\varepsilon) \tag{4.15}$$

where $\delta\overline{\psi}$ is the solution of

$$\delta y' = \overline{F}_y(t,\overline{\psi}(t))\delta y + \delta F(t,\overline{\psi}(t)) \tag{4.16}$$

satisfying the initial condition

$$\delta\overline{\psi}(\tau) = \delta w. \tag{4.17}$$

Moreover,

$$|E(t,\delta w,\alpha,\varepsilon)|/\varepsilon \to 0 \tag{4.18}$$

as $\varepsilon \to 0$ uniformly for t in $[t_0+\varepsilon\delta t_0, t_1+\varepsilon\delta t_1]$, for all $(\delta t_0, \delta t_1, \delta w)$ in a compact set K and all α in P^r.

REMARK 4.2. The reader is reminded that δF depends on $F_1,\ldots,F_r$ and α. The function $\delta\overline{\psi}$ therefore also depends on these quantities.

<u>Proof</u>. By virtue of (4.11) we can write

$$G(t,y,\varepsilon\alpha,\varepsilon) = \overline{F}(t,y) + \varepsilon\delta F(t,y). \tag{4.19}$$

Hence we can write (4.13) as

$$\frac{dy}{dt} = G(t,y,\varepsilon\alpha,\varepsilon) + g(t,y,\alpha,\varepsilon). \tag{4.20}$$

If we let $\beta = (\varepsilon\alpha, \varepsilon)$ and let

$$\hat{G}(t,y,\beta) = G(t,y,\varepsilon\alpha,\varepsilon),$$

we see that the function $\hat{G}$ has the properties that are imposed on the function G of Lemmas 4.1 and 4.2. Therefore these lemmas and their consequences are applicable to the system

$$\frac{dy}{dt} = G(t,y,\varepsilon\alpha,\varepsilon). \tag{4.21}$$

Unfortunately, we have no information about the regularity of g as a function of α and ε, so that we cannot apply these results to (4.20). We therefore proceed in two steps. First we analyze (4.21) and certain related perturbations of the initial data and end points. We then compare these perturbed solutions with those of (4.20), which we can do because of the "smallness" of g in an appropriate sense.

Consider (4.21) with initial condition

$$y(\tau) = \eta = \bar{\eta}+\varepsilon\delta w, \tag{4.22}$$

where $\bar{\eta} = \bar{\psi}(\tau)$. We denote the solution of this initial value problem by

$$\theta = \theta(\cdot,\eta,\beta) \qquad \beta = (\varepsilon\alpha,\varepsilon).$$

Since τ will remain fixed throughout our discussion we do not indicate the dependence of θ on τ in our notation. From (4.19) we see that for $\varepsilon = 0$ the system reduces to (4.10). The unique solution of (4.10) on the interval $[t_0,t_1]$ is $\bar{\psi}$. Thus

$$\bar{\psi}(t) = \theta(t,\bar{\eta},0).$$

It follows from Lemma 4.2 and Remark 4.1 that there exists an $\varepsilon_0 > 0$ such that for all $0 \leq \varepsilon \leq \varepsilon_0$ the function $\theta(\cdot,\eta,\beta)$ with $\beta = (\varepsilon\alpha,\varepsilon)$ is a solution of (4.21) subject to initial conditions

(4.22) on $[t_0+\varepsilon\delta t_0, t_1+\varepsilon\delta t_1]$ for all $(\delta t_0,\delta t_1,\delta w)$ in K and all α in P^r. Moreover, the function θ is differentiable with respect to η and β and the derivatives θ_η and θ_β are continuous functions of (t,η,β) on the domain of definition of θ. From the properties of θ_β and of G, from (4.7), and from (4.6) it follows that θ_ε is a continuous function of (t,η,ε) on its domain of definition. From Corollary 4.1 we get that

$$\theta'_\eta(t,\bar{\eta},0) = G_y(t,\bar{\psi}(t),0)\theta_\eta(t,\bar{\eta},0) \tag{4.23}$$

$$\theta'_\varepsilon(t,\bar{\eta},0) = G_y(t,\bar{\psi}(t),0)\theta_\varepsilon(t,\bar{\eta},0) + G_\varepsilon(t,\bar{\psi}(t),0), \tag{4.24}$$

where θ_ε and G_ε are as in (4.7) and the prime denotes differentiation with respect to t. Note that θ_η is a matrix and that θ_ε is a vector. From (4.19) we get that

$$G_y(t,\bar{\psi}(t),0) = \bar{F}_y(t,\bar{\psi}(t))$$

$$G_\varepsilon(t,\bar{\psi}(t),0) = \delta F(t,\bar{\psi}(t)).$$

Substitution of the right side of these equations in (4.23) and (4.24) yields:

$$\theta'_\eta(t,\bar{\eta},0) = \bar{F}_y(t,\bar{\psi}(t))\theta_\eta(t,\bar{\eta},0) \tag{4.25}$$

and

$$\theta'_\varepsilon(t,\bar{\eta},0) = \bar{F}_y(t,\bar{\psi}(t))\theta_\varepsilon(t,\bar{\eta},0) + \delta F(t,\bar{\psi}(t)). \tag{4.26}$$

The matrix θ_η and the vector θ_ε also satisfy the initial conditions

$$\theta_\eta(\tau,\bar{\eta},0) = I \qquad \theta_\varepsilon(\tau,\bar{\eta},0) = 0. \tag{4.27}$$

We now compare $\theta(t,\eta,\beta)$ with $\bar{\psi}(t) = \theta(t,\bar{\eta},0)$. Let

$$\begin{aligned}\Pi(\varepsilon_0) = \{(t,\delta t_0,\delta t_1,\delta w,\alpha,\varepsilon):\ & t \in [t_0+\varepsilon\delta t_0, t_1+\varepsilon\delta t_1];\\ & (\delta t_0,\delta t_1,\delta w) \in K;\ \ \alpha \in P^r,\ \ 0 < \varepsilon \le \varepsilon_0\}.\end{aligned} \tag{4.28}$$

Since θ_ε and θ_η exist and are continuous in all of their arguments and since $\eta = \bar{\eta}+\varepsilon\delta w$, we have

$$\theta(t,\eta,\beta)-\bar{\psi}(t) = \varepsilon[\theta_\eta(t,\bar{\eta},0)+o(1)]\delta w$$

$$+ \varepsilon[\theta_\varepsilon(t,\bar{\eta},0)+o(1)],$$

where the term $o(1)$ tends to zero as $\varepsilon \to 0$, uniformly on $\Pi(\varepsilon_0)$. Thus we may write

$$\theta(t,\eta,\beta)-\bar{\psi}(t) = \varepsilon[\theta_\eta(t,\bar{\eta},0)\delta w+\theta_\varepsilon(t,\bar{\eta},0)] + E_2(t,\delta w,\alpha,\varepsilon),$$

where $|E_2|/\varepsilon \to 0$ as $\varepsilon \to 0$, uniformly on $\Pi(\varepsilon_0)$.

If we let

$$\delta\bar{\psi}(t) = \theta_\eta(t,\bar{\eta},0)\delta w + \theta_\varepsilon(t,\bar{\eta},0),$$

then the preceding equation can be written as

$$\theta(t,\eta,\beta) = \bar{\psi}(t) + \varepsilon\delta\bar{\psi}(t) + E_2(t,\delta w,\alpha,\varepsilon). \tag{4.29}$$

From (4.25), (4.26), and (4.27) we see that $\delta\bar{\psi}$ is a solution of the system (4.16) subject to the initial conditions (4.17). Therefore, by the variation of parameters formula,

$$\delta\bar{\psi}(t) = \Psi(t,\tau)[\delta w + \int_\tau^t \Psi^{-1}(s,\tau)\delta F(s,\bar{\psi}(s))ds], \tag{4.30}$$

where Ψ is the fundamental matrix solution of the linear homogeneous system

$$\frac{dy}{dt} = \bar{F}_y(t,\bar{\psi}(t))y$$

satisfying the initial condition $\Psi(\tau,\tau) = I$.

Recall that we let $\mathscr{I}\times\mathscr{Y}$ denote a compact interval in $E^1 \times E^{n+1}$ such that the graph of $\bar{\psi}$ is interior to $\mathscr{I}\times\mathscr{Y}$. It follows from (4.29), from the properties of E_2, and from (4.30), that there exists an $\varepsilon_0' < \varepsilon_0$ such that for $\varepsilon \le \varepsilon_0'$ the graph of θ is interior to $\mathscr{I} \times \mathscr{Y}$ for all $(t, \delta t_0, \delta t_1, \delta w, \alpha, \varepsilon)$ in $\Pi(\varepsilon_0')$. We now relabel ε_0' as ε_0. All previous statements involving $\Pi(\varepsilon_0)$ are still valid. This procedure of taking $\varepsilon_0' < \varepsilon_0$ and relabeling ε_0' as ε_0 with the result that all previous statements involving $\Pi(\varepsilon_0)$ or $\varepsilon \le \varepsilon_0$ are valid for the new value of ε_0, will occur several times in the proof. Henceforth when we do this we shall merely state that we take $\varepsilon_0' < \varepsilon_0$ and relabel ε_0' as ε_0.

We now return to the perturbed differential equation (4.13), or its equivalent, (4.20). We consider (4.20) and the initial condition (4.14). Since the right hand side of (4.20) is $C^{(1)}$ in y for fixed t, α, ε, there exists an ε_0' such that for $0 \le \varepsilon \le \varepsilon_0'$, for all α in P^r, and all δw in the compact set K', which is the projection of K into the δw-space E^{n+1}, the following statement is true. The differential equation (4.20) subject to initial condition (4.14) has a unique solution

$$\psi = \psi(\cdot, \delta w, \alpha, \varepsilon) \tag{4.31}$$

defined on a maximal interval $(\omega_-(\delta w, \alpha, \varepsilon),\ \omega_+(\delta w, \alpha, \varepsilon))$ such that the graph of ψ is interior to $\mathscr{I} \times \mathscr{Y}$. We now take ε_0' to be less than ε_0 and relabel ε_0' as ε_0.

It follows from (2.13) and from (4.12) that for (t,y) in the compact set $\mathscr{I} \times \mathscr{Y}$ which contains the graph of $\bar{\psi}$ in its interior, the right hand side of (4.20) is bounded by an integrable function in $L_1[\mathscr{I}]$ for all α in P^r and all $0 < \varepsilon \le \varepsilon_0$. From this observation it follows that the set of solutions of ψ given by (4.31) is an equicontinuous family on $\mathscr{I}$, where we extend $\psi(\cdot, \delta w, \alpha, \varepsilon)$ to be continuous on all of $\mathscr{I}$ by defining it to be the appropriate constants outside of the interval $(\omega_-(\delta w, \alpha, \varepsilon),\ \omega_+(\delta w, \alpha, \varepsilon))$.

We now compare $\psi(t, \delta w, \alpha, \varepsilon)$ with $\theta(t, \bar{\eta} + \varepsilon\delta w, \beta)$, where

$\bar{\eta} = \bar{\psi}(\tau)$ and $\beta = (\varepsilon\alpha, \varepsilon)$. In order to simplify the notation we shall suppress the dependence on $(\delta w, \alpha, \varepsilon)$ and write

$$\psi(t) = \psi(t, \delta w, \alpha, \varepsilon)$$
$$\theta(t) = \theta(t, \bar{\eta} + \varepsilon\delta w, \beta), \qquad \beta = \varepsilon(\alpha, 1)$$

where the arguments $(\delta w, \alpha, \varepsilon)$ in the two functions have the same values. Since ψ is a solution of (4.20) and θ is a solution of (4.21), and both have the same initial value (4.22), we have

$$|\theta(t) - \psi(t)| \leq \int_{\tau}^{t} |G(s, \theta(s), \varepsilon\alpha, \varepsilon) - G(s, \psi(s), \varepsilon\alpha, \varepsilon)| ds + \left| \int_{\tau}^{t} g(s, \psi(s), \alpha, \varepsilon) ds \right| \tag{4.32}$$

for all t in $(\omega_{-}(\delta w, \alpha, \varepsilon), \omega_{+}(\delta w, \alpha, \varepsilon))$.

Since the family of solutions ψ is an equicontinuous family, it follows from (2.14) that the second term on the right in (4.32) is less than ε^2. From (4.19), (4.12) and the fact that $0 < \varepsilon \leq \varepsilon_0'$, it follows that there exists a constant C such that

$$|G(s, \theta(s), \varepsilon\alpha, \varepsilon) - G(s, \psi(s), \varepsilon\alpha, \varepsilon)| \leq C\mu(s) |\theta(s) - \psi(s)|.$$

Hence

$$|\theta(t) - \psi(t)| \leq C \int_{\tau}^{t} \mu(s) |\theta(s) - \psi(s)| ds + \varepsilon^2.$$

From Gronwall's Inequality (Lemma IV.4.2) we get that for all t in $(\omega_{-}(\delta w, \alpha, \varepsilon), \omega_{+}(\delta w, \alpha, \varepsilon))$

$$|\theta(t) - \psi(t)| \leq \varepsilon^2 \exp\{C \int_{\mathscr{I}} \mu(s) ds\}. \tag{4.33}$$

We saw that $\theta(t, \eta, \beta)$ with $\eta = \bar{\psi}(\tau) + \varepsilon\delta w$ and $\beta = (\varepsilon\alpha, \varepsilon)$ is defined on all of $\Pi(\varepsilon_0)$ and that the graph of each function $\theta(\cdot, \eta, \beta)$ is interior to $\mathscr{I} \times \mathscr{Y}$. It therefore follows from (4.33) that $\psi(t, \delta w, \alpha, \varepsilon)$ is defined on all of $\Pi(\varepsilon_0')$ for some $\varepsilon_0' < \varepsilon_0$

and that the graph of each function $\psi(\cdot,\delta w,\alpha,\varepsilon)$ is interior to $\mathcal{J}\times\mathcal{Y}$. In particular note that $\psi(\cdot,\delta w,\alpha,\varepsilon)$ is defined on $[t_0-\varepsilon\delta t_0, t_1+\varepsilon\delta t_1]$ for $0 < \varepsilon \leq \varepsilon_0'$. We now again relabel ε_0' as ε_0.

It also follows from (4.33) that

$$\psi(t) = \theta(t) + E_1(t,\delta w,\alpha,\varepsilon), \tag{4.34}$$

where $|E_1(t,\delta w,\alpha,\varepsilon)|/\varepsilon \to 0$ as $\varepsilon \to 0$ uniformly on $\Pi(\varepsilon_0)$.

From the equality

$$\psi(t) - \overline{\psi}(t) = (\psi(t)-\theta(t)) + (\theta(t)-\overline{\psi}(t))$$

and from (4.29) and (4.34) it follows that

$$\psi(t) = \overline{\psi}(t) + \varepsilon\delta\overline{\psi}(t) + E_1(t,\delta w,\alpha,\varepsilon) + E_2(t,\delta w,\alpha,\varepsilon).$$

If we set

$$E(t,\delta w,\alpha,\varepsilon) = E_1(t,\delta w,\alpha,\varepsilon) + E_2(t,\delta w,\alpha,\varepsilon)$$

in the preceding we get (4.15). Since E_1 and E_2 satisfy (4.18) uniformly on $\Pi(\varepsilon_0)$, the same holds for E. Since we have already shown that $\delta\overline{\psi}$ is a solution of (4.16) satisfying (4.17), the lemma is established.

We conclude this section by listing some consequences of Lemma 4.3. We assume that $\overline{F}$ and $\overline{\psi}$ are such that the mapping

$$t \to \overline{F}(t,\overline{\psi}(t)) \tag{4.35}$$

is continuous at $t = t_i$, $i = 0,1$. Let

$$\overline{F}_i = \overline{F}(t_i,\overline{\psi}(t_i)) \quad i = 0,1.$$

It follows from the continuity of (4.35) at $t = t_i$, $i = 0,1$ and the relation

$$\overline{\psi}(t_i+\varepsilon\delta t_i) = \overline{\psi}(t_i) + \int_{t_i}^{t_i+\varepsilon\delta t_i} \overline{F}(s,\overline{\psi}(s))ds$$

that

$$\overline{\psi}(t_i+\varepsilon\delta t_i) = \overline{\psi}(t_i) + \varepsilon\delta t_i\overline{F}_i + o(\varepsilon).$$

From the continuity of $\delta\overline{\psi}$ we have

$$\delta\overline{\psi}(t_i+\varepsilon\delta t_i) = \delta\overline{\psi}(t_i) + o(1),$$

where the $o(1)$ term is as $\varepsilon \to 0$. From the last two relations and from (4.15) we get

$$\psi(t_i+\varepsilon\delta t_i) = \overline{\psi}(t_i) + \varepsilon[\delta\overline{\psi}(t_i)+\delta t_i F_i] + o(\varepsilon), \tag{4.36}$$

where the $o(\varepsilon)$ term depends on $(t,\delta w,\alpha,\varepsilon)$ and is $o(\varepsilon)$ uniformly on $\Pi(\varepsilon_0)$.

Let

$$\delta y_i = \delta\overline{\psi}(t_i) + \delta t_i\overline{F}_i \qquad i = 0,1. \tag{4.37}$$

From (4.30) we get

$$\delta y_i = \Psi(t_i,\tau)[\delta w + \int_{\tau}^{t_i}\Psi^{-1}(s,\tau)\delta F(s,\overline{\psi}(s))ds]+\delta t_i\overline{F}_i \tag{4.38}$$

$$i = 0,1.$$

From (4.36) and (4.37) we obtain the following relation, which will be of great importance in the sequel:

$$e(\psi)-e(\overline{\psi}) = \varepsilon(\delta t_0,\delta y_0,\delta t_1,\delta y_1) + \rho^*(\delta t_0,\delta t_1,\delta w,\alpha,\varepsilon), \tag{4.39}$$

where $\rho^*/\varepsilon \to 0$ as $\varepsilon \to 0$, uniformly for $(\delta t_0,\delta t_1,\delta w)$ in K and α in P^r.

5. A Convex Set of Variations

DEFINITION 5.1. By a *variation* we shall mean a four-tuple $(\delta t_0, \delta t_1, \delta w, \delta F)$, where δt_0 and δt_1 are real numbers, δw is a vector in E^{n+1} and δF is an element in $\mathrm{co}(\mathscr{F}) - \bar{F}$.

We shall denote the set of all variations by $\mathscr{V}$. We define addition of variations and multiplication by a real number in the obvious way. Under these definitions $\mathscr{V}$ is a convex set. For every finite set of elements $\delta F_1, \ldots, \delta F_m$ in $\mathrm{co}(\mathscr{F}) - \bar{F}$ we define a convex set $\mathscr{V}(\delta F_1, \ldots, \delta F_m)$ in $\mathscr{V}$ as follows:

$$\mathscr{V}(\delta F_1, \ldots, \delta F_m) = \{(\delta t_0, \delta t_1, \delta w, \delta F) : \delta t_0 \in E^1,\ \delta t_1 \in E^1,$$
$$\delta w \in E^{n+1},\ \delta F \in \mathrm{co}[\delta F_1, \ldots, \delta F_m]\}.$$

We identify $\mathscr{V}(\delta F_1, \ldots, \delta F_m)$ with $E^1 \times E^1 \times E^{n+1} \times P^m$ in the obvious way and we put the metric topology of $E^{2+(n+1)+m}$ on $\mathscr{V}(\delta F_1, \ldots, \delta F_m)$.

For each of the functions δF_i, $i = 1, \ldots, m$, there exist functions $F_{i1}, F_{i2}, \ldots, F_{ir_i}$ in $\mathscr{F}$ and a vector α_i in P^{r_i} such that

$$\delta F_i = \sum_{j=1}^{r_i} \alpha_i^j F_{ij} - \bar{F}.$$

Any element δF in $\mathrm{co}[\delta F_1, \ldots, \delta F_m]$ has the form

$$\delta F = \sum_{i=1}^{m} \beta^i \delta F_i \qquad \beta^i \geq 0 \qquad \sum_{i=1}^{m} \beta^i = 1 \tag{5.1}$$

and therefore can be written as

$$\delta F = \sum_{i=1}^{m} \beta^i \sum_{j=1}^{r_i} \alpha_i^j F_{ij} - \bar{F}.$$

If we set $r = r_1 + \ldots + r_m$ and relabel the functions F_{ij}, $i = 1, \ldots, m$, $j = 1, \ldots, r_i$, as $F_1, \ldots, F_r$ we obtain that every F in $\mathrm{co}[\delta F_1, \ldots, \delta F_m]$ can be written as

$$\delta F = \sum_{i=1}^{r} \alpha^i F_i - \overline{F}, \tag{5.2}$$

where α is in P^r and depends on β. Note that the functions $F_1,\ldots,F_r$ are the same for all δF in $\mathrm{co}[\delta F_1,\ldots,\delta F_m]$. Equation (5.2) shows that the designation of an element in $\mathrm{co}[\delta F_1,\ldots,\delta F_m]$ by the symbol δF is consistent with our previous use of this symbol. We also call attention to the fact that the mapping $\beta \to \alpha = \alpha(\beta)$ from P^m to P^r is continuous. Thus, if $\{\beta_n\}$ is a sequence of points in P^m converging to β in P^m then $\alpha_n = \alpha(\beta_n)$ converges to $\alpha = \alpha(\beta)$ in P^r.

Since the family $\mathscr{F}$ is quasiconvex it follows that for every compact interval $\mathscr{I} \subset \mathscr{I}_0$, every compact interval $\mathscr{Y} \subset \mathscr{Y}_0$, every $0 < \varepsilon < 1$, and every $\beta \in P^m$, there exists a function $g(\cdot,\cdot,\alpha(\beta),\varepsilon)$ defined on $\mathscr{I} \times \mathscr{Y}$ such that

$$\overline{F} + \varepsilon\delta F + g \in \mathscr{F},$$

where δF is given by (5.1), or equivalently by (5.2). The function g satisfies the conditions of Definition 2.2 with $\alpha = \alpha(\beta)$ for all β in P^m. In particular, if $\beta_n \to \beta$ then

$$g(\cdot,y,\alpha(\beta_n),\varepsilon) \to g(\cdot,y,\alpha(\beta),\varepsilon) \tag{5.3}$$

in measure on $\mathscr{I}$ for all y in $\mathscr{Y}$ and $0 < \varepsilon < 1$.

Let $\delta F_1,\ldots,\delta F_m$ be given and let $\sum$ be a compact subset of $\mathscr{V}(\delta F_1,\ldots,\delta F_m)$. Since the topology of $\mathscr{V}(\delta F_1,\ldots,\delta F_m)$ is the euclidean topology, there is a compact set K in $E^{2+(n+1)}$ and a compact subset B of P^m such that $(\delta t_0,\delta t_1,\delta w,\delta F)$ is an element of $\sum$ if and only if $(\delta t_0,\delta t_1,\delta w)$ is in K and $\beta \in B$, where β is as in (5.1). Also note that since $\beta \to \alpha(\beta)$ is continuous, the set

$$A = \{\alpha\colon \alpha \in P^r,\ \ \alpha = \alpha(\beta),\ \ \beta \in B\}$$

is a compact subset of P^r. We also point out that variations belonging to $\sum$ have the property that δF is given by (5.2) with $\alpha \in A$.

It follows from Lemma 4.3 that there exists an $\varepsilon_0 > 0$ such that if $\varepsilon < \varepsilon_0$ then for every variation in $\sum$ there exists a trajectory $\psi = \psi(\cdot,\delta w,\alpha(\beta),\varepsilon)$ with the properties described in Lemma 4.3. This enables us to define a mapping h_ε from $\sum$ to $E^{2+2(n+1)}$ for each $0 < \varepsilon < \varepsilon_0$ as follows:

$$h_\varepsilon(\delta t_0,\delta t_1,\delta w,\delta F) = (e(\psi)-e(\overline{\psi}))/\varepsilon. \tag{5.4}$$

From (4.39) we get that

$$h_\varepsilon(\delta t_0,\delta t_1,\delta w,\delta F) = (\delta t_0,\delta y_0,\delta t_1,\delta y_1) + \rho_\varepsilon(\delta t_0,\delta t_1,\delta w,\alpha(\beta)),$$

where $\rho_\varepsilon \to 0$ as $\varepsilon \to 0$, uniformly for $(\delta t_0,\delta t_1,\delta w,\beta)$ in $\sum$. From (4.38) we see that we may write

$$(\delta t_0,\delta y_0,\delta t_1,\delta y_1) = L(\delta t_0,\delta t_1,\delta w,\delta F), \tag{5.5}$$

where L is a linear map defined on all of $\mathscr{V}$. Thus we may write

$$h_\varepsilon = L + \rho_\varepsilon. \tag{5.6}$$

LEMMA 5.1. The mappings h_ε and ρ_ε are continuous on $\sum$ for each fixed ε in the interval $0 < \varepsilon < \varepsilon_0$. Moreover,

$$\lim_{\varepsilon\to 0} \rho_\varepsilon(\delta t_0,\delta t_1,\delta w,\alpha(\beta)) = 0,$$

uniformly for $(\delta t_0,\delta t_1,\delta w,\beta)$ in $\sum$.

The validity of the last statement in the lemma was pointed out immediately after the definition of h_ε in (5.4). It follows immediately from the definition of L (see (4.38) and (5.1)) that L is continuous on $\sum$. Therefore to prove the lemma it suffices to show that h_ε is continuous on $\sum$.

The components δt_0 and δt_1 of the map h_ε are clearly continuous. Since $\overline{\psi}$ is fixed, it follows that we must show that for fixed ε, and fixed $i = 0,1$,

$$\psi(t_i+\varepsilon\delta t_i,\delta w,\alpha(\beta),\varepsilon) \to \psi(t_i+\varepsilon\delta t_i',\delta w',\alpha(\beta'),\varepsilon)$$

as $(\delta t_0,\delta t_1,\delta w,\beta) \to (\delta t_0',\delta t_1',\delta w',\beta')$ in $\sum$. To simplify notation let

$$\psi(t) = \psi(t,\delta w,\alpha(\beta),\varepsilon)$$
$$\tilde{\psi}(t) = \psi(t,\delta w',\alpha(\beta'),\varepsilon).$$

Then

$$|\psi(t_i+\varepsilon\delta t_i) - \tilde{\psi}(t_i+\varepsilon\delta t_i')| \le |\psi(t_i+\varepsilon\delta t_i)-\tilde{\psi}(t_i+\varepsilon\delta t_i)| + \int_{\varepsilon\delta t_i}^{\varepsilon\delta t_i'}|\tilde{\psi}'(s)|ds. \tag{5.7}$$

The function $\tilde{\psi}$ is a solution of (4.13), or equivalently, (4.20), with δF given by (5.2) and with $\alpha = \alpha(\beta')$. Thus

$$\tilde{\psi}'(t) = (1-\varepsilon)\overline{F}(t,\tilde{\psi}(t)) + \varepsilon\sum_{i=1}^{r}\alpha^i(\beta')F_i(t,\tilde{\psi}(t)) + g(t,\tilde{\psi}(t),\alpha(\beta'),\varepsilon). \tag{5.8}$$

Similarly,

$$\psi'(t) = (1-\varepsilon)\overline{F}(t,\psi(t)) + \varepsilon\sum_{i=1}^{r}\alpha^i(\beta)F_i(t,\psi(t)) + g(t,\psi(t),\alpha(\beta),\varepsilon). \tag{5.9}$$

From (5.8), (4.12), and (2.13), we get that

$$|\tilde{\psi}'(s)| \le 2\mu(s),$$

where μ is independent of β, ε, and δw. Hence for each $0 < \varepsilon < \varepsilon_0$, the integral on the right in (5.7) tends to zero as

$(\delta t_0, \delta t_1, \delta w, \beta) \to (\delta t_0', \delta t_1', \delta w', \beta')$.

To complete the proof we must show that for fixed $0 < \varepsilon < \varepsilon_0$, the first term on the right in (5.7) tends to zero as $(\delta t_0, \delta t_1, \delta w, \beta) \to (\delta t_0', \delta t_1', \delta w', \beta')$. Since

$$\psi(\tau) = \overline{\psi}(\tau) + \varepsilon\delta w \quad \tilde{\psi}(\tau) = \overline{\psi}(\tau) + \varepsilon\delta w'$$

it follows that for all t in $[t_0+\varepsilon\delta t_0, t_1+\varepsilon\delta t_1]$,

$$|\psi(t)-\tilde{\psi}(t)| \leq \varepsilon|\delta w-\delta w'| + \int_\tau^t |\psi'(s)-\tilde{\psi}'(s)|ds. \tag{5.10}$$

From (5.8) and (5.9) we get, after setting $\alpha' = \alpha(\beta')$ and $\alpha = \alpha(\beta)$, that

$$\begin{aligned}\psi'(s)-\tilde{\psi}'(s) &= (1-\varepsilon)\{\overline{F}(s,\psi(s))-F(s,\tilde{\psi}(s))\} \\ &+ \varepsilon\sum_{i=1}^{r}\alpha^i(F_i(s,\psi(s))-F_i(s,\tilde{\psi}(s))) \\ &+ \{g(s,\psi(s),\alpha,\varepsilon)-g(s,\tilde{\psi}(s),\alpha,\varepsilon)\} \\ &+ \varepsilon\sum_{i=1}^{r}(\alpha_i-\alpha_i')F_i(s,\tilde{\psi}(s)) \\ &\quad + \{g(s,\tilde{\psi}(s),\alpha,\varepsilon)-g(s,\tilde{\psi}(s),\alpha',\varepsilon)\}.\end{aligned}$$

It now follows from (4.12), (2.13), (5.3), and the convergence of $\alpha = \alpha(\beta)$ to $\alpha' = \alpha(\beta')$ that

$$|\psi'(s)-\tilde{\psi}'(s)| \leq \mu(s)|\psi(s)-\tilde{\psi}(s)| + \omega(s,\alpha,\alpha',\varepsilon),$$

where $\omega(\cdot,\alpha,\alpha',\varepsilon) \to 0$ in measure on $\mathscr{I}$ as $\alpha \to \alpha'$ and $|\omega(s,\alpha,\alpha',\varepsilon)| \leq \mu(s)$ for all α, α', and ε. Substitution of the last inequality into (5.10) gives

$$\begin{aligned}|\psi(t)-\tilde{\psi}(t)| &\leq \varepsilon|\delta w-\delta w'| + \int_{\mathscr{I}}|\omega(s,\alpha,\alpha',\varepsilon)|ds \\ &+ \int_\tau^t \mu(s)|\psi(s)-\tilde{\psi}(s)|ds.\end{aligned}$$

Hence by Gronwall's Inequality (Lemma IV.4.2)

$$|\psi(t_i+\varepsilon\delta t_i)-\tilde{\psi}(t_i+\varepsilon\delta t_i)| \leq$$

$$\left\{\varepsilon|\delta w-\delta w'| + \int_{\mathcal{I}} |\omega(s,\alpha,\alpha',\varepsilon)|ds\right\}\exp\{\int_{\mathcal{I}} \mu(s)ds\}.$$

Since $\omega(\cdot,\alpha,\alpha',\varepsilon) \to 0$ in measure on $\mathcal{I}$ as $\alpha \to \alpha'$ and $|\omega(s,\alpha,\alpha',\varepsilon)| \leq \mu(s)$, it follows that

$$\psi(t_i+\varepsilon\delta t_i) \to \tilde{\psi}(t_i+\varepsilon\delta t_i) \qquad i = 0,1$$

as $(\delta t_0,\delta t_1,\delta w,\beta) \to (\delta t_0',\delta t_1',\delta w',\beta')$ for all $0 < \varepsilon < \varepsilon_0$, as required.

6. The Separation Lemma

The proof of the Separation Lemma will involve the Separation Theorem for convex sets in E^n, a corollary of the Brouwer fixed-point theorem, and an elementary observation about convex sets in E^n. We begin with a review of some of this material.

If $\mathcal{S}$ is a set in E^{n+1} by $a+\mathcal{S}$ we mean the set of all vectors x of the form $a+s$ with s in $\mathcal{S}$. If $\mathcal{S}_1$ and $\mathcal{S}_2$ are two sets by their <u>linear span</u> we mean the set of all vectors s of the form $s = s_1+s_2$ with s_1 in $\mathcal{S}_1$ and s_2 in $\mathcal{S}_2$. By a linear variety in E^{n+1} we mean a set of the form $a+\mathcal{S}$, where $\mathcal{S}$ is a vector subspace of E^{n+1}. The <u>dimension</u> of a linear variety is the dimension of the vector subspace $\mathcal{S}$. A subspace of dimension n is also called a <u>hyperplane through the origin</u>. A <u>hyperplane</u> in E^{n+1} is a linear variety of dimension n. It consists of all vectors y such that $\langle\nu,y\rangle = \gamma$, where ν is a fixed non-zero vector in E^{n+1} and γ is a scalar. If $\gamma = 0$ the hyperplane passes through the origin. The equation $\langle\nu,y\rangle = \gamma$ is called the equation of the hyperplane.

Two sets A and B are said to be <u>separated</u> by the hyperplane with equation $\langle \nu,y\rangle = \gamma$ if $\langle \nu,a\rangle \leq \gamma$ for all a in A and $\langle \nu,b\rangle \geq \gamma$ for all b in B.

For a set A, by the <u>carrier plane</u> of A we mean the linear variety of lowest dimension containing A. Note that since every set A is always contained in E^{n+1}, it follows that every set A has a carrier plane. The relative topology of A is the topology induced on A by its carrier plane.

The principal result concerning the separation of convex sets is the following:

LEMMA 6.1. Let A and B be two convex sets in E^{n+1}. Then A and B can be separated by a hyperplane if either (i) the carrier planes of A and B are such that their linear span is not all of E^{n+1}; or (ii) the relative interiors of A and B are disjoint.

If the carrier planes of A and B are such that their linear span is not all of E^{n+1}, then their linear span is contained in a hyperplane, say, $\langle \nu,y\rangle = \gamma$. This hyperplane will serve as the separating hyperplane, since $\langle \nu,a\rangle = \langle \nu,b\rangle = \gamma$ for all a in A and b in B. A proof of the statement that convex sets A and B with disjoint relative interiors can be separated by a hyperplane can be found in ([52], Theorem 11.3).

Let Z be a proper subset of a euclidean space E^d and let $Z^{\perp}$ denote the orthogonal complement of Z in E^d. Let $\mathcal{S}$ be a compact and convex subset of E^d. Then for every s in $\mathcal{S}$ there is a unique $y(s)$ in $Z^{\perp}$ and $z(s)$ in Z such that $s = y(s) + \bar{z}(s)$. The set $(y(s) + Z) \cap \mathcal{S}$ is non-empty, convex, and compact. It therefore has a unique element of minimum norm, which we denote by y_s. We can therefore define a mapping Λ from $\mathcal{S}$ into the boundary of $\mathcal{S}$ as follows:

$$\Lambda(s) = s_y. \tag{6.1}$$

LEMMA 6.2. The mapping Λ is continuous on $\mathcal{S}$.

We first note that if $\mathcal{S} \subset Z$ then $\Lambda(s)$ is constant on $\mathcal{S}$ and so is continuous. Let us then suppose that $\mathcal{S}$ is not contained in Z and that Λ is not continuous on $\mathcal{S}$. Then there would exist a point s_0 in $\mathcal{S}$ and a sequence $\{s_n\}$ of distinct points in $\mathcal{S}$ such that $s_n \to s_0$ and $\lim \Lambda(s_n) \neq \Lambda(s_0)$. Since $\mathcal{S}$ is compact there is a subsequence of $\{s_n\}$, which we again label as $\{s_n\}$, such that $\Lambda(s_n)$ converges to some point s^* in $\mathcal{S}$, with $s^* \neq \Lambda(s_0)$. Let $s_n = y_n+z_n$ and let $s_0 = y_0+z_0$, where y_n and y_0 are in $Z^{\perp}$ and z_n and z_0 are in Z. Then $\Lambda(s_n) = y_n+z_n'$ and $\Lambda(s_0) = y_0+z_0'$ for some z_n' and z_0' in Z. Also, $y_n \to y_0$, and consequently $s^* = y_0+z_0^*$ for some z_0^* in Z.

Let $\varepsilon = |s^*|-|\Lambda(s_0)| > 0$. Then there exists an integer k_0 such that whenever $k > k_0$ and $j > j_0$, $|\Lambda(s_k)-\Lambda(s_j)| < \varepsilon/4$. Let k be an integer greater than k_0 and such that $|\Lambda(s_k)-s^*| < \varepsilon/4$. Since $\mathcal{S}$ is convex, the line segment $[\Lambda(s_0),\Lambda(s_k)]$ lies in $\mathcal{S}$. Since the points $\{s_n\}$ are distinct, so are the points $\{y_n\}$. Hence $y_k \notin y_0+Z$. Since $y_n \to y_0$ it follows that there exist an integer $j > k_0$ such that y_j+Z intersects the segment $[\Lambda(s_0),\Lambda(s_k)]$ at a point whose distance from $\Lambda(s_0)$ is less than $\varepsilon/4$. Hence $|\Lambda(s_j)| < |\Lambda(s_0)| + \varepsilon/4$. Therefore

$$\begin{aligned}\varepsilon = |s^*|-|\Lambda(s_0)| &\leq |s^*-\Lambda(s_k)|+|\Lambda(s_k)-\Lambda(s_j)|+|\Lambda(s_j)|-|\Lambda(s_0)| \\ &\leq \varepsilon/4.\end{aligned}$$

This contradiction proves the lemma.

The Brouwer fixed point theorem states that if f is a continuous mapping of the closed unit ball B_n of E^n into itself, then there is a point x in B_n such that $f(x) = x$. If $\mathcal{S}$ is a convex set of dimension d in E^n, then $\mathscr{S}$ is homeomorphic to the closed unit ball in E^d. Hence the following is true.

LEMMA 6.3. If f is a continuous mapping of a convex compact set $\mathscr{S}$ in E^n into itself, then there is a point s in $\mathscr{S}$ such that $f(s) = s$.

In Section 5 we noted that the linear map L defined by (5.5) and (4.38) is defined on all of $\mathscr{V}$. Let

$$\mathscr{K} = L(\mathscr{V}).$$

Then $\mathscr{K}$ is a set in $E^{2+2(n+1)}$ containing the origin. Moreover, since $\mathscr{V}$ is convex the set $\mathscr{K}$ is convex.

Let $\mathscr{N}$ be the manifold with upper boundary $\mathscr{M}$ with respect to which $\overline{\psi}$ is an $\mathscr{F}$-$\mathscr{N}$ extremal (see Section 2). Let $\mathscr{M}_T$ denote the tangent plane to $\mathscr{M}$ at $e(\overline{\psi})$. Let $\mathscr{N}_T$ denote the tangent half plane to $\mathscr{N}$ at $e(\overline{\psi})$. That is, $\mathscr{N}_T$ consists of the points in the tangent plane to the manifold $\mathscr{N}_1$ (see Section 2, page 243) at $e(\overline{\psi})$ that are images of points in $\mathscr{N}$ under the standard homeomorphism between $\mathscr{N}_1$ and its tangent plane at $e(\overline{\psi})$. Figure 2 illustrates in schematic form the various sets for the problem with $\hat{y} = (y^0, y^1)$, initial point fixed and terminal set a one dimensional manifold $\mathscr{T}$.

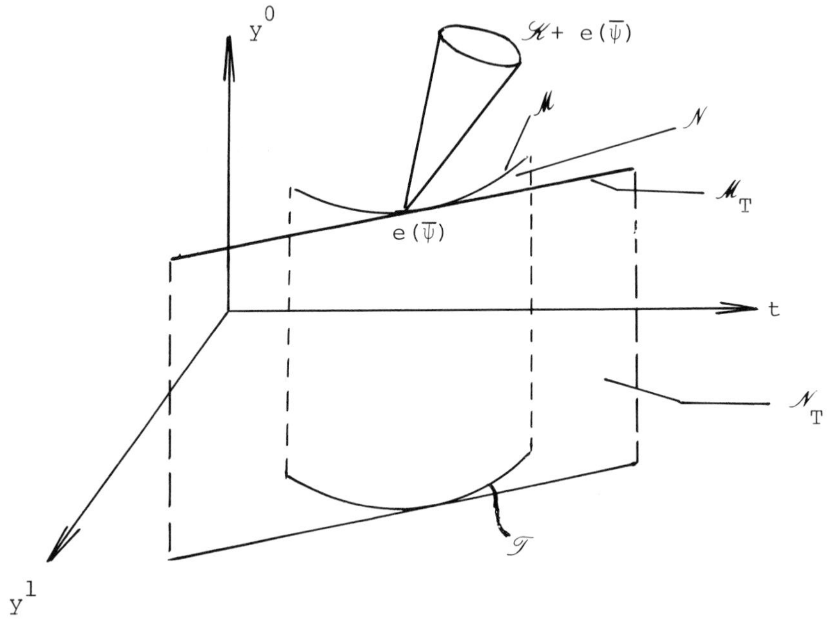

Figure 2

We now state and prove the principal result of this section.

LEMMA 6.4. (Separation Lemma). The sets $\mathscr{K}$ and $\mathscr{N}_T - e(\bar{\psi})$ can be separated by a hyperplane through the origin.

<u>Proof</u>. There is no loss of generality in assuming that $e(\bar{\psi}) = 0$, since this can always be accomplished by a translation of the origin of coordinates.

Suppose the lemma were false. Then since $\mathscr{K}$ and $\mathscr{N}_T$ are both convex it follows from Lemma 6.1 that the following statements are true: (i) The carrier plane of $\mathscr{K}$ and the carrier plane of $\mathscr{N}_T$ are such that their linear span is the entire space. (ii) There exists a point q that is a relative interior point of both $\mathscr{K}$ and $\mathscr{N}_T$.

Since q is a relative interior point of $\mathscr{N}_T$ there exists a relatively open ball $\mathscr{O}_T$ in the carrier plane of $\mathscr{N}_T$ such that $\mathrm{cl}(\mathscr{O}_T)$ is compact and is contained in the relative interior of $\mathscr{N}_T$. Therefore

$$\mathscr{O}_T \cap \mathscr{M}_T = \emptyset.$$

Let the dimension of the carrier plane of $\mathscr{K}$ be m, where $0 \leq m \leq 2+2(n+1)$. Then there exists a simplex $\mathscr{K}_m$ of dimension m such that $\mathscr{K}_m$ has $(m+1)$ vertices, q is in the relative interior of $\mathscr{K}_m$ and $\mathscr{K}_m$ is in the relative interior of $\mathscr{K}$. Let $q_1,\ldots,q_{m+1}$ denote the vertices of $\mathscr{K}_m$. Then there exist variations $p_1,\ldots,p_{m+1}$ in $\mathscr{V}$ such that $q_i = L(p_i)$. Let

$$p_i = (\delta t_{0i}, \delta t_{1i}, \delta w_i, \delta F_i) \qquad i = 1,\ldots,m+1. \tag{6.2}$$

Let $\mathscr{S}_m$ denote the compact simplex in $\mathscr{V}(\delta F_1,\ldots,\delta F_{m+1})$ whose vertices are the variations p_i in (6.2). Since $\mathscr{S}_m$ is a compact subset of $\mathscr{V}(\delta F_1,\ldots,\delta F_{m+1})$ the mapping h_ε defined in (5.6) is defined on $\mathscr{S}_m$ for all sufficiently small $\varepsilon > 0$ and has the properties stated in Lemma 5.1.

Let $\mathscr{O}$ be the neighborhood of $e(\overline{\psi})$ in $E^{2+2(n+1)}$ that occurs in the definition of $\mathscr{F}$-$\mathscr{N}$ extremality. Since $\mathscr{N}$ is a differentiable manifold with upper boundary $\mathscr{M}$, where $\mathscr{M}$ is a differentiable manifold, since by the definition of $\mathscr{F}$-$\mathscr{N}$ extremality $e(\overline{\psi})$ belongs to $\mathscr{M}$, and since we are supposing that $e(\overline{\psi}) = 0$, the following is true. There exists a homeomorphism h^* from a neighborhood $\mathscr{O}_T'$ of $e(\overline{\psi}) = 0$ in $\mathscr{N}_T$ onto a neighborhood $\mathscr{O}'$ of 0 in $\mathscr{N}$ of the form

$$h^*(y) = y + r^*(y), \tag{6.3}$$

where

$$\frac{|r^*(y)|}{|y|} \to 0 \quad \text{as} \quad y \to 0, \quad y \in \mathcal{O}'_T. \tag{6.4}$$

There is no loss of generality in assuming $\mathcal{O}'_T$ and $\mathcal{O}'$ to be such that

$$\mathcal{O}'_T \subset \mathcal{O} \qquad \text{and} \qquad \mathcal{O}' \subset \mathcal{O} \quad . \tag{6.5}$$

Figure 3 illustrates some of the preceding definitions in schematic fashion.

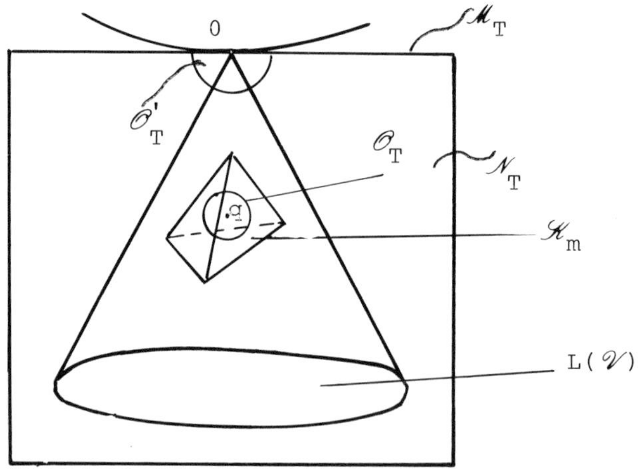

Figure 3

Let $\mathcal{S} = \mathcal{S}_m \times \mathrm{cl}(\mathcal{O}_T)$. Denote points in $\mathcal{S}_m$ by σ and points in $\mathrm{cl}(\mathcal{O}_T)$ by η. Since $\mathrm{cl}(\mathcal{O}_T)$ is compact it follows that for sufficiently small $\varepsilon > 0$

$$\varepsilon[\mathrm{cl}(\mathcal{O}_T)] \subseteq \mathcal{O}'_T. \tag{6.6}$$

Hence we can define a mapping γ_ε from $\mathcal{S}$ into $E^{2+2(n+1)}$ as follows:

$$\gamma_\varepsilon(\sigma,\eta) = h_\varepsilon(\sigma) - \frac{1}{\varepsilon}h^*(\varepsilon\eta), \tag{6.7}$$

where h_ε is defined in (5.4) and h^* is defined in (6.3). From (6.3) and (5.6) we get

$$\gamma_\varepsilon(\sigma,\eta) = L(\sigma)-\eta + \rho_\varepsilon(\sigma) - \frac{1}{\varepsilon} r^*(\varepsilon\eta).$$

Let

$$L_1(\sigma,\eta) = L(\sigma)-\eta \tag{6.8}$$
$$R(\sigma,\eta,\varepsilon) = \rho_\varepsilon(\sigma) - \frac{1}{\varepsilon} r^*(\varepsilon\eta).$$

Then

$$\gamma_\varepsilon(\sigma,\eta) = L_1(\sigma,\eta) + R(\sigma,\eta,\varepsilon). \tag{6.9}$$

From the compactness of $\mathcal{S}_m$ and cl $\mathcal{O}_T$, and from (6.4) and Lemma 5.1 it follows that for each fixed (σ,η) in $\mathcal{S}$

$$\lim_{\varepsilon\to 0} R(\sigma,\eta,\varepsilon) = 0, \tag{6.10}$$

uniformly for (σ,η) in $\mathcal{S}$.

Since q belongs to the relative interior of $\mathcal{K}_m$ there is a σ_0 in the interior of $\mathcal{S}_m$ such that $L(\sigma_0) = q$. Since $q \in \mathcal{O}_T$, we have that (σ_0,q) is in $\mathcal{S}$ and

$$L_1(\sigma_0,q) = L(\sigma_0)-q = 0.$$

Thus the origin of $E^{2+2(n+1)}$ is in the image of $\mathcal{S}$ under L. Since the linear span of the carrier plane of $\mathcal{K}_m$ and $\mathcal{O}_T$ is all of $E^{2+2(n+1)}$ it follows that $L_1(\mathcal{S})$ contains a ball B_δ with center at the origin and radius δ. From (6.10) we get that there exists an $\varepsilon_0 > 0$ such that if $0 < \varepsilon < \varepsilon_0$ then

$$|R(\sigma,\eta,\varepsilon)| < \delta. \tag{6.11}$$

The mapping L_1 will in general not be one-one on $\mathcal{S}$. We now determine a subset $\mathcal{S}^*$ of $\mathcal{S}$ such that the restriction of L_1 to $\mathcal{S}^*$ is 1-1 and has a continuous inverse.

Since $\text{cl}(\mathcal{O}_T) \subset \mathcal{N}_T$ and $\mathcal{N}_T$ is contained in the tangent plane to a certain manifold at the origin, it follows that the carrier plane Π_1 of $\mathrm{cl}(\mathcal{O}_T)$ is a linear space. Let Π_2 denote the linear space spanned by the variations $p_1,\ldots,p_{m+1}$ defined in (6.2). Thus

$$\Pi_2 = \{v: \ v = \sum_{i=1}^{m+1} \alpha^i p_i\}.$$

Clearly, $\mathcal{S} \subseteq \Pi_2 \times \Pi_1$. The mapping L_1 defined on $\mathcal{S}$ in (6.8) can be extended to a linear mapping L_2 defined on all of $\Pi_2 \times \Pi_1$ by the formula

$$L_2(v,q) = \sum_{i=1}^{m+1} \alpha^i L(p_i) - q \qquad (v \in \Pi_2,\ q \in \Pi_1).$$

Let Z denote the kernel of L_2 and let $Z^{\perp}$ denote the orthogonal complement of Z relative to $\Pi_2 \times \Pi_1$. Let Λ be defined relative to Z as in (6.1) and let

$$\mathcal{S}^* = \Lambda(\mathcal{S}).$$

Then $\mathcal{S}^* \subseteq \mathcal{S}$ and $L_1(\mathcal{S}^*) = L_1(\mathcal{S})$. Let L^* denote the restriction of L_1 to $\mathcal{S}^*$. Then

$$L^*(\mathcal{S}^*) = L_1(\mathcal{S}). \tag{6.12}$$

It follows from the definitions of Λ and $\mathcal{S}^*$ that if s_1 and s_2 are distinct elements of $\mathcal{S}^*$ and if $s_1 = y_1+z_1$, $s_2 = y_2+z_2$, where $y_i \in Z^{\perp}$ and $z_i \in Z$, $i = 1,2$, then $y_1 \neq y_2$. Hence the mapping L^* is 1-1 on $\mathcal{S}^*$, and L^{*-1} is defined on $L^*(\mathcal{S}^*)$. Since $\mathcal{S}$ is compact and Λ is continuous, $\mathcal{S}^* = \Lambda(\mathcal{S})$ is compact. Since L^* is the restriction of a continuous map, it is continuous on $\mathcal{S}^*$. Hence L^{*-1} is continuous on $L^*(\mathcal{S}^*)$.

Let ε_0 be as in (6.11). For each $0 < \varepsilon < \varepsilon_0$ we define a mapping H_ε on $L^*(\mathcal{S}^*)$ as follows

$$H_\varepsilon(p) = -\gamma_\varepsilon(L^{*-1}(p)) + p. \tag{6.13}$$

If we let $s = L^{*-1}(p)$, then $s \in \mathcal{S}^*$ and therefore $s \in \mathcal{S}$. Hence the mapping H_ε is well defined. Since L^{*-1} is continuous and γ_ε is continuous for fixed ε, it follows that for each $0 < \varepsilon < \varepsilon_0$ the map H_ε is continuous. By virtue of (6.9)

$$\gamma_\varepsilon(L^{*-1}(p)) = L_1(L^{*-1}(p)) + R(s,\varepsilon).$$

But $L_1 = L^*$ on $\mathcal{S}^*$ so

$$\gamma_\varepsilon(L^{*-1}(p)) = p + R(s,\varepsilon).$$

Hence from the preceding and from (6.11) we get

$$|H_\varepsilon(p)| = |R(s,\varepsilon)| < \delta. \tag{6.14}$$

In the paragraph preceding the inequality (6.11) we saw that the ball in $E^{2+2(n+1)}$ of radius δ and center at the origin is contained in $L_1(\mathcal{S})$. From this fact, from the inequality in (6.14), and from (6.12) it follows that H_ε maps $L^*(\mathcal{S}^*)$ into itself. Since $\mathcal{S}$ is compact and convex and L_1 is linear, $L_1(\mathcal{S})$ is compact and convex. By (6.12), the same is true for $L^*(\mathcal{S}^*)$. Thus for each $0 < \varepsilon < \varepsilon_0$, H_ε is a continuous map of the compact convex set $L^*(\mathcal{S}^*)$ into itself. Hence by Lemma 6.3 the mapping H_ε has a fixed point $\bar{p}(\varepsilon)$. Thus

$$-\gamma_\varepsilon(L^{*-1}(\bar{p}(\varepsilon))) + \bar{p}(\varepsilon) = \bar{p}(\varepsilon)$$

and so

$$\gamma_\varepsilon(L^{*-1}(\bar{p}(\varepsilon))) = 0. \tag{6.15}$$

Let $\bar{s}(\varepsilon) = L^{*-1}(\bar{p}(\varepsilon))$. Since $\bar{s}(\varepsilon) \in \mathcal{S}^*$, we have $\bar{s}(\varepsilon) = (\bar{\sigma}(\varepsilon),\bar{\eta}(\varepsilon))$ for some $(\bar{\sigma}(\varepsilon),\bar{\eta}(\varepsilon))$ in $\mathcal{S}$. Thus (6.15) becomes

$$\gamma_\varepsilon(\bar{\sigma}(\varepsilon),\bar{\eta}(\varepsilon)) = 0.$$

From (6.7) we then get

$$h_{\varepsilon}(\bar{\sigma}(\varepsilon)) = \frac{1}{\varepsilon} h^*(\varepsilon\bar{\eta}(\varepsilon)).$$

If we denote the trajectory corresponding to $\bar{\sigma}(\varepsilon)$ by ψ_{ε}, then from (5.4) and the last equality we get

$$e(\psi_{\varepsilon}) = h^*(\varepsilon\bar{\eta}(\varepsilon)). \tag{6.16}$$

The point $\bar{\eta}(\varepsilon)$ is in $\mathcal{O}_T$. It therefore follows from (6.6) that for ε sufficiently small, $\varepsilon\bar{\eta}(\varepsilon)$ is in $\mathcal{O}'_T$. Hence by the definition of h^*, the right hand side of (6.16) is a point lying in a relatively open set $\mathcal{O}'$ in $\mathcal{N}$. By (6.5) it also lies in the neighborhood $\mathcal{O}$ that occurs in the definition of $\mathcal{F}-\mathcal{N}$ extremality. Thus

$$e(\psi_{\varepsilon}) = \mathcal{O} \cap \mathcal{E} \cap \mathcal{N}. \tag{6.17}$$

On the other hand, since cl $\mathcal{O}_T$ is contained in the relative interior of $\mathcal{N}_T$ and since $\bar{\eta}(\varepsilon) \in \mathcal{O}_T$, it follows that $\bar{\eta}(\varepsilon)$ is in the relative interior of $\mathcal{N}_T$. Hence $\varepsilon\bar{\eta}(\varepsilon)$ cannot belong to $\mathcal{M}_T$. Consequently $h^*(\varepsilon\bar{\eta}(\varepsilon))$ cannot belong to $\mathcal{M}$. Therefore $e(\psi_{\varepsilon})$ does not belong to $\mathcal{M}$. This assertion and (6.17) contradict the assumed $\mathcal{F}-\mathcal{N}$ extremality of $\bar{\psi}$. Hence our assumption that $\mathcal{K}$ and $\mathcal{N}_T$ cannot be separated is incorrect, and the lemma is proved.

7. Analytic Consequences of the Separation Lemma

The necessary conditions of Theorem 3.1 are implicit in Lemma 6.4, the Separation Lemma. We now show this and thereby complete the proof of Theorem 3.1.

Let Π denote the hyperplane through the origin that separates $\mathcal{K}$ and $\mathcal{N}_T - e(\bar{\psi})$. Let c be a normal to Π at the origin and let c point into the half space containing $\mathcal{N}_T - e(\bar{\psi})$. Then

$$\langle c,\zeta\rangle \leq 0 \leq \langle c,\nu\rangle \tag{7.1}$$

for all $\zeta \in \mathscr{K}$ and $\nu \in \mathscr{N}_T - e(\overline{\psi})$. Let ν' be a vector in $\mathscr{M}_T - e(\overline{\psi})$. Then since $\mathscr{M}_T - e(\overline{\psi}) \subset \mathscr{N}_T - e(\overline{\psi})$, we have $\langle c,\nu'\rangle \geq 0$. But $\mathscr{M}_T - e(\overline{\psi})$ is a subspace, so that $-\nu' \in \mathscr{M}_T - e(\overline{\psi})$ and so $\langle c,-\nu'\rangle \geq 0$. Hence for all $\nu' \in \mathscr{M}_T - e(\overline{\psi})$ we have $\langle c,\nu'\rangle = 0$. In other words, c is orthogonal to $\mathscr{M}_T$ at $e(\overline{\psi})$.

It follows from (5.5) and from the definition of $\mathscr{K}$ as the image of $\mathscr{V}$ under L that a typical element ζ in $\mathscr{K}$ has the form

$$\zeta = (\delta t_0, \delta y_0, \delta t_1, \delta y_1).$$

If we write c in the form

$$c = (c_0^0, c_0, c_1^0, c_1),$$

where $c_i = (c_i^1, \ldots, c_i^n)$, $i = 0,1$, then from (7.1) we have that for every $\zeta \in \mathscr{K}$,

$$c_0 \delta t_0 + \langle c_0, \delta y_0\rangle + c_1^0 \delta t_1 + \langle c_1, \delta y_1\rangle \leq 0.$$

If we substitute the expressions for δy_i, $i = 0,1$ given in (4.38) we get that for arbitrary scalars δt_i, $i = 0,1$, arbitrary δw in E^{n+1} and arbitrary δF in $co(\mathscr{F}) - \overline{F}$

$$\sum_{i=0}^{1} (c_i^0 + \langle c_i, \overline{F}_i\rangle)\delta t_i + \langle c_i, \Psi(t_i,\tau)\{\delta w + \int_\tau^{t_i} \Psi^{-1}(s,\tau)\delta F(s,\overline{\psi}(s))ds\}\rangle \leq 0. \tag{7.2}$$

Here Ψ is the fundamental matrix solution of the linear homogeneous system

$$\frac{dy}{dt} = \overline{F}_y(t,\overline{\psi}(t))y \tag{7.3}$$

satisfying the initial conditions $\Psi(\tau,\tau) = I$ and

$$\overline{F}_i = \overline{F}(t_i, \overline{\psi}(t_i)) \qquad i = 0,1. \tag{7.4}$$

If in (7.2) we take $\delta t_0 = 0$, $\delta w = 0$, $\delta F = 0$ and $\delta t_i = \pm 1$ we get

$$c_1^0 + \langle c_1, \overline{F}_1 \rangle = 0. \tag{7.5}$$

If in (7.2) we then take $\delta t_0 = \pm 1$, $\delta w = 0$, $\delta F = 0$, $\delta t_1 = 0$, we get

$$c_0^0 + \langle c_0, \overline{F}_0 \rangle = 0. \tag{7.6}$$

If we now take $\delta F = 0$, δw arbitrary, and make use of (7.5) and (7.6) we get

$$\langle c_0, \Psi(t_0,\tau)\delta w \rangle + \langle c_1, \Psi(t_1,\tau)\delta w \rangle = 0.$$

Therefore

$$\langle \Psi^*(t_0,\tau)c_0 + \Psi^*(t_1,\tau)c_1, \delta w \rangle = 0$$

for all δw in E^{n+1}, and therefore

$$\Psi^*(t_0,\tau)c_0 = -\Psi^*(t_1,\tau)c_1. \tag{7.7}$$

Finally, if we take $\delta w = 0$, then for arbitrary δF in $co(\mathscr{F}) - \overline{F}$ we get

$$\langle \Psi^*(t_0,\tau)c_0, \int_\tau^{t_0} \Psi^{-1}\delta\overline{F}ds \rangle$$
$$+ \langle \Psi^*(t_1,\tau)c_1, \int_\tau^{t_1} \Psi^{-1}\delta\overline{F}ds \rangle \le 0,$$

where we have set $\delta\overline{F}(s) = \delta F(s, \overline{\psi}(s))$. If we use (7.7) this inequality can be written

$$\langle \Psi^*(t_1,\tau)c_1, \int_{t_0}^{t_1} \Psi^{-1}\delta\overline{F}ds \rangle \le 0,$$

which in turn can be written

$$\int_{t_0}^{t_1} \langle \Psi^{*-1}(s,\tau)\Psi^*(t_1,\tau)\; c_1,\; \delta F(s,\overline{\psi}(s))\rangle\, ds \le 0. \tag{7.8}$$

Let

$$\eta(t) = \Psi^{*-1}(t,\tau)\Psi^*(t_1,\tau)c_1$$
$$\eta_1 = \Psi^*(t_1,\tau)c_1$$

Then

$$\eta(t) = \Psi^{*-1}(t,\tau)\eta_1 \tag{7.9}$$

and

$$\eta(t_1) = c_1 \qquad \eta_1 = \eta(\tau). \tag{7.10}$$

Upon differentiating the identity

$$\Psi(t,\tau)\Psi^{-1}(t,\tau) = I$$

and using (7.3) we get

$$0 = \Psi'\Psi^{-1} + \Psi(\Psi^{-1})' = (\overline{F}_y\Psi)\Psi^{-1} + \Psi(\Psi^{-1})'$$
$$= \overline{F}_y + \Psi(\Psi^{-1})'.$$

Hence $(\Psi^{-1})' = -\Psi^{-1}\overline{F}_y$, which is equivalent to the relation

$$(\Psi^{*-1})' = -\overline{F}^*_y(\Psi^{*-1}).$$

Thus Ψ^{*-1} is a fundamental matrix of solutions for the linear system

$$z' = -\overline{F}^*_y(t,\overline{\psi}(t))z$$

and satisfies the initial condition $\Psi^{*-1}(\tau,\tau) = I$. From this observation and from (7.9) and (7.10) it follows that

$$\eta'(t) = -\overline{F}^*_y(t,\overline{\psi}(t))\eta(t) \qquad \text{a.e.}$$

From this relation (3.4) of Theorem 3.1 follows by taking the transpose of both sides.

Relation (3.3) in Theorem 3.1 is an immediate consequence of (3.1) and (3.2).

To see that $\eta(t) \neq 0$ for all t in $[t_0,t_1]$ we first note that if $\eta(t_2)$ were zero for some t_2 in $[t_0,t_1]$, then from (7.9) we would have that $\eta_1 = 0$. But then by (7.10) this would mean that $c_1 = 0$. From (7.5) we then get $c_1^0 = 0$, and from (7.7) we get that $c_0 = 0$. We then use (7.6) to get that $c_0^0 = 0$. Hence $c = 0$, which cannot be.

Using (7.9) and (7.10) we can write (7.8) as

$$\int_{t_0}^{t_1} \langle \eta(s), \delta F(s,\bar{\psi}(s)) \rangle ds \leq 0 \tag{7.11}$$

for all δF in $co(\mathscr{F})-\bar{F}$. If in (7.11) we now take $\delta F = F-\bar{F}$, where F is an arbitrary element of $\mathscr{F}$, we get that

$$\int_{t_0}^{t_1} \langle \eta(s), \bar{F}(s,\bar{\psi}(s)) \rangle ds \geq \int_{t_0}^{t_1} \langle \eta(s), F(s,\bar{\psi}(s)) \rangle ds,$$

for all F in $\mathscr{F}$. This is precisely (3.5) of Theorem 3.1.

We have already shown that $c = (c_0^0, c_0, c_1^0, c_1)$ is orthogonal to $\mathscr{M}$ at $e(\bar{\psi})$. We shall show that (3.6) is orthogonal to $\mathscr{M}$ at $e(\bar{\psi})$ by showing that the vector (3.6) is precisely c. From (7.10) we have $c_1 = \eta(t_1)$. From (7.9) we have that

$$\eta(t_0) = \Psi^{*-1}(t_0,\tau)c_1.$$

From this and from (7.7) we get that

$$\eta(t_0) = -c_0. \tag{7.12}$$

From (7.5), (7.4), (7.10) and the definition of $\bar{H}$ we get

$$c_1^0 = -\langle c_1, \bar{F}_1 \rangle = -\langle \eta(t_1), \bar{F}(t_1,\bar{\psi}(t_1)) \rangle = -\bar{H}(t_1,\bar{\psi}(t_1),\eta(t_1)).$$

Similarly, from (7.6), (7.4), (7.11) and the definition of $\bar{H}$ we get

$$c_0^0 = \langle -c_0, \bar{F}_0 \rangle = \langle \eta(t_0), \bar{F}(t_0, \bar{\psi}(t_0)) \rangle = \bar{H}(t_0, \bar{\psi}(t_0), \eta(t_0)).$$

This establishes the transversality condition.

8. Proofs of Corollaries V.3.1 and V.3.2

The notion of point of density of a measurable set and the notion of approximate continuity of a measurable function will be used in our proofs. We shall review these definitions and summarize some facts about these concepts that we shall use. For proofs and further discussion the reader is referred to Natanson ([46], p. 260-262).

Let E be a measurable set on the line, let x_0 be an arbitrary point, and for $h > 0$ let $I(h)$ denote the interval $[x_0-h, x_0+h]$. The point x_0 is said to be a point of density of E if

$$\lim_{h \to 0} \text{meas}(E \cap I(h))/2h = 1.$$

For a measurable set E it is true that almost all points of E are points of density.

Let f be a real valued function defined on an interval $[a,b]$ and let x_0 be an interior point of $[a,b]$. Then f is said to be approximately continuous at x_0 if there is a measurable subset E of $[a,b]$ such that x_0 is a point of density of E and the restriction of f to $E \cup \{x_0\}$ is continuous at x_0; i.e.

$$\lim_{\substack{x \to x_0 \\ x \in E}} f(x) = f(x_0).$$

A real valued measurable function f defined on a closed interval $[a,b]$ is approximately continuous at almost all points of $[a,b]$.

The definition of approximate continuity and the statement that a measurable function is approximately continuous almost everywhere extends to mappings from the line to E^n by consideration of the real valued component mappings.

We now prove Corollary V.3.1. It follows from Assumption V.(3.1)-(ii), the continuity of ϕ, the continuity of $\hat{\lambda}$, and the measurability of u that the mapping $\mathscr{A}$ defined by

$$\mathscr{A}(t) = H(t,\phi(t),u(t),\hat{\lambda}(t)) \tag{8.1}$$

is measurable on $[t_0,t_1]$. Hence (8.1) is approximately continuous on $[t_0,t_1]$.

Suppose that the conclusion of Corollary V.3.1 were false. Then there would be a set E of positive measure on which V.(3.5) failed to hold. Let t_2 be a point of E at which the function $\mathscr{A}$ defined in (8.1) is approximately continuous. Since V.(3.5) fails at t_2, there is a point z in $\mathscr{C}$ such that

$$H(t_2,\phi(t_2),u(t_2),\hat{\lambda}(t_2))-H(t_2,\phi(t_2),z,\hat{\lambda}(t_2))<0. \tag{8.2}$$

Since $\mathscr{A}$ is approximately continuous on $[t_0,t_1]$, so is the mapping γ defined by

$$\gamma(t) = H(t,\phi(t),u(t),\hat{\lambda}(t)) - H(t,\phi(t),z,\hat{\lambda}(t)). \tag{8.3}$$

Moreover, t_2 is a point of approximate continuity of γ. It therefore follows from (8.2) that there is a measurable set E_1 with $\operatorname{meas}(E_1) > 0$ such that for t in E_1

$$\gamma(t) < 0.$$

Now define a control v as follows:

$$\begin{aligned} v(t) &= u(t) && t \notin E_1 \\ &= z && t \in E_1. \end{aligned}$$

Then

$$\int_{t_0}^{t_1} [H(t,\phi(t),u(t),\hat{\lambda}(t))-H(t,\phi(t),v(t),\hat{\lambda}(t))]dt$$

$$= \int_{E_1} \gamma(t)dt < 0,$$

which contradicts V.(3.3). This proves Corollary V.3.1.

We now prove Corollary V.3.2. Let T denote the set of points t in $[t_0,t_1]$ at which V.(3.5) holds, at which u is approximately continuous, and at which λ and ϕ are differentiable. Then meas $T = t_1-t_0$.

Let t and t_2 be points of T. Let $\Delta t = t-t_2$, let $\Delta\phi = \phi(t)-\phi(t_2)$, let $\Delta\hat{\lambda} = \hat{\lambda}(t)-\hat{\lambda}(t_2)$, and for $0 \leq s \leq 1$ let

$$P(s;t_2,t) = (t_2+s\Delta t,\phi(t_2)+s\Delta\phi,u(t),\hat{\lambda}(t_2)+s\Delta\hat{\lambda}). \tag{8.4}$$

From (8.1) and V.(3.5) we get

$$\begin{aligned}\mathscr{A}(t)-\mathscr{A}(t_2) &= H(t,\phi(t),u(t),\hat{\lambda}(t))-H(t_2,\phi(t_2),u(t_2),\hat{\lambda}(t_2)) \\ &\leq H(t,\phi(t),u(t),\hat{\lambda}(t))-H(t_2,\phi(t_2),u(t), \hat{\lambda}(t_2)) \\ &= H(P(1;t_2,t))-H(P(0;t_2,t)).\end{aligned}$$

If we now apply the mean value theorem to the function $s \to H(P(s;t_2,t))$ defined for $0 \leq s \leq 1$ and write $P(s)$ instead of $P(s;t_2,t)$, we get that there is a θ in the open interval $(0,1)$ such that

$$\begin{aligned}\mathscr{A}(t)-\mathscr{A}(t_2) \leq H_t(P(\theta))\Delta t &+ \langle H_x(P(\theta)),\Delta\phi\rangle \\ &+ \langle H_p(P(\theta)),\Delta\hat{\lambda}(t)\rangle.\end{aligned} \tag{8.5}$$

Since ϕ and λ are continuous on $[t_0,t_1]$ they are bounded on $[t_0,t_1]$. Since $\hat{\lambda} = (\lambda^0,\lambda)$, where λ^0 is a constant, $\hat{\lambda}$ is bounded on $[t_0,t_1]$. By assumption, u is bounded on $[t_0,t_1]$. Hence there is a closed ball B in $(E^1 \times E^n \times E^m \times E^{n+1})$ such that for all t, t_2 in $[t_0,t_1]$ and all $0 \leq s \leq 1$, the points $P(s;t_2,t)$ are in B. It then follows from the continuity of $\hat{f}_t$ and $\hat{f}_x$ on $\mathscr{R}\times\mathscr{U}$ and the continuity of $\hat{\lambda}$ that there is a constant $K_1 > 0$ such that for all t,t_2 in $[t_0,t_1]$ and all $0 \leq s \leq 1$

$$|H_t(P(s;t_2,t))| \le K_1 \qquad |H_x(P(s;t_2,t))| \le K_1$$
$$|H_p(P(s;t_2,t))| \le K_1. \tag{8.6}$$

From V.(3.2) we get that

$$\Delta\phi = \int_{t_2}^{t} H_p(s,\phi(s),u(s),\hat{\lambda}(s))ds$$

$$\Delta\lambda = -\int_{t_2}^{t} H_x(s,\phi(s),u(s),\hat{\lambda}(s))ds.$$

Since ϕ, $\hat{\lambda}$ and u are bounded on $[t_0,t_1]$ and since H_x and H_p are continuous, it follows that there is a constant K_2 such that for all t, t_2 in $[t_0,t_1]$

$$|\Delta\phi| \le K_2\Delta t \qquad |\Delta\hat{\lambda}| \le K_2\Delta t. \tag{8.7}$$

From (8.5), (8.6) and (8.7) we now get that there exists a constant K such that for all t, t_2 in T,

$$\mathscr{A}(t) - \mathscr{A}(t_2) \le K|t_2 - t|. \tag{8.8}$$

By arguments similar to those used in obtaining (8.5) we get

$$\begin{aligned}\mathscr{A}(t) - \mathscr{A}(t_2) &\ge H(t,\phi(t),u(t_2),\hat{\lambda}(t)) - H(t_2,\phi(t_2),u(t_2),\hat{\lambda}(t_2)) \\ &= H_t(P'(\theta'))\Delta t + \langle H_x(P'(\theta')),\Delta\phi\rangle \\ &\quad + \langle H_p(P'(\theta')),\Delta\hat{\lambda}\rangle,\end{aligned} \tag{8.9}$$

where $0 < \theta' < 1$ and

$$P'(s) = P'(s;t_2,t) = (t_2+s\Delta t,\phi(t_2)+s\Delta\phi,u(t_2),\hat{\lambda}(t_2)+\Delta\hat{\lambda}).$$

From this we conclude by arguments used to establish (8.8) that

$$\mathscr{A}(t) - \mathscr{A}(t_2) \ge -K|t_2 - t|.$$

Upon combining the last inequality with (8.8) we see that $\mathscr{A}$ satisfies the Lipschitz condition

$$|\mathscr{H}(t)-\mathscr{H}(t_2)| \leq K|t-t_2|$$

for all t, t_2 in T. Since meas $T = t_1-t_0$, the set T is dense in $[t_0,t_1]$. Hence there exists a function h that is continuous on all of $[t_0,t_1]$ and such that $h(t) = \mathscr{H}(t)$ for all t in T. Thus

$$h(t) = H(t,\phi(t),u(t),\hat{\lambda}(t)) \quad \text{a.e.}$$

as asserted in Corollary V.3.2. The function h also satisfies a Lipschitz condition with the same constant as $\mathscr{H}$ does.

Since h is Lipschitzian it is absolutely continuous. Let T_1 denote those points of T at which h is differentiable. Then meas $T_1 = t_1-t_0$. Let t_2 now be a point of T_1 and suppose $t_2 \neq t_1$. Then since $h'(t_2)$ exists, we have

$$h'(t_2) = \lim_{t\to t_2} \frac{h(t)-h(t_2)}{t-t_2},$$

where the limit is taken over those values of $t > t_2$ such that $t \in T_1$. Since meas $T_1 = t_1-t_0$, the set of such points t is not empty and has t_2 as a limit point. Since for such t,

$$\frac{h(t)-h(t_2)}{t-t_2} = \frac{\mathscr{H}(t)-\mathscr{H}(t_2)}{t-t_2}, \tag{8.10}$$

and since $\Delta t = t-t_2 > 0$, it follows from (8.5) that

$$\frac{h(t)-h(t_2)}{\Delta t} \leq H_t(P(\theta)) + \langle H_x(P(\theta)), \frac{\Delta\phi}{\Delta t}\rangle + \langle H_p(P(\theta)), \frac{\Delta\hat{\lambda}}{\Delta t}\rangle.$$

We now let $t \to t_2$ so that $\Delta t \to 0$. From the continuity of ϕ and λ, from (8.4), and from the fact that t_2 is a point of density of u it follows that

$$P(\theta) \to \Pi(t_2) = (t_2, \phi(t_2), u(t_2), \hat{\lambda}(t_2))$$

as $t \to t_2$. From the fact that $\phi'(t_2)$ and $\lambda'(t_2)$ exist and from the continuity of H_t, H_x and H_p we get

$$h'(t_2) \leq H_t(\Pi(t_2)) + \langle H_x(\Pi(t_2)), \phi'(t_2)\rangle$$
$$+ \langle H_p(\Pi(t_2)), \lambda'(t_2)\rangle .$$

If we now use V.(3.2) we get

$$h'(t_2) \leq H_t(\Pi(t_2)). \tag{8.11}$$

From (8.10) and (8.9) we get

$$\frac{h(t)-h(t_2)}{\Delta t} \geq H_t(P'(\theta')) + \langle H_x(P'(\theta')), \frac{\Delta\phi}{\Delta t}\rangle$$
$$+ \langle H_p(P'(\theta')), \frac{\Delta\hat{\lambda}}{\Delta t}\rangle .$$

By arguments similar to those used in the preceding paragraph we get that

$$h'(t_2) \geq H_t(\Pi(t_2)).$$

Combining this result with (8.11) gives

$$h'(t_2) = H_t(t_2, \phi(t_2), u(t_2), \hat{\lambda}(t_2)).$$

Since t_2 is an arbitrary point of T_2 and meas $T_2 = t_1 - t_0$, we have shown that

$$h'(t) = H_t(t, \phi(t), u(t), \hat{\lambda}(t)) \quad \text{a.e.}$$

If the function u is piecewise continuous then $\mathcal{A}$ is continuous at all points, except possibly at those points at which u is discontinuous. Since u is piecewise continuous at a point τ of discontinuity of u we have that

$$u(\tau+0) = \lim_{t\to\tau^+} u(t)$$

and

$$u(\tau-0) = \lim_{t\to\tau^-} u(t)$$

exist and are finite. Since $\mathscr{C}$ is closed, $u(\tau+0)$ and $u(\tau-0)$ belong to $\mathscr{C}$. If $t < \tau$, then

$$H(t,\phi(t),u(t),\hat{\lambda}(t)) \geq H(t,\phi(t),u(\tau+0),\hat{\lambda}(t)).$$

If we now let $t \to \tau$ we get

$$H(\tau,\phi(\tau),u(\tau-0),\hat{\lambda}(\tau)) \geq H(\tau,\phi(\tau),u(\tau+0),\hat{\lambda}(\tau)).$$

By similar arguments we get

$$H(\tau,\phi(\tau),u(\tau+0),\hat{\lambda}(\tau)) \geq H(\tau,\phi(\tau),u(\tau-0),\hat{\lambda}(\tau)).$$

It therefore follows that $\mathscr{A}$ is continuous at all points. Thus $\mathscr{A}(t) = h(t)$ everywhere, and so $\mathscr{A}$ is absolutely continuous. Moreover, $\mathscr{A}'(t) = h'(t)$ so that

$$\mathscr{A}'(t) = H_t(t,\phi(t),u(t),\hat{\lambda}(t)) \quad \text{a.e.}$$

BIBLIOGRAPHICAL NOTES

Chapter I

1. The production planning problem of Section 2 was formulated and solved by Arrow and Karlin in Chapter 4 of [1]. They give references to previous work with discrete time versions of the problem.

2. For an early treatment of the flight mechanics problem see Leitmann [34], where references to still earlier work are given.

3. The paper in the engineering literature that stimulated work on Example 5 and the consequent interest in optimal control theory in this country is McDonald [37]. Bushaw [14] solved the time optimal problem posed in [37] on the assumption that the optimal control was "bang-bang".

Chapter III

1. Example 2.2 is a modification of an example used to show the nonexistence of a minimizing function in the calculus of variation. The original example is due to Weierstrass (see [12], p. 418-419).

2. Example 2.4 is due to Roxin [53].

3. Theorem 5.1 with the assumption that each set $\mathcal{Q}^+(t,x)$ is convex replaced by the more stringent assumption that each set $\mathcal{Q}(t,x)$ is convex is essentially due to Filippov [22] and to Roxin [53].

4. A wealth of examples in which Theorem 5.1 finds application will be found in Athans and Falb [2].

5. In [16] and [17] Cesari introduced the sets $\mathcal{Q}^+(t,x)$ and the Cesari property, which he called "property (Q)". Theorem 4.1 in the equivalent form of Corollary 4.1 is essentially due to Cesari, [17] and [18]. The proof in these notes is different from Cesari's, as is the use of the weak Cesari property. It was first given by us in [7].

6. The condition (6.2) for equi-absolute continuity is due to Cesari [18]. Corollary 6.1, however, goes back to de la Vallee Poussin. (See Natanson [46], p. 159.)

7. Theorem 7.1 is a generalization due to McShane and Warfield [44] of an implicit function theorem for measurable functions introduced by Filippov in his paper [22]. The latter result is known as Filippov's Lemma. Another generalization of Filippov's lemma was given by Castaing [15].

8. Cesari and his students have studied in great detail conditions on f^0 and f that guarantee the Cesari property and have explored very thoroughly the relationships among various classical conditions in the calculus of variations and the Cesari property. These results are summarized in [19], where reference to other work is given.

9. Theorem 8.1 was given by us in [8].

Chapter IV

1. A generalization of Theorem 2.1 is given in [6].

2. The concept of relaxed trajectories goes back to L. C. Young [60] who introduced it in problems of the calculus of variations under the name of "generalized curves". A more exhaustive treatment which included the study of generalized curves in the Bolza problem was later given by E. J. McShane in a series of three papers [40], [41], [42]. Relaxed controls and relaxed trajectories were introduced into control theory independently and in different forms by Warga [58] and Gamkrelidze [24]. In [43] McShane improved his earlier treatment of generalized curves and applied it to control problems with constraints that are not necessarily compact. In this connection, also see Cesari [18]. Warga [59] and L. C. Young in his book [60] have treatments of the relaxed problem different from the one used in this text. Our definition is that of Gamkrelidze [24].

3. Our treatment of the Chattering Lemma is based upon that of Gamkrelidze [25].

4. The properties of the attainable set for linear systems whose control set is a cube in E^m were given by LaSalle [32]. The properties of the attainable set for nonlinear systems under hypotheses similar to those of Theorem 5.3 were first given by Roxin [53].

5. The underlying idea in the proofs of Lemma 6.1 and Theorem 6.1 is taken from the elegant proof of Lindenstrauss [36] of Liapunov's Theorem. Lemma 6.1 is taken from Hermes and LaSalle ([27], Theorem 8.2).

6. Theorem 6.3 was first stated by Neustadt [47]. The extension to the non-compact case was given by Olech [49] and by Jacobs [30]. The proof in the text is different from these proofs.

7. The first results on bang-bang control for the linear time optimal problem are to be found in LaSalle [31] and [32], Bellman, Glicksberg and Gross [3], and Gamkrelidze [23]. Previous writers, notably Bushaw [14] and McDonald [37] had assumed that the optimal control must be bang-bang. For a version of the bang-bang principle that is sharper than the one given in Theorem 6.3 see Sonneborn and Van Vleck [56].

Chapter V

1. For a derivation of the maximum principle along the lines given in Section 1 under less restrictive hypotheses on W see Berkovitz [5] and Mirică [45].

2. Example 3.1 goes back to Bolza ([12], p. 116-117).

3. The multiplier rule for variational problems with differential equation side conditions goes back to Euler and Lagrange. It was not until the early 1900's, however, that proofs of the multiplier rule without gaps were finally given by Kneser and Hilbert.

A short history of the development of the multiplier rule up to the year 1909 is given in Bolza [12], pp. 566-568. Further historical remarks can be found in Bliss [9].

4. The Weierstrass condition in the generality given here was first proved by McShane [39]. Prior to [39], the Weierstrass condition was established under the assumption that the multiplier rule held with a unique set of multipliers (ψ^0, ψ) with $\psi^0 = -1$. McShane did away with this requirement. In [39] McShane introduced a convex set of variations. This idea was later exploited and developed further by Pontryagin and his co-workers in their proof of the maximum principle [50], [51].

5. Exercise 5.2 was treated by Berkovitz [4]. References to other work will be found there.

Chapter VI

1. The proof of the maximum principle given here is essentially that of Gamkrelidze [25].

BIBLIOGRAPHY

1. K. J. Arrow, S. Karlin and H. Scarf, Studies in the Mathematical Theory of Inventory and Production, Stanford University Press, Stanford, California, 1958.

2. M. Athans and P. Falb, Optimal Control, McGraw-Hill, New York, 1966.

3. R. Bellman, I. Glicksberg, and O. Gross, On the "bang-bang" control problem, Quart. Appl. Math., 14(1956), 11-18.

4. L. D. Berkovitz, Variational methods in problems of control and programming, J. Math. Anal. Appl., 3(1961), 145-169.

5. L. D. Berkovitz, Necessary conditions for optimal strategies in a class of differential games and control problems, SIAM J. Control 5(1967), 1-24.

6. L. D. Berkovitz, An existence theorem for optimal controls, J. Optimization Theory Appl., 6(1969), 77-86.

7. L. D. Berkovitz, Existence theorems in problems of optimal control, Studia Math., 44(1972), 275-285.

8. L. D. Berkovitz, Existence theorems in problems of optimal control without property (Q), in Techniques of Optimization, A. V. Balakrishnan ed., Academic Press, New York and London, 1972, 197-209.

9. G. A. Bliss, The problem of Lagrange in the calculus of variations, Amer. J. Math., 52(1930), 673-741.

10. G. A. Bliss, Lectures on the Calculus of Variations, The University of Chicago Press, Chicago, 1946.

11. V. G. Boltyanskii, R. V. Gamkrelidze and L. S. Pontryagin, The theory of optimal processes I, the maximum principle. Izv. Akad. Nauk SSSR. Ser Mat., 24(1960), 3-42. English translation in Amer. Math. Soc. Transl. Ser. 2, 18(1961), 341-382.

12. O. Bolza, Vorlesungen uber Variationsrechnung, Reprint of 1909 edition, Chelsea Publishing Co., New York.

13. A. E. Bryson and Y. C. Ho, Applied Optimal Control, Blaisdell 1969, Waltham, Toronto, London.

14. D. Bushaw, Optimal discontinuous forcing terms, Contributions to the Theory of Nonlinear Oscillations IV, Annals of Math Study 41, S. Lefschetz ed., Princeton University Press, Princeton, (1958), 29-52.

15. C. Castaing, Sur les multi-applications mesurables, Rev. Francaise Automat. Informat. Recherche Opérationnelle 1(1967), 91-126.

16. L. Cesari, Existence theorems for optimal solutions in Pontryagin and Lagrange problems, SIAM J. Control 3(1966), 475-498.

17. L. Cesari, Existence theorems for weak and usual optimal solutions in Lagrange problems with unilateral constraints I, Trans. Amer. Math. Soc.,124(1966), 369-412.

18. L. Cesari, Existence theorems for optimal controls of the Mayer type, SIAM J. Control 6(1968), 517-552.

19. L. Cesari, Closure, lower closure, and semicontinuity theorems in optimal control, SIAM J. Control 9(1971), 287-315.

20. N. Dunford and J. T. Schwartz, Linear Operators Part I: General Theory, Interscience, New York, 1958.

21. H. G. Eggleston, Convexity, Cambridge University Press, Cambridge, 1958.

22. A. F. Filippov, On certain questions in the theory of optimal control, SIAM J. Control 1(1962), 76-89. Orig. Russ. article in Vestnik Moskov. Univ. Ser. Mat. Mech. Astr.,2(1959), 25-32.

23. R. V. Gamkrelidze, Theory of time-optimal processes for linear systems, Izv. Akad. Nauk. SSSR. Ser Mat., 22(1958), 449-474 (Russian).

24. R. V. Gamkrelidze, On sliding optimal regimes, Dokl. Akad. Nauk SSSR. 143(1962), 1243-1245. Translated as Soviet Math. Dokl., 3(1962), 390-395.

25. R. V. Gamkrelidze, On some extremal problems in the theory of differential equations with applications to the theory of optimal control, SIAM J. Control 3(1965), 106-128.

26. E. Hille and R. S. Phillips, Functional Analysis and Semi-Groups, Revised Ed., American Mathematical Society, Providence, 1957.

27. H. Hermes and J. P. LaSalle, Functional Analysis and Time Optimal Control, Academic Press, New York-London, 1969.

28. M. R. Hestenes, Calculus of Variations and Optimal Control Theory, John Wiley, New York-London-Sydney, 1966.

29. J. G. Hocking and G. S. Young, Topology, Addison-Wesley, Reading, Mass., 1961.

30. M. Q. Jacobs, Attainable sets in systems with unbounded controls, J. Differential Equations 4(1968), 408-423.

31. J. P. LaSalle, Study of the Basic Principle Underlying the Bang-Bang Servo, Goodyear Aircraft Corp. Report GER-5518 (July 1953). Abstract 247t. Bull. Amer. Math. Soc. 60(1954), 154.

32. J. P. LaSalle, The time optimal control problem, Contributions to the Theory of Nonlinear Oscillations Vol. 5, Annals of Math Study No. 45 Princeton University Press, Princeton (1960), 1-24.

33. E. B. Lee and L. Markus, Foundations of Optimal Control Theory, John Wiley, New York, London, Sydney, 1967.

34. G. Leitmann, On a Class of Variational problems in rocket flight, Jour. Aero/Space Sciences 26(1959), 586-591.

35. G. Leitmann, An Introduction to Optimal Control, McGraw-Hill, New York, 1966.

36. J. Lindenstrauss, A short proof of Liapounoff's convexity theorem, J. Math. Mech., 15(1966), 971-972.

37. D. McDonald, Non linear techniques for improving servo performance, National Electronics Conference 6(1950), 400-421.

38. E. J. McShane, Integration, Princeton University Press, Princeton, 1944.

39. E. J. McShane, On multipliers for Lagrange problems, Amer. J. Math., 61(1939), 809-819.

40. E. J. McShane, Necessary conditions in generalized-curve problems in the calculus of variations, Duke Math. J., 7(1940), 1-27.

41. E. J. McShane, Existence theorems for Bolza problems in the calculus of variations, Duke Math. J., 7(1940), 28-61.

42. E. J. McShane, Generalized curves, Duke Math. J., 6(1940), 513-536.

43. E. J. McShane, Relaxed controls and variational problems, SIAM J. Control 5(1967), 438-485.

44. E. J. McShane and R. B. Warfield, Jr., On Filippov's implicit functions lemma, Proc. Amer. Math. Soc., 18(1967), 41-47.

45. S. Mirică, On the admissible synthesis in optimal control theory and differential games, SIAM J. Control 7(1969), 292-316.

46. I. P. Natanson, Theory of Functions of a Real Variable, Eng. Trans. by Leo Boron, revised ed., F. Ungar, New York, 1961.

47. L. W. Neustadt, The existence of optimal controls in the absence of convexity conditions, J. Math. Anal. Appl., 7(1963), 110-117.

48. L. W. Neustadt, Optimization: A Theory of Necessary Conditions, Princeton University Press, Princeton, New Jersey, To Appear.

49. C. Olech, Extremal solutions of a control system, J. Differential Equations 2(1966), 74-101.

50. L. S. Pontryagin, V. G. Boltyanskii, R. V. Gamkrelidze, E. F. Mischenko, The Mathematical Theory of Optimal Processes, (Translated by K. N. Trirogoff, L. W. Neustadt, editor) John Wiley, 1962.

51. L. S. Pontryagin, Optimal regulation processes, Uspehi Mat. Nauk (N.S.) 14(1959), 3-20. English translation in Amer. Math. Soc. Transl. Ser. 2,18(1961), 321-339.

52. R. T. Rockafellar, Convex Analysis, Princeton University Press, Princeton, 1970.

53. E. Roxin, The existence of optimal controls, Mich. Math. J., 9(1962), 109-119.

54. W. Rudin, Real and Complex Analysis, McGraw-Hill, New York, St. Louis, San Francisco, Toronto, London, Sydney, 1966.

55. G. Scorza-Dragoni, Un teorema sulle funzioni continue rispetto ad una e misurabile rispetto ad un'altra variabile, Rend. Sem. Mat. Univ. Padova 17(1948), 102-106.

56. L. M. Sonneborn and F. S. Van Vleck, The bang-bang principle for linear control systems, SIAM J. Control 2(1964), 151-159.

57. M. M. Vainberg, Variational Methods for the Study of Nonlinear Operators, Eng. Trans. by A. Feinstein, Holden-Day, San Francisco, London, Amsterdam, 1964.

58. J. Warga, Relaxed variational problems, J. Math. Anal. Appl., 4(1962), 111-128.

59. J. Warga, Optimal Control of Differential and Functional Equations, Academic Press, New York, 1972.

60. L. C. Young, Generalized curves and the existence of an attained absolute minimum in the calculus of variations, Compt. Rend. Soc. Sci. et Lettres. Varsovie, Cl III 30(1937), 212-234.

61. L. C. Young, Lectures on the Calculus of Variations and Optimal Control Theory, W. B. Saunders Co., Philadelphia, London, Toronto, 1969.

INDEX